Chemical and Physical Behavior of Human Hair

Clarence R. Robbins

Chemical and Physical Behavior of Human Hair

Third Edition

With 111 Figures

Springer-Verlag
New York Berlin Heidelberg London Paris
Tokyo Hong Kong Barcelona Budapest

Clarence R. Robbins
Research and Development Division
Colgate-Palmolive Company
Piscataway, NJ 08854-5596
USA

Library of Congress Cataloging-in-Publication Data

Robbins, Clarence R.
Chemical and physical behavior of human hair/C.R. Robbins.—
3rd ed.
p. cm.
Includes bibliographical references and index.
ISBN 0-387-94191-6 (New York).—ISBN 3-540-94191-6 (Berlin)
1. Hair. I. Title.
[DNLM: 1. Hair. 2. Hair—analysis. WR 450 R632c 1994]
QP88.3.R62 1994
612.7'99—dc20
DNLM/DLC
for Library of Congress 93-38111

Printed on acid-free paper.

The first edition was published by Van Nostrand Reinhold Company in 1979 under the same title.

Production managed by Jim Harbison; manufacturing supervised by Vincent Scelta.
Typeset by Asco Trade Typesetting Ltd, Hong Kong.
Printed and bound by Braun-Brumfield, Inc., Ann Arbor, MI.
Printed in the United States of America.

9 8 7 6 5 4 3 2 1

ISBN 0-387-94191-6 Springer-Verlag New York Berlin Heidelberg
ISBN 3-540-94191-6 Springer-Verlag Berlin Heidelberg New York

To my wife Gene, for 32 years of hope by trying to make every day a good one. To my father, an example for me and an inspiration to many. To my daughter Laurie, my son Mark and his wife, Kim, my mother Ethel, my brother John and his family, my mother-in-law Daun, my father-in-law Gordon, and my many other relatives and friends who help to make life meaningful for me.

Preface to the First Edition

In this book, I have tried to provide a reasonably up to date and complete account of the chemical and physical properties and behavior of human hair. Chapter 1 describes the structure of human hair, from the microscopic level down to the molecular level, Chapter 2 discusses its chemical composition, and Chapters 3 through 7 emphasize the organic and physical chemistry of the reactions and interactions of hair with different cosmetic or potential cosmetic ingredients. The last chapter describes the physical behavior of hair before, during, and after cosmetic treatments.

The principal audience is obviously the cosmetic and fiber industries and their suppliers, but it is my hope that the examples of physico-chemical concepts involving reaction mechanisms and structure elucidation might stimulate the academic world to become more actively involved in hair research and perhaps to offer some of these examples of classical applied science in organic and biochemistry courses.

In many instances, reference is made to the wool literature, because it is more complete, and the two fibers are so similar. I have also taken the liberty to express my own opinions and generalizations on many of the subjects herein. I caution the reader not to be content with the generalizations I have made, but to examine the original data for himself. The purpose of this book is to serve as a convenient starting point, not to be the final word.

Clarence R. Robbins

Preface to the Second Edition

Ten years have elapsed since I wrote the first English edition of this book, and progress in hair science has made a second edition necessary. Since 1978, at least two new major scientific cosmetic journals have appeared, and four International Hair Science Symposia and one International Symposium on Forensic Hair Comparisons have been held.

Thus, continuing studies in hair science have been numerous in the past ten years, so much so that a large proportion of the text had to be completely rewritten. In addition a number of omissions that were kindly pointed out by reviewers have been included in this edition, as well as corrections sent to me by readers.

I sincerely hope that this new edition fulfills the original purposes: to provide a reasonably up to date and complete account of the chemical and physical properties and behavior of human hair, to serve as a convenient starting point for hair research—and, as indicated in the first edition, not to be the final word.

CLARENCE R. ROBBINS

Preface to the Third Edition

More than five years have elapsed since I completed the 2nd English edition of this book and fifteen years since the first edition. During that interim, this book has been translated into Japanese and a Russian translation is well on its way. In addition, many advances have been made in both our understanding of the structure of human hair, of the chemical and physical mechanisms of how cosmetics work, and in improved hair products in the marketplace.

For example, the cuticle of human hair is largely a protective barrier that does not participate directly in the tensile strength of human hair. The structure of the cortex is now known to consist of intermediate filamentous type proteins, that are surrounded by highly crosslinked sulfur rich proteins. The intercellular regions are largely nonkeratinous and have been demonstrated to be highly important to the penetration and to the swelling behavior of the fiber. The mechanism of chemisorption onto keratin fibers has been further elucidated leading to better conditioning shampoos than ever before, and although, there are still many mysteries surrounding the biochemical mechanism of hair growth, we are at least beginning to understand some of the fundamental perplexities in this vast area of research.

Thus, this new edition portrays human hair from these new perspectives and hopefully fulfills the original purpose: to provide a reasonably up to date and complete account of the chemical and physical properties and behavior of human hair. It is also my hope that this edition will provide a convenient starting point for hair research and serve as a stimulus to academia by offering examples of classical applied science in an area of interest to most lay persons—and as indicated in the first edition, not to be the final word.

CLARENCE R. ROBBINS

Contents

1

Morphological and Macromolecular Structure

Human hair is a keratin-containing appendage that grows from large cavities or sacs called follicles. Hair follicles extend from the surface of the skin through the stratum corneum and the epidermis into the dermis (Figure 1–1).

Hair serves protective, sensory, and sexual attractiveness functions. It is characteristic of all mammals, and in humans grows over a large percentage of the body surface. Regardless of the species of origin or body site, human hair grows in three distinct stages and has certain common structural characteristics. For example, hair fibers grow in a cyclical pattern consisting of three distinct stages called anagen (growing stage), catagen (transition stage), and telogen (resting stage) (Figure 1–2).

Morphologically, a fully formed hair fiber contains three and sometimes four different units or structures. At its surface, hair contains a thick protective covering consisting of layers of flat overlapping scalelike structures called the cuticle (Figure 1–3). The cuticle layers surround the cortex, which contains the major part of the fiber mass. The cortex, the second unit, consists of spindle-shaped cells that are aligned along the fiber axis. Cortical cells contain the fibrous proteins of hair. Thicker hairs often contain one or more loosely packed porous regions called the medulla, located near the center of the fiber. The fourth unit is the intercellular cement that glues or binds the cells together, forming the major pathway for diffusion into the fibers.

Although this book is concerned with hair fibers in general, the primary

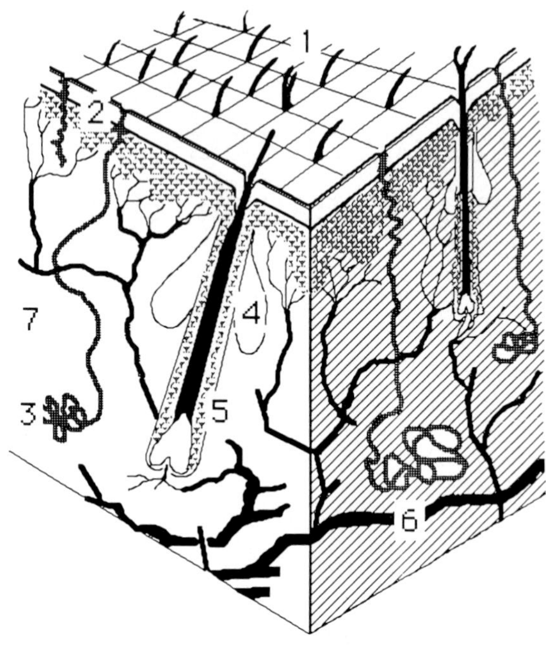

FIGURE 1–1. A section of human skin illustrating a hair fiber in its follicle as it grows, how it is nourished and as it emerges through the skin. (Published with permission of Academic Press, Inc. [119].)

focus is on human scalp hair, and this first chapter is concerned primarily with the growth and morphology of this unique structure.

General Structure and Hair Growth

The schematic diagram of Figure 1–4 illustrates an active human hair bulb and fiber inside the follicle or sac that originates in the subcutaneous tissue of the skin. The dermal papilla, located near the center of the bulb, is important to the growth of hairs. Basal layers, which produce hair cells, nearly surround the bulb. Melanocytes, which produce hair pigment, also exist within the bulb. Blood vessels (see Figure 1–1) carry nourishment to the growing hair fiber deep within the skin at the base of the bulb.

The human hair fiber can be divided into three distinct zones along its axis. The zone of biological synthesis and orientation resides at and around the bulb of the hair. The next zone in an outward direction along

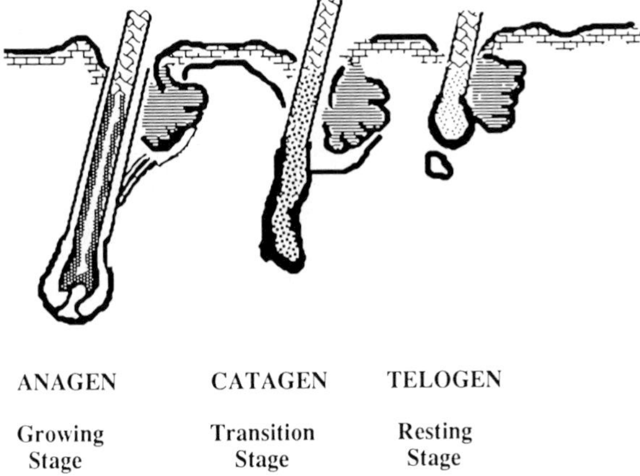

FIGURE 1–2. Schematic illustrating the three stages of growth of the human hair fiber.

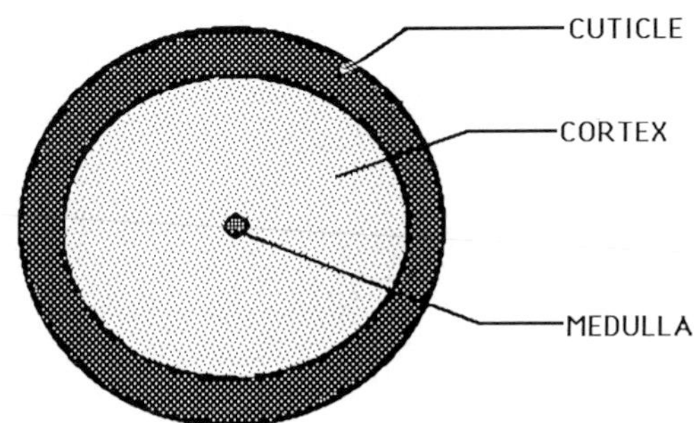

FIGURE 1–3. Schematic diagram of a cross section of a human hair fiber.

the hair shaft is the zone of keratinization, where stability is built into the hair structure via the formation of cystine linkages [1]. The third zone that eventually emeges through the skin surface is the region of the permanent hair fiber; the permanent hair fiber consists of dehydrated cornified cuticle, cortical, and sometimes medullary cells, and intercellular cement.

The major emphasis in this book is on the chemistry, structure, and the physics of the permanent zone of the human hair fiber. As indicated, the primary focus is on human scalp hair as opposed to hair of other parts of the body.

Kaswell [2] has suggested that the diameter of human hair fibers varies from 15 to 110 μm. Randebrook [3] provides a somewhat larger estimate, 40 to 120 μm. Figure 1–5 illustrates the range in fiber diameters and cross-sectional shapes of hairs from five Caucasian adults. For a more complete

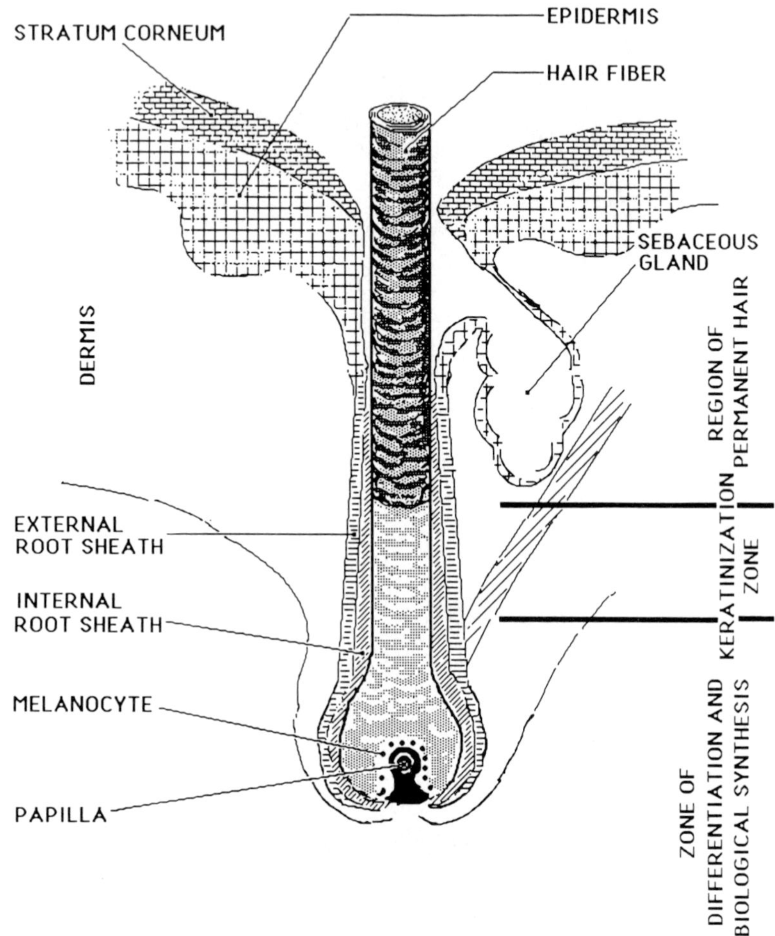

FIGURE 1–4. Pilosebaceous unit illustrating a hair follicle with its fiber and the different zones of growth and structural organization as the fiber emerges through the scalp. (Published with permission of Academic Press, Inc. [120].)

discussion of hair fiber diameter, see Chapter 8 (this volume) and also the review by Bogaty [4] and the references therein.

Three distinct regions containing different types of cells are generally apparent in cross sections of human hair fibers (see Figures 1–3 and 1–6). These three cellular regions and the intercellular matter are described in summary form in the introductory section. After brief discussions of functions of hair, hair growth/hair loss, and treatments for hair loss, the remainder of this chapter focuses on these three types of cells and the intercellular binding material of human scalp hair.

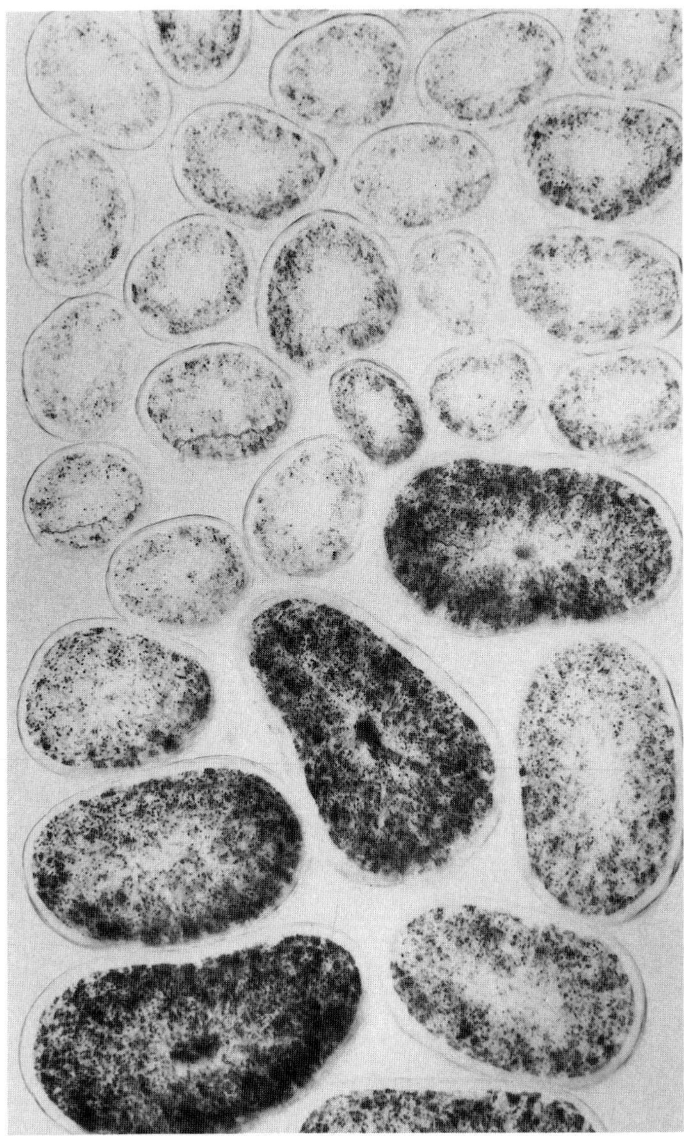

FIGURE 1–5. Light micrograph of scalp hair fiber cross sections, illustrating varying fiber cross-sectional size, shape, and pigmentation. Note lack of pigment in the cuticle.

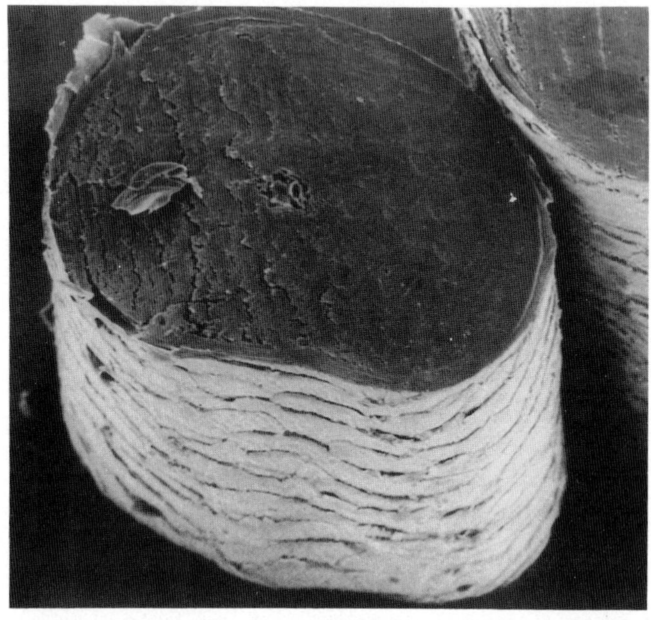

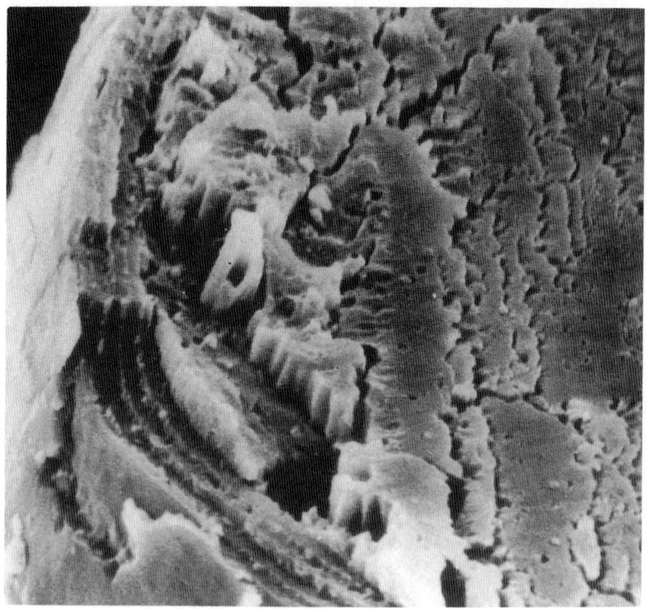

FIGURE 1–6. Treated hair fibers cross-sectioned with a microtome. Top: Cuticle, cortex, and medulla. Bottom: Cuticle layers.

Functions of Hair

Human scalp hair provides both protective and cosmetic or adornment functions. Scalp hair protects the head from the elements by functioning as a thermal insulator. Hair also protects the scalp against sunburn, other effects of light radiation, and mechanical abrasion.

Hair on parts of the body other than the scalp provides related protective and adornment functions. The adornment function of eyebrows is obvious. However, eyebrows also inhibit sweat and prevent extraneous matter from running into the eyes. In addition, eyebrows protect the bony ridges above the eyes, and assist in communication and in the expression of emotion.

Eyelashes are also important to adornment. Eyelashes protect the eyes from sunlight and foreign objects, and they assist in communication. Nasal hairs filter inspired air and retard the flow of air into the respiratory system, thus allowing air to be warmed or cooled as it enters the body. Hair on other parts of the anatomy serves related functions. In addition, a general function of all hairs is as sensory receptors, because all hairs are supplied with sensory nerve endings. The sensory receptor function can enhance hair in its protective actions.

Hair Growth

General Features of Hair Growth

Mitosis or equational cell division occurs near the base of the bulb at the lowermost region of the zone of differentiation and biological synthesis (see Figure 1–4). This is the primary region of protein synthesis and of hair growth. The basal layers, which produce hair cells, and the melanocytes, which produce hair pigment, are located within the bulb. The dermal papilla, near the center of the bulb, is believed to play a role in controlling the growth cycle of hairs. The newly formed cells migrate upward in the follicle as they move away from the base of the bulb, and they elongate as they move into a region of elongation (in the zone of differentiation and biological synthesis). As the cells continue to move upward into the keratogenous zone, dehydration begins. Disulfide bonds form through a mild oxidative process, and ultimately the permanent hair fiber is formed. Melanocytes in the lower portion of the bulb produce melanin pigment. This pigment is incorporated into the cortical cells by a phagocytosis mechanism [5] that occurs in the zone of differentiation and biological synthesis.

Other important structures associated with each hair fiber are sebaceous glands (the oil-producing glands of the epidermis), arrector muscles, nerve endings, and blood vessels that feed and nourish the dividing and growing cells of the hair near the bulb (see Figures 1–1 and 1–4).

As mentioned earlier, hair fibers grow in three distinct stages (see Figure

1–2), and these stages are controlled by androgens, hormones that stimulate the activity of male sex glands and male characteristics which are produced by the adrenals and the sex glands. The three stages are as follows.

1. The ***anagen*** stage, or the growing stage, is characterized by intense metabolic activity in the hair bulb. For scalp hair, this activity generally lasts for 2 to 6 years, producing hairs that grow to approximately 100 cm in length (~3 ft); however, human scalp hair longer than 150 cm (5 ft) is frequently observed in long hair contests (Figure 1–7).

FIGURE 1–7. "Three Women," by Belle Johnson. Photograph taken about 1900. Hair generally grows to a maximum length of about 3 ft; however, specimens more than 5 ft in length have been documented. (Reprinted with permission of the Massillon Museum, Massillon, OH.).

Terminal hair does grow at slightly different rates on different regions of the head. For example, hair grows at about 16 cm/year (~6.2 in./year) on the vertex or the crown area of the scalp, but at a slightly slower rate (~14 cm/year) in the temporal area and generally at slower rates on other body regions (e.g., ~10 cm/year) in the beard region.

2. The *catagen* stage, the transition stage, lasts for only a few weeks. During catagen, metabolic activity slows down, and the base of the bulb migrates upward in the skin toward the epidermal surface.
3. The *telogen* stage, or the resting stage, also lasts only a few weeks. At this stage, growth has completely stopped and the base of the bulb has atrophied to the point at which it approaches the level of the sebaceous canal. A new hair then begins to grow beneath the telogen follicle, pushing the old telogen fiber out. The telogen fiber is eventually shed.

In humans, prenatal hairs originate from the malpighian layer or the stratum germinativum of the epidermis, usually in the third or fourth month of fetal life. Prenatal hairs, sometimes called lanugo, are either lightly pigmented or contain no pigment and are usually shed before birth or soon thereafter. This prenatal or infant hair generally grows to a limit of about 15 cm and is then replaced by children's hair. Children's hair or prepubertal hair, sometimes called primary terminal hair, is longer and more coarse than infant hair and generally grows to a maximum length of about 60 cm.

Soon after the onset of puberty and the consequent hormonal changes, hair grows longer and more coarse, producing what is called secondary terminal hair. In addition to these changes in scalp hair, hair in the axillary, pubic, and beard (for males) areas becomes longer and more coarse at the onset of puberty. Secondary terminal scalp hair is generally longest and most coarse in the late teens (Figure 1–8), generally growing to a maximum length of about 100 cm.

In the mid- to late twenties, hormonal changes induce slow gradual shortening of anagen for scalp hair, which causes these hair fibers to grow shorter and finer. Ultimately, in many persons this results in the transition of terminal hairs to vellus hairs, producing the condition known as baldness. Vellus hairs grow on those "hairless" regions of the body including

TABLE 1–1. Human scalp hair and age.

Hair type	Approximate age (years)	Approximate maximum length (cm)	Approximate maximum diameter (μm)
Infant hair (lanugo)	<1	15	20
Children's (primary terminal)	1 to 12	60	60
Adult hair (secondary terminal)	>13	100	100
Vellus hair	>30	0.1	4

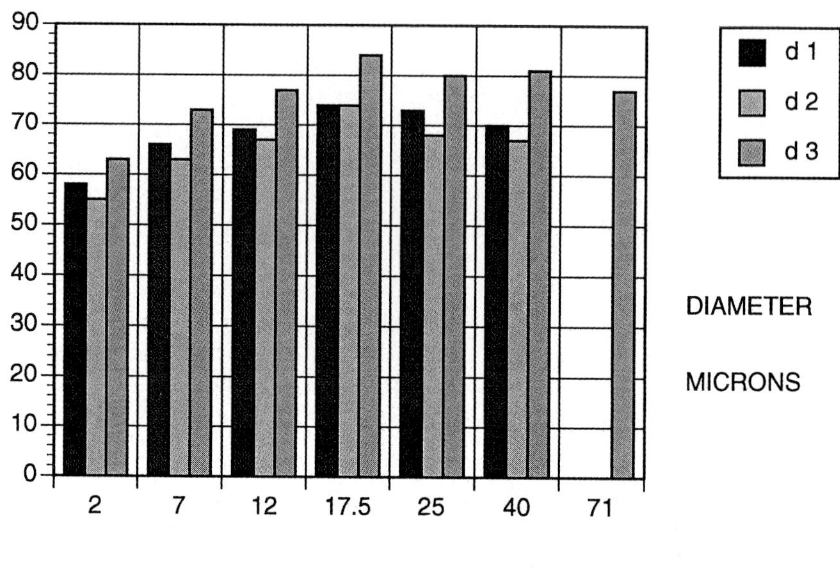

FIGURE 1–8. Hair fiber diameter and age. Data from 3 separate studies (d1, d2, d3).

TABLE 1–2. Differences between terminal and vellus hairs.

Terminal hairs	Vellus hairs
Long hairs (~1.0 m or longer)	Short hairs (~1 mm)
Thick hairs (30–120 μm)	Thin hairs (4 μm or less)
Generally one hair per pilosebaceous unit	More than one hair per pilosebaceous unit
Usually pigmented	Unpigmented
Longer life cycle (3–6 years in anagen)	Shorter life cycle (in telogen ~90% of time)

the bald scalp, the nose, and many other areas of the body that appear hairless.

As indicated earlier, terminal hairs is a term that is normally applied to those long thick hairs that occur on children and adults in contrast to lanugo and vellus hairs. Terminal hairs, at some stage of development, grow on the scalp, eyelash area, eyebrow area, axillary and pubic areas, trunk and limbs, and beard and mustache areas of males. Vellus hairs grow on virtually all other areas of the human body except the following hairless areas: the palms of the hands, the soles of the feet, the undersurface of the fingers and toes, the margin of the lips, the areolae of the nipples, the umbilicus, the immediate vicinity of the urogenital and anal openings, the nail regions, and scar tissue.

Conditions of Excessive Hair Growth

Scalp hair at maturity normally grows to a length of about 3 ft; however, in long hair contests, lengths greater than 5 ft are frequently observed. In March of 1988, Dianne Witt of Massachusetts had the longest scalp hair on record (Guiness Book of Records), which was measured at more than 10 ft or more than 300 cm long. Miss Witt's hair appears to be growing at a normal rate of about 6 in. per year; therefore, because of some condition, that probably involves interference with the ability of testosterone to control the anagen to telogen cycle, her hair has remained in anagen phase for more than 20 years (see the section entitled A Mechanism for Hair Growth/Hair Loss and Changes in Hair Size later in this chapter).

Hypertrichosis is a condition in which an excessive growth of terminal hair occurs often on the limbs, trunk, or face. Hypertrichosis may be localized or diffuse. The most common type is called essential hirsutism or idiopathic hypertrichosis of women. In this condition, terminal hairs grow on women in those areas where hairiness is considered a secondary sex characteristic of males, for example, the trunk, the limbs, or the beard or mustache areas. This condition is generally not caused by an endocrinologic abnormality, but rather is believed to be linked to the transport of testosterone from the endocrine glands to the site of action (Figure 1–9).

Endocrinopathic hirsutism is a rare condition that results from excessive synthesis of hormones with androgenic properties. This abnormality produces masculinization of females. One symptom of this condition is excessive growth of terminal hairs in regions that are normally "hairless" in females. Classic examples of this disease are oftentimes exhibited in circus side shows.

Hair Loss (Alopecia)

The transition of terminal hairs to vellus hairs occurs gradually and at different rates for different persons. This phenomenon tends to occur in a more diffuse pattern among women than among men; thus, the term male pattern baldness is used for the patterns of balding that either begin in the crown of the scalp and move forward or begin in the frontal area of the scalp and recede to create a characteristic baldness pattern (Figure 1–10).

Nonetheless, it has been shown by Venning and Dawber [6] that many women do develop patterned balding much in the same manner as men. In fact, about 13% of premenopausal women and 37% of postmenopausal women show some degree of "male pattern baldness."

The normal scalp will contain about 175 to 300 terminal hairs per square centimeter [7], and it loses about 50 to 100 hairs per day [8–10] as normal hair fallout. This fallout provides a normal anagen period of about 3 to 6 years for a total of about 100,000 to 110,000 hairs on the

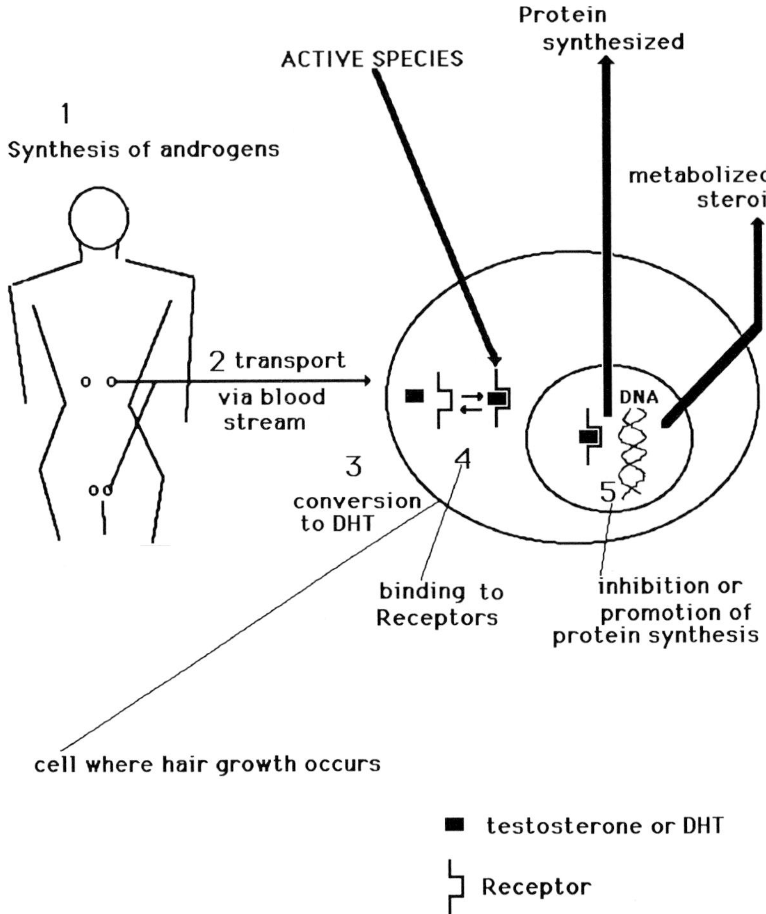

FIGURE 1–9. Mechanism suggested for hair growth/hair loss that is stimulated or retarded by an androgen combined with a protein receptor to form the active species that controls hair growth.

scalp. Shedding rates normally decrease both during and after pregnancy. However, some women report thinning of scalp hair during pregnancy. Furthermore, in a normal scalp, the proportion of follicles in anagen peaks to nearly 90% in the spring (March) in temperate climates and falls steadily to a low of about 80% in the late fall (November) when the telogen count is highest [11]. This effect is accompanied by increasing hair fallout in the fall. As baldness approaches, the anagen period decreases and thus the percentage of hairs in anagen also decreases [12] from the norm of about 80% to 90% in anagen. The remainder of hairs are in catagen and telogen.

Anagen/telogen ratios are sometimes used as a criterion of the balding condition, that is, as the balding process progresses, the ratio of anagen

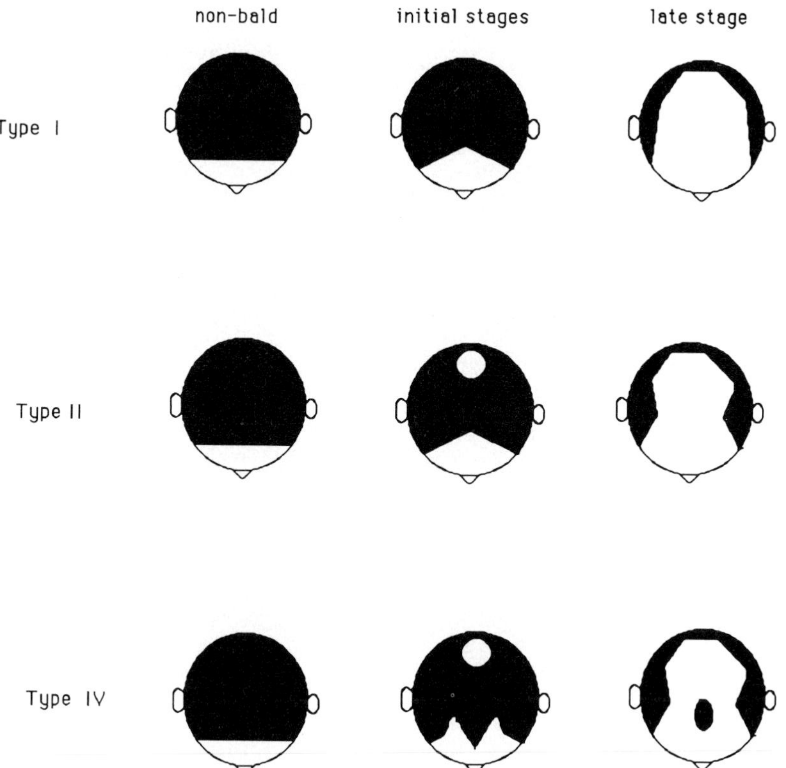

FIGURE 1–10. Examples of baldness classifications and stages of baldness.

hairs to telogen hairs decreases. These ratios may be determined by plucking hairs and microscopically evaluating the roots (Figures 1–11 and 1–12) or by the phototrichogram method (Figure 1–13) in which a small area of the scalp is shaved, photographed, and rephotographed 3 to 5 days later. Comparison of the two photographs reveals those hairs that have grown, (anagen hairs) and those hairs that have not grown (telogen hairs), providing an accurate determination of anagen/telogen ratios.

There is also a small but significant reduction in the number of hair follicles per square centimeter in male pattern alopecia. A bald area will contain about two-thirds to three-fourths the number of follicles of a normal scalp area, which will contain approximately 460 follicles per square centimeter [13]. A single pilosebaceous unit (see Figure 1–4) often consists of more than one fiber and may contain either terminal or vellus hairs.

Alopecia or hair loss may occur over any body region such as the scalp, face, trunk, or limbs. Obviously, alopecia of the scalp (baldness) has received the most attention. In most forms of baldness, progressive miniaturization of hair follicles results in a transition of terminal hairs to vellus

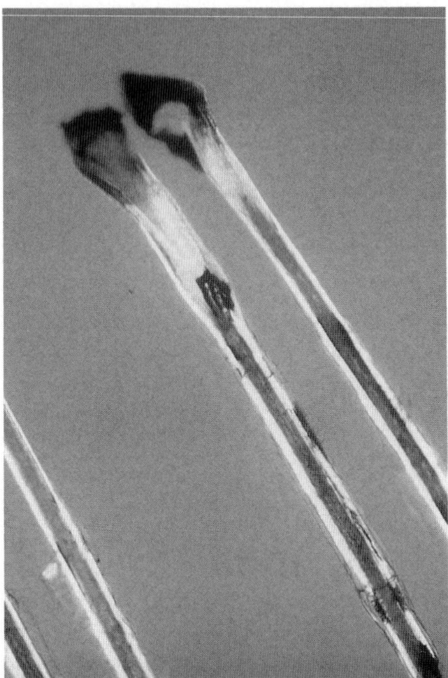

FIGURE 1–11. A light micrograph of plucked hair fibers in the anagen stage.

hairs [14] as opposed to the common misconception portrayed by the term "hair loss." In a normal healthy scalp, approximately 80% to 90% of the hairs are in anagen, 1% to 2% in catagen, and 10% to 20% in telogen [15].

The most common form of hair loss is genetically involved and is linked to androgens, thus the term androgenetic alopecia. Androgenetic alopecia or common baldness is a normal aging phenomenon and occurs in both sexes. To the extent that it is a cosmetic concern or problem, it occurs in about 40% of men and in about 10% of women.

Alopecia areata, another form of hair loss, is believed to be related to the immune system (e.g., autoimmunity). This disease generally occurs as patchy baldness on an otherwise normal scalp, although sometimes hair of other body regions is affected. When the entire scalp is involved, the condition is called alopecia totalis. If terminal hair loss occurs over the entire body, which is rare, it is called alopecia universalis. Emotional stress has also been shown to be one of the initiating causes of areata. Topical application of steroids is sometimes used to treat this condition.

Alopecia induced by physical stress has been termed trichotillomania. This condition occurs from physically pulling or twisting a localized area of hair until noticeable thinning develops. This type of hair loss sometimes

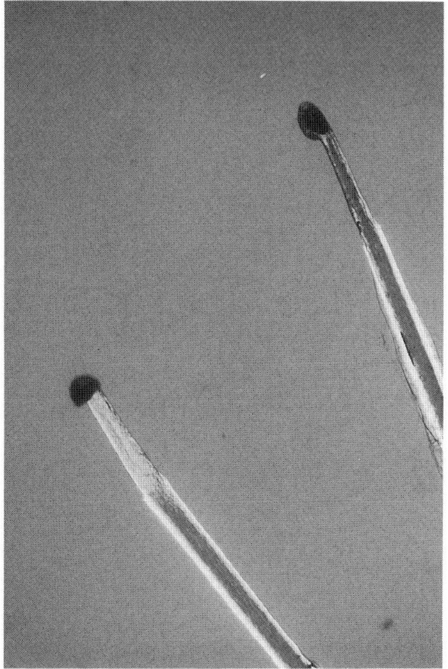

FIGURE 1–12. A light micrograph of plucked hair fibers in the telogen stage.

occurs in children who unconsciously pull or twist a region of hair. A similar type of hair loss may also occur in adults. Telogen effluvium is a term used to describe a sudden but diffuse hair loss caused by an acute physical or psychological stress. This condition usually lasts only a few months and is reversible. Drugs used in chemotherapy often induce alopecia; however, this type of hair loss is also usually reversible.

A Mechanism for Hair Growth/Hair Loss and Changes in Hair Size

The rate of the cyclic activity of the pilosebaceous unit or the rate of conversion from anagen to telogen stages determines whether there is hair growth or hair loss. This activity also controls the changes in hair size that occur during different stages of the life of mammals. At different ages of humans, therefore, such as birth, puberty, and maturity, hairs grow in different sizes (see Figure 1–8), and all these changes involve androgens.

More than 40 years ago, Hamilton [16] demonstrated that androgens are a factor in male pattern baldness. For example, long-term injections of testosterone induce a rapid transformation of terminal hairs to vellus hairs

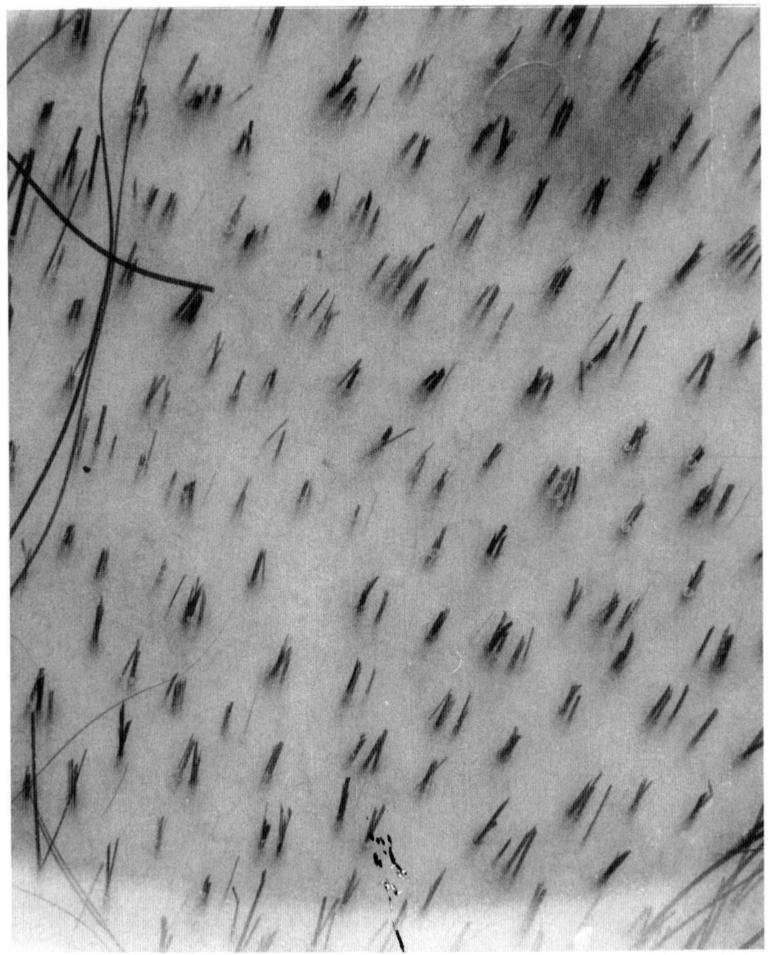

FIGURE 1–13 A, B. Enlarged photographs of the scalp. (A) Immediately after shaving. (B) Three days after shaving, illustrating the hairs that have grown (in anagen phase) and those that have not grown (telogen).

in the frontal scalp of stump-tailed macaques [14]. Thus, testosterone, an androgen produced by the adrenals and the sex glands, was shown to play a critical role in controlling the growth patterns of human scalp hair fibers.

Estrogens have been shown to provide positive effects on hair growth when taken internally or applied topically. Systemic estrogen probably prolongs the anagen phase of hair growth by suppressing androgen production [18]. However, both estrogens and antiandrogens when applied topically have also been shown to be capable of suppressing hair loss [19]. Antiandrogens, substances that are capable of blocking androgen function, in-

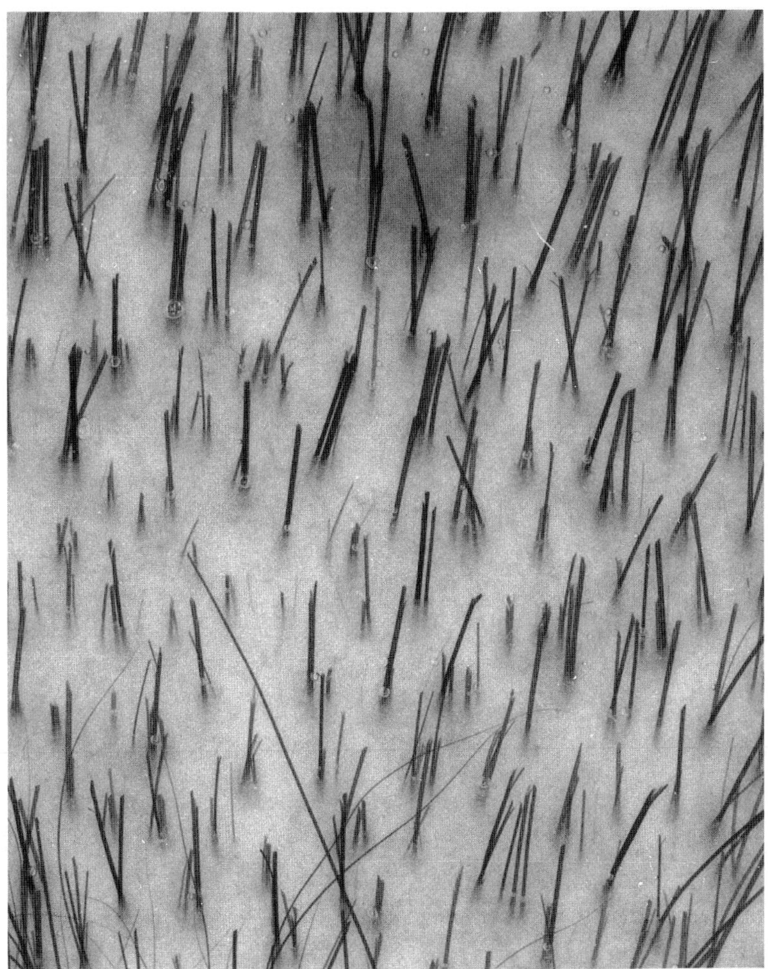

FIGURE 1–13. (*continued*)

clude spironolactone, cyproterone acetate, and progesterone (an inhibitor of 5α-reductase; Figure 1–14), an important enzyme in the conversion of androgens to the most active form of testosterone. The topical application of estrogens and antiandrogens probably causes a local inhibition of the androgen function and may ultimately provide the best solution to hair growth, as shown by the proposed mechanism.

Chemical cures for baldness and the search for a greater understanding of the mechanism of this phenomenon not only center around androgens but also involve drugs known to be capable of inducing hypertrichosis, such as streptomycin, minoxidil, estradiol, and oxandrolone. The latter three drugs have shown promise in reversing the symptoms of male pattern bald-

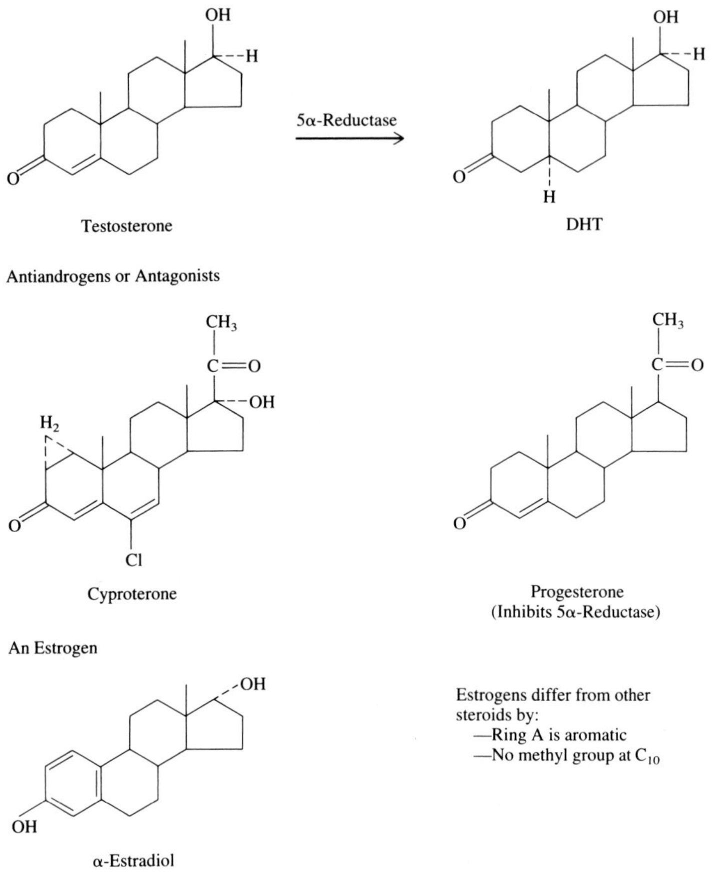

Testosterone

5α-Reductase →

DHT

Antiandrogens or Antagonists

Cyproterone

Progesterone
(Inhibits 5α-Reductase)

An Estrogen

α-Estradiol

Estrogens differ from other steroids by:
—Ring A is aromatic
—No methyl group at C_{10}

FIGURE 1–14. Chemical structures of the active androgens, testosterone and dihydrotestosterone (DHT), some examples of antiandrogens, and an estrogen.

ness. Minoxidil (6-amino-1,2-dihydro-1-hydroxy-imino-4-piperidino pyrimidine) and diazoxide have also demonstrated hypertrichosis, and the former vasodilator (minoxidil) has been shown to be capable of reversing male pattern alopecia in clinical trials during treatment periods. However, with minoxidil, best results are obtained under occlusion [13] and in subjects whose condition of balding has not progressed for an extended period of time. For a thorough review of baldness, current treatments, and regulatory issues, see Gerstein [17].

Normal androgen control of hair growth, that is, control of the anagen/telogen cycle and the subsequent alteration of hairs to different sizes, may be considered as occurring in five distinct stages or steps (see Figure 1–9).

Step 1: The synthesis or the production of androgens by the adrenals and the ovaries or the testes

Step 2: The transport of these hormones in the bloodstream on carrier proteins such as sex hormone binding globulin (SHBG) to peripheral tissues such as the pilosebaceous apparatus, and the subsequent dissociation from the binding proteins

Step 3: The conversion in the hair follicle to the more active hormone 5α-dihydrotestosterone (DHT)

Step 4: The transport of testosterone and DHT into hair cells and the binding of these steroids inside the cells to specific receptor proteins to form the active species involved in protein synthesis

Step 5: The inhibition or the promotion of protein synthesis in the nucleus that controls the hair cycle, and, finally, the metabolic degradation of the steroid and clearing of that species from the activated/inactivated hair cells.

Thus, any agent or process that either enhances or interferes with any of these five steps will lead to either less or greater production of longer, coarser hairs. Interference in the transport process of step 2 may result in terminal hairs produced where vellus hairs are normally produced. For example, the second step involving the transport of testosterone on carrier proteins is the effect seen in most hirsuite women [20]. What is observed is a reduction in transport proteins (SHBG) and the concomitant increase in the level of free unbound testosterone in the bloodstream. Thus, interference with the transport mechanism causes thick terminal hairs to be produced in body regions where they are not normally produced.

Transport of not only testosterone but also that of other androgens capable of being synthesized into testosterone is important, as suggested by Sawaya et al. [21] in showing that the enzyme 3β-hydroxysteroid dehydrogenase, which converts other androgens into testosterone (Figure 1–15), shows greater activity in samples of balding scalp as compared to normal hairy scalp. In addition, balding men show increased activity of the enzyme 5α-reductase in the pilosebaceous units and in the skin of the frontal scalp, although men with a deficiency of this enzyme do not develop baldness [22].

After testosterone has been transported to the pilosebaceous unit or synthesized in the vicinity of the bulb or the sebaceous gland [23], it is converted into its active form, DHT, by the enzyme 5α-reductase (step 3; see also Figure 1–9). Hair root cells contain androgen receptors, but the latest thinking is that these receptors are intranuclear rather than intracellular [24,25]. In addition, Sawaya et al. [26] have shown a greater androgen-binding capacity (i.e., DHT) in the nuclei of sebaceous glands taken from patients with bald scalps than from patients with normal hairy scalps. Thus, for balding persons DHT migrates into hair cells in the bulb and binds to a specific DNA receptor to form the active species that influences the hair cycle, that is, the form that decreases protein synthesis and shortens the anagen period.

Another important finding is the fact that pilosebaceous units that grow

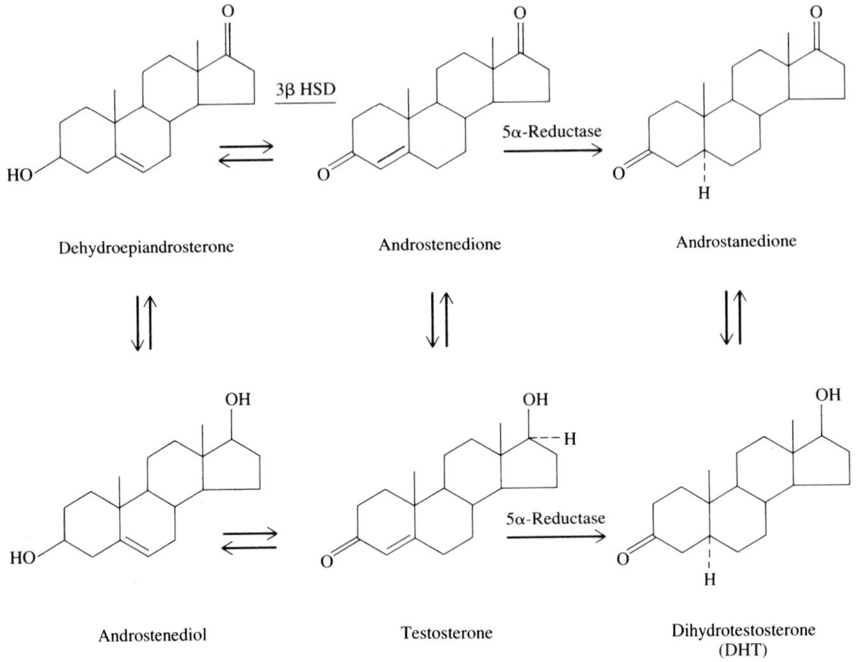

3β HSD

5α-Reductase

Dehydroepiandrosterone

Androstenedione

Androstanedione

5α-Reductase

Androstenediol

Testosterone

Dihydrotestosterone (DHT)

FIGURE 1–15. The conversion of androgens to testosterone and DHT.

thick terminal hairs when surgically transplanted to a region that is hairless will continue to produce thick terminal hairs [27]. Furthermore, in some cases thick terminal hairs will begin to grow, sometimes in isolation, in regions that are normally hairless such as areas of facial hair in women. In other words, the response to the androgens are dependent on the specific pilosebaceous unit, which is often but not always regionally dependent. These findings suggest that specific pilosebaceous units, depending on body region or stage of life, may be programmed by means of different receptor proteins to respond to androgens in a way that either induce baldness or grow hair.

These conclusions, particularly with regard to different receptor proteins for stimulating or retarding hair growth, could explain the fact that in males thick terminal hairs begin to grow in the axilla, the mons pubis, and the beard areas at puberty in spite of increased levels of testosterone, and yet at a later time in life increased levels of testosterone in the scalp help to cause male pattern baldness. Hamilton [28,29] has shown that when injected intramuscularly with testosterone propionate normal persons exhibit an increased growth of coarse sternal hairs although eunuchs castrated before age 20 when similarly injected show even less growth of beard hair than eunuchs castrated after age 21 [29].

TABLE 1–3. Growth of male beard hair.

Group examined	Average weight of beard hair (mg/24 h) for ages 30–80 years
Normal control persons	32
Castrated after age 21	14
Castrated before age 20	8

The foregoing situation of testosterone injection of normal persons suggests that this androgen can somehow induce or promote hair growth in the skin of adults in the thorax tissues, as opposed to the effect in the scalp where this same hormone inhibits hair growth. The data from eunuchs suggest that not only will a decreased testosterone level in some body regions (e.g., the beard area in males) decrease hair growth, but further that if two of the glands that produce this hormone are removed before maturation of the local tissue responsible for hair growth, then hair growth will be further inhibited in that tissue. In other words, hair growth is dependent on the local tissue as well as on the androgen level.

Once again, we see evidence for the programing of specific pilosebaceous units to respond to androgens to either induce baldness or to grow hair, and also a specific local tissue response. These different hair growth effects could occur through differences in the receptor proteins that are produced by the local tissue immediately surrounding the bulb, and by the effect of that active species, which is a combination of androgen and receptor protein on the synthesis of proteins by DNA (see Figure 1–9).

To summarize hair growth, human scalp hair grows at an average rate of approximately 6 in. per year. The life cycle of a hair fiber consists of three stages—anagen (growth stage), catagen (transition stage), and telogen (resting stage when the hair is shed)—and this life cycle is partially controlled by androgens and the local tissue. Testosterone and DHT are the primary androgens that determine not only whether hairs increase or decrease in size with age but also other aspects of hair growth and hair loss; however, the response by the local tissue tends to be a regional response and determines whether the androgen induces hair growth or shortens the hair cycle leading to baldness (see Figure 1–9).

Differences in anagen can vary from a few months to 8 years. For normal terminal scalp hairs, 3 to 6 years of anagen is an average growth time, producing hairs approximately 1 m long (~3 ft) before shedding. Human hair generally grows in a mosaic pattern; thus, in any given area of the scalp, one finds hairs in various stages of their life cycle. In a normal healthy scalp, most hairs are in anagen (about 80%–90%), although there are seasonal changes in hair growth, with maximum shedding (telogen) in August and September. In male pattern baldness (alopecia hereditaria) and in all forms of hair loss, there is a more rapid turnover to telogen; thus, a greater percentage of hairs are in telogen. In addition, baldness is characterized by vellus hairs more than by hair loss, although a small reduction in the number of follicles per unit area does occur. For additional details regarding the biological synthesis and formation of human hair, see the references by Barth [20], Sawaya [21,26], Gerstein [17], Montagna and Ellis [30], and Mercer [31], and the July supplement to the *Journal of Investigative Dermatology* (1993).

Surgical Treatment of Hair Loss

Several surgical procedures have been used for treatment of hair loss. Although these procedures may be used for most forms of alopecia, they are used primarily for treatment of androgenetic alopecia or even for hair loss from tissue injury such as burns, particularly in cases where extensive baldness exists. These procedures are based on the fact that hairs growing actively in one region of the scalp, such as the occipital region, will continue to grow actively when moved to a bald region, confirming the role of the local tissue in the hair growth process. Those treatments currently used include:

Hair transplantation
Scalp reduction
Transposition flap
Soft tissue expansion

In the most common form of hair transplantation, small skin plugs containing 15 to 20 growing terminal hairs each are surgically removed and placed into a smaller cylindrical hole in the balding region of the scalp. Usually several sessions of transplantation are required, involving the placement of 50 or more plugs per session, and the placement or angling of the plugs is important to the end cosmetic effect. Elliptical grafts or even smaller minigrafts may be employed [32]. Within 2 to 4 weeks after transplantation, the donor hairs usually fall out but are replaced by new hairs, generally within 3 months.

Oftentimes, in cases where the bald area is rather large, scalp reduction is done in conjunction with hair transplantation. This method involves surgical excision of a strip of the bald skin to reduce the total hairless area. Repeated scalp reductions can be performed together with transplantation to provide better coverage for a very bald person.

The transposition flap method [33] involves moving a flap of skin that contains a dense area of hair to a bald area. This method is sometimes employed together with minigraft implantation along the frontal hairline to provide a cosmetic effect that is more natural in appearance.

Soft tissue expansion is a more recent development for treatment of alopecia. In this procedure, soft silicone bags are inserted under the skin in the hair-bearing area of the scalp, usually in the region of the occipital area of the scalp. The bags are then slowly filled with salt water during a 2- to 4-month period. After expansion of the hair-bearing skin, the bags are removed, the bald area of the scalp is excised, and flaps are created with the expanded hair-bearing skin.

The Cuticle

The cuticle is a chemically resistant region surrounding the cortex in animal hair fibers, see (see Figures 1–3, 1–5, and 1–6). Its chemical resistance is illustrated via an experiment by Geiger [34]. When isolated cuticle material and whole wool fiber are completely reduced and alkylated, the alkali solubility [35] of the cuticle material is approximately one-half that of whole fiber (85%). Cuticle cells are generally isolated from keratin fibers by shaking in formic acid [36,37], by enzymatic digestion [34,38,39], or by shaking in water [40].

The cuticle consists of flat overlapping cells (scales) (Figures 1–6, 1–16, and 1–17. The cuticle cells are attached at the proximal end (root end), and they point toward the distal end (tip end) of the hair fiber, like shingles on a roof. The shape and orientation of the cuticle cells are responsible for the differential friction effect in hair (see Chapter 8). Each cuticle cell is approximately 0.5 to 1.0 μm thick and approximately 45 μm long. The cuticle in human hair is generally 5 to 10 scales thick [41,42], whereas in different wool fibers the cuticle is only 1 to 2 scales thick [43]. The number of scale layers can serve as a clue to the species of origin in forensic studies.

The cuticle of human hair contains smooth unbroken scale edges at the root end near the scalp (see Figure 1–17). However, cuticle damage evidenced by broken scale edges that can usually be observed several centimeters away from the scalp is caused by weathering and mechanical damage from the effects of normal grooming actions, such as combing, brushing, and shampooing (Figures 1–17 and 1–18). In certain long hair fibers (25 cm or longer), progressive surface damage may be observed (illustrated by Figure 1–18). Stage 1 shows intact smooth scale edges and scale surfaces; stage 2 contains broken scale edges; in stage 3, the scales have been partially removed, and in stage 4 the hair splits indicating extensive cortical damage. Garcia et al. have described this phenomenon of hair degradation in some detail [44]; however, see Chapter 5 for additional information and references.

The cuticle of both human hair and wool fiber has been shown to contain a higher percentage of cystine than whole fiber [45] and more of the other amino acids that are generally not found in α-helical polypeptides [46]. Analysis of the cuticle of wool fiber by the polarizing microscope shows negligible birefringence [1]. The cuticle of human hair also demon-

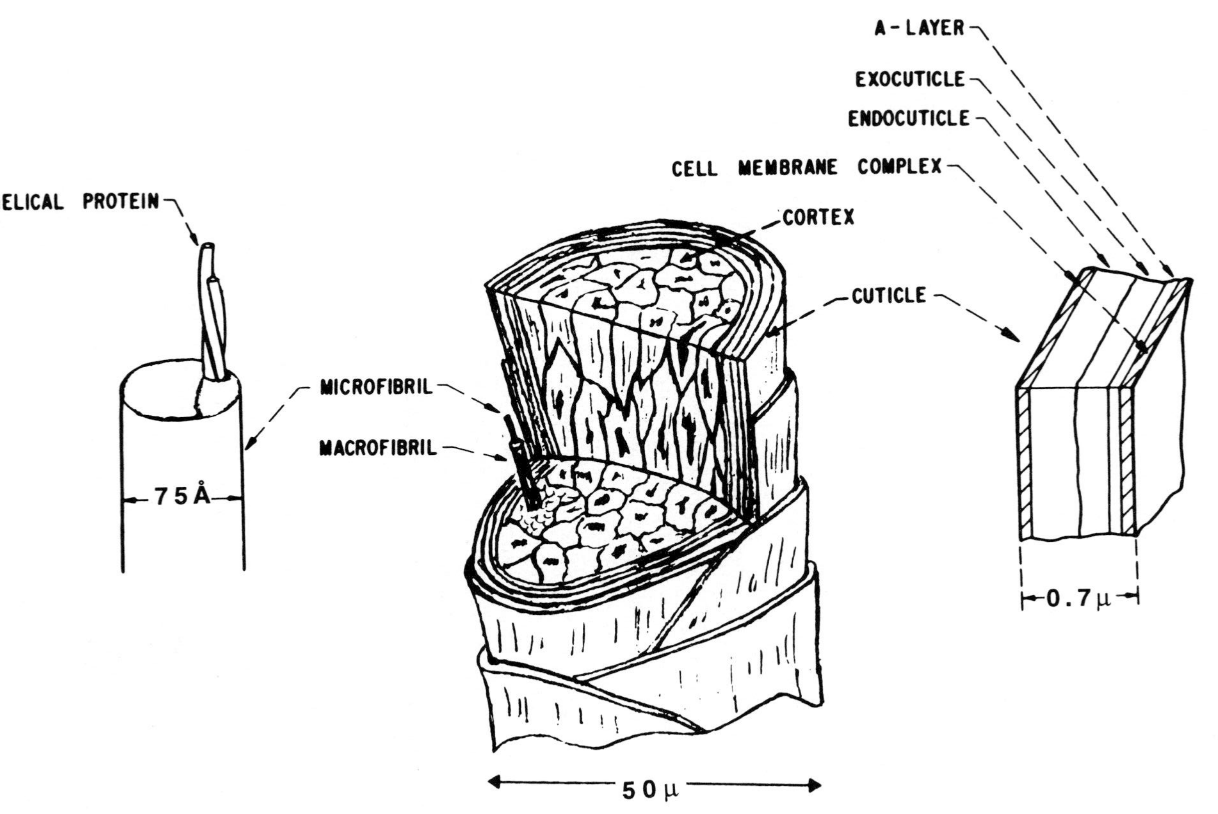

FIGURE 1–16. Stereogram of the hair fiber structure illustrates substructures of the cuticle and the cortex.

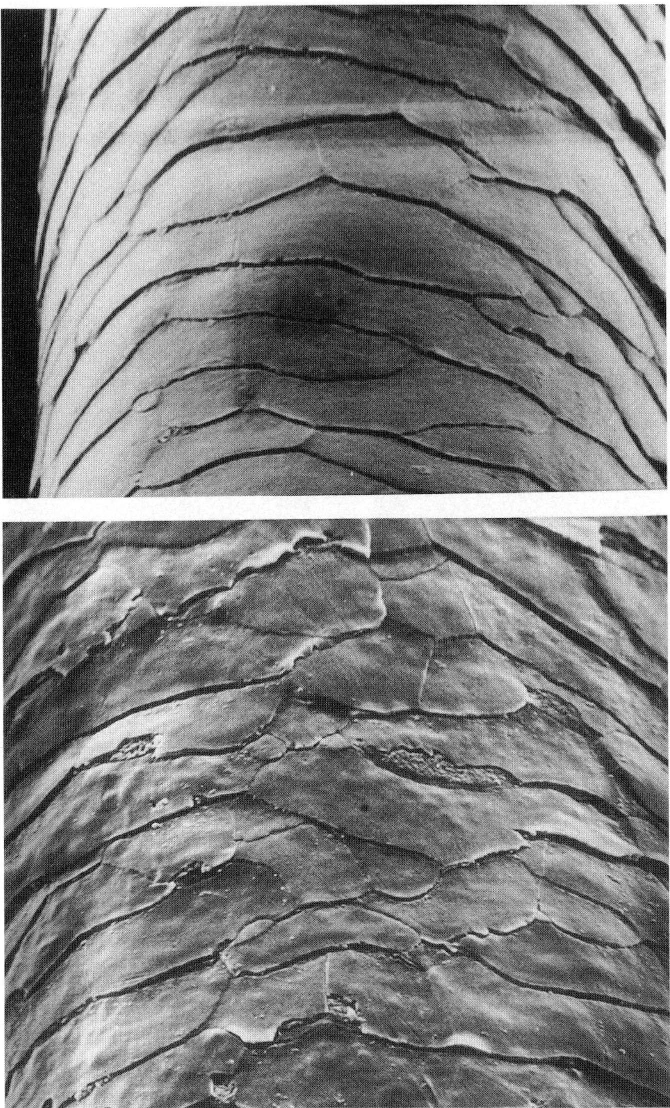

FIGURE 1–17. Cuticle scales of human hair. Top: Near root end. Note smooth scale edges. Bottom: Near tip end of another fiber. Note worn and broken scale edges.

strates negligible birefringence. Astbury and Street [47] have provided X-ray evidence confirming that, in contrast to the cortex, the cuticle of hairs is not a highly organized system at the molecular level.

The schematic diagram of Figure 1–19 illustrates the structure of cuticle cells. Each cuticle cell contains a thin outer membrane, the epicuticle. Different estimates of the thickness of this membrane have been cited (25–

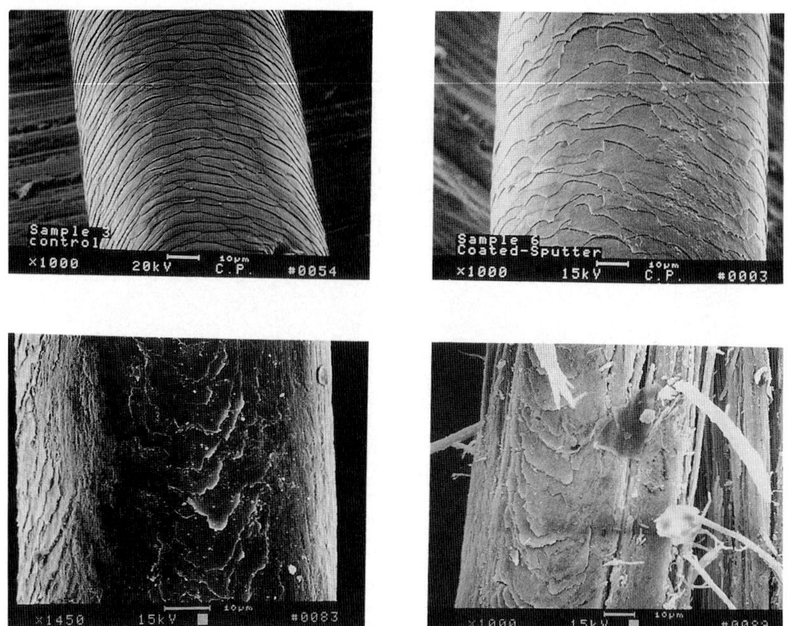

FIGURE 1–18. Top left: Stage 1; note smooth cuticle edges. Top right: Stage 2; note broken cuticle scale edges. Bottom left: Stage 3; note complete removal of cuticle (central area). Bottom right: Stage 4; split hair.

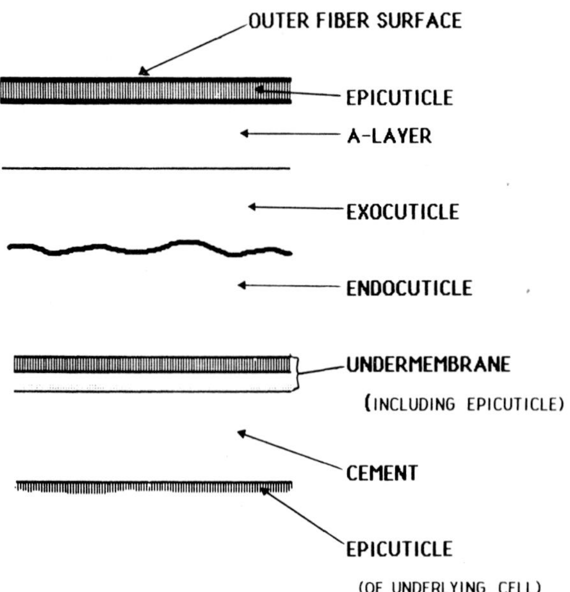

FIGURE 1–19. Schematic diagram of the proposed structure of a cuticle cell in cross section.

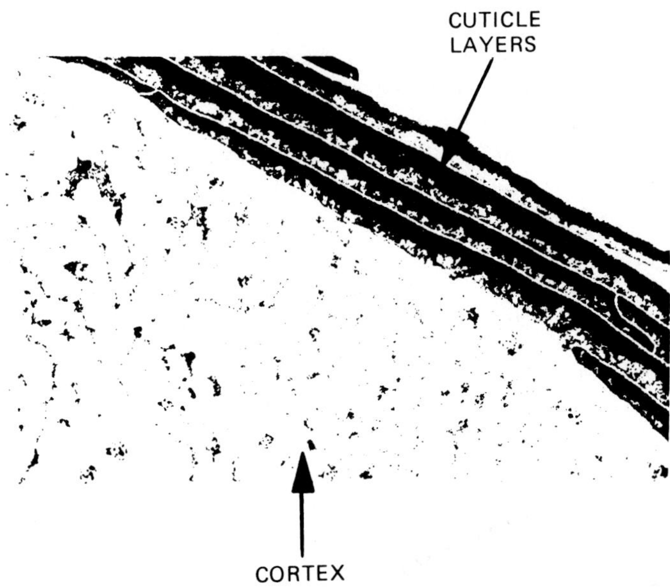

FIGURE 1–20. Transmission electron micrograph (TEM) of a cross section of a hair fiber treated with silver methenamine illustrates high and low sulfur layers of cuticle cells (stained, high-sulfur regions).

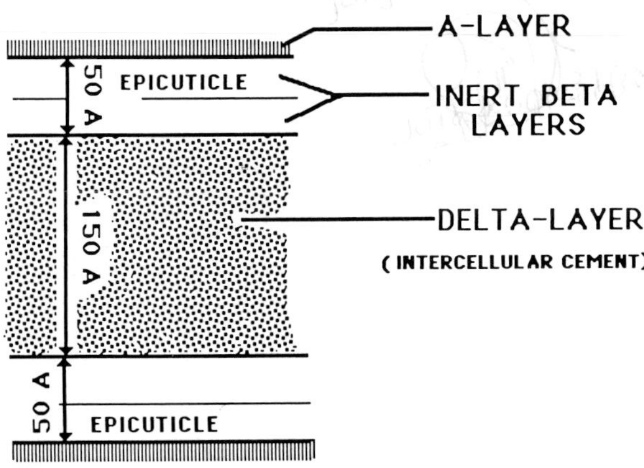

FIGURE 1–21. Schematic illustrating cell membrane complex in animal hairs. (From Fraser et al. [41].)

250 Å; however, 25 Å is probably the most common estimate [48,49]). Beneath the cuticle cell membranes are three major layers: the A layer, a resistant layer with a high cystine content (>30%); the exocuticle, sometimes called the B layer, also rich in cystine (~15%); and the endocuticle, low in cystine content (~3%) [100] (Figure 1–20). A portion of the undermembrane of Figure 1–19 is also epicuticle. For more details of the

intercellular structures, see Figure 1–21. Thus, the cuticle of human hair is a laminar structure similar to the cuticle of wool fiber, and the different layers of the cuticle have been described for merino wool [50] and for human hair [44,51–53]. See Chapter 2 for a more complete description of the amino acid composition of the cuticle and its different component parts.

Epicuticle

The surface of mammalian hairs is covered with a thin material called epicuticle [54] that is approximately 25 Å thick [48,49] (see Figure 1–19). The epicuticle was first observed in wool fibers by Allworden [55], who noted that sacs or bubbles form at the surface of the fibers during treatment with chlorine water (Figure 1–22). Allworden sacs form as a result of diffusion of chlorine water into cuticle cells and the ensuing reactions. Chlorine water degrades proteins beneath the epicuticle, producing water-soluble species too bulky to diffuse out of the semipermeable membrane. Swelling then results from osmotic forces, producing the characteristic Allworden sacs [56].

It was suggested at one time that the epicuticle is a continuous membrane covering the entire fiber [57]. However, Leeder and Bradbury [36] isolated single cuticle cells from wool fiber and demonstrated that single cuticle cells undergo the Allworden reaction, thus proving that a membrane surrounds each cuticle scale. What has been described as a continuous epicuticle may be a cell membrane complex, consisting of epicuticle cell membrane complex and intercellular binding material that could produce the appearance of a continuous sheath.

Leeder et al. [58] have defined the epicuticle as a chemically resistant proteinaceous membrane that remains on keratin fiber surfaces after strongly bound lipids have been removed with potassium *t*-butoxide in anhydrous butanol. Thus, the epicuticle is a proteinaceous layer about 25 Å thick covered by strongly bound structural lipid, which Leeder calls the F layer. The F layer represents the outermost fiber surface of Figure 1–19. After removal of these surface lipids, wool fiber still undergoes the Allworden reaction [59], confirming that these surface lipids are only a portion of the cell membranes of the cuticle.

Swift and Holmes [49] agree that the epicuticle of hair contains both lipid and fibrous protein layers, is approximately 25 Å thick, and is cell membrane material but does not have sufficient contrast with its surroundings to allow microscopic identification. Leeder et al. [60] have provided evidence that the cell membrane lipids of wool fiber do not consist of phospholipids that normally form bilayers in living tissue. Leeder recommends that further research is necessary to clarify the structure of keratin cell membranes.

Of several methods described for isolation of epicuticle, one by Langer-

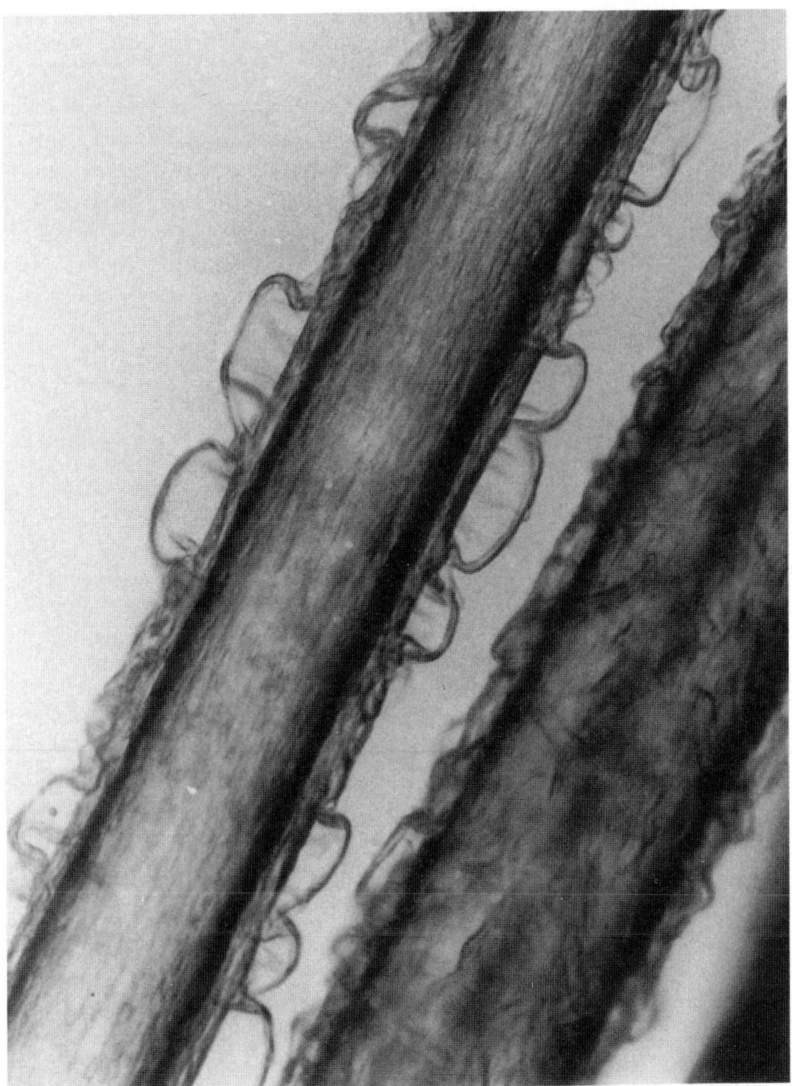

FIGURE 1–22. Allworden sacs formed at the surface of hair fibers during reaction with chlorine water.

malm and Philip [48] involves dissolving the bulk of the fiber from the membrane material with dilute sodium sulfide. Another method, that of Lindberg et al. [57], is treatment of intact fibers with chlorine water or bromine water followed by neutralization and shaking. Neither of these procedures produces pure epicuticle, but they probably provide a portion of the cell membranes, including part of or all the epicuticle.

Swift and Holmes [49] have described a relatively nondestructive method involving extraction with hot ethanol for removing epicuticle from human hair fibers. Hair fibers extracted extensively with hot ethanol are less resistant to enzymatic degradation than ether-extracted hair and do not undergo the characteristic Allworden reaction with chlorine water. It has been suggested therefore that extraction of hair with hot ethanol removes either a portion of or the entire epicuticle. Chemical analysis of epicuticle substance removed from hair by hot ethanol extraction indicates both protein and fatty acids (20–30%) [61,62]. Qualitatively similar results have been reported by Leeder and Bradbury for analysis of epicuticle isolated from merino wool [36].

The Cortex

The cortex constitutes the major part of the fiber mass of human hair and consists of cells and intercellular binding material. The intercellular binding material, or the cell membrane complex, is described later in this chapter.

Cortical Cells

Cortical cells are generally 1 to 6 μm thick and approximately 100 μm long [3] (see Figures 1–16, 1–23, and 1–24), although considerable variation in their size and shape has been reported [63]. Figures 1–23 and 1–24 are scanning electron migrographs (SEMs) of split hairs with separated cortical cells appearing like splintered wood. Human hair contains a symmetrical cortex, unlike wool fiber, and most of the cells are of the same general type with regard to the ratio of fibrillar to nonfibrillar matter (highly crystalline, fibrillar; less organized, nonfibrillar).

Most wool fibers contain two or even three types of cortical cells that are sometimes segregated into distinct regions (Figure 1–25) which can be readily observed in cross section [64]. These cell types are called orthocortex, paracortex, and mesocortex. Orthocortical cells contain less matrix between the intermediate filaments and a lower sulfur content (~3%); paracortical cells are smaller in diameter, and they have smooth and rounded borders and a high sulfur content (~5%) [65].

Mesocortical cells contain an intermediate cystine content [66]. Morphologically, the cortical cells of human scalp hair are similar and resemble orthocortical cells of wool more closely, but they contain a high sulfur content. Kassenbeck [65], however, has shown that cortical cells adjacent to the cuticle are more flat and contain a lower sulfur content than the remaining cortical cells that comprise the bulk of the cortex; he calls these heterotype cortical cells. The morphology of human hair cortical cells therefore may depend on the sulfur content but probably depends as well

FIGURE 1–23. Scanning electron micrograph (SEM) of a split hair. Note cortical cell fragments.

on other functional groups of these cells that determine the state of swelling at the time hardening occurs in the zone of keratinization.

Kassenbeck [65] suggests that the biological function of crimped animal hairs is to trap large volumes of air in the hair coat to provide thermal insulation. Some animals have both summer and winter fur.

Summer fur begins to grow rapidly in the spring, producing long and coarse hairs that are less crimped to inhibit the formation of air pockets and to permit cooling. Winter fur begins to grow in the autumn, yielding short, stiff, crimped hairs to trap large volumes of air in the coat for thermal insulation. Perhaps the seasonal effect on anagen/telogen ratios for human scalp hair is related to the summer/winter effects on hair growth in fur-bearing animals. Kassenbeck [65] further explains that the growth rate of animal hair and the morphological structures of both the cuticle and cortex are relevant to the hair shape and to the cooling and insulation functions.

Cortical cells also contain pigment granules and nuclear remnants. The

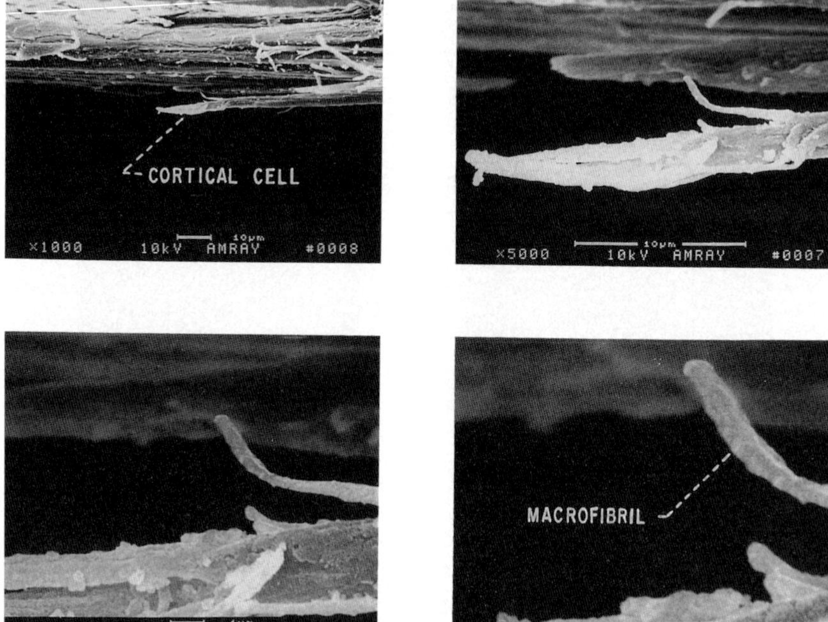

FIGURE 1–24. Scanning electron micrograph (SEM) of a split hair shows details of cortical structure.

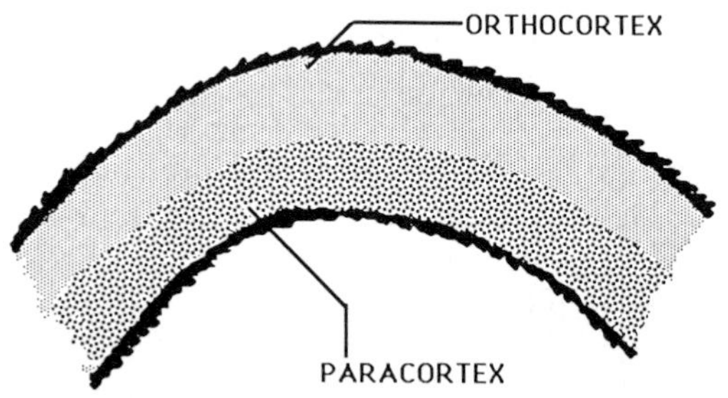

FIGURE 1–25. Schematic of a wool fiber illustrates orthocortex and paracortex regions of the cortex in relationship to crimp.

nuclear remnants are small, elongated cavities near the center of the cells. The pigment granules are small, oval, or spherical particles of approximately 2,000 to 8,000 Å (0.2–0.8 μm) in diameter [67] that are dispersed throughout the cortical cells. Both these structures constitute only a small fraction of the cortex. Pigment granules generally do not occur in the cuticle of scalp hair; however, pigment granules have been observed in

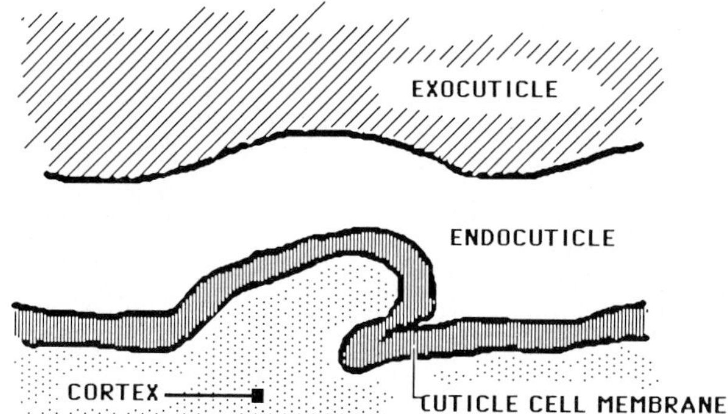

FIGURE 1–26. Schematic illustrates Piper's interlocking scheme for linking cuticle to cortex.

the cuticle and the medulla of beard hair, especially in heavily pigmented hair [68].

Birbeck and Mercer [69] suggested that pigment granules enter the cortical cells by a phagocytosis mechanism in the zone of differentiation and biological synthesis. Piper [70] has presented evidence that cortical cells are linked to adjacent cuticle cells via complex interlocking structures by means of a mechanism involving phagocytosis (Figure 1–26).

Cortical cells may be isolated from human hair by procedures involving either shaking in formic acid [36,37] or enzymatic digestion [34,38,39]. Another procedure involves shaking hair fibers in water to strip the cuticle cells from the hair, providing cortex with intact cell membranes free of cuticle [40]. In addition to nuclear remnants and pigment granules, the cortical cells of human hair contain highly important spindle-shaped fibrous structures called macrofibrils or macrofilaments (see Figures 1–16 and 1–24).

Macrofibrils

The spindle-shaped macrofibrils in human hair are approximately 0.1 to 0.4 μm in width or diameter [3] and comprise a major portion of the cortical cells (see Figures 1–24 and 1–27). Each macrofibril consists of intermediate filaments originally called microfibrils (highly organized fibrillar units) and the matrix, a less organized structure that surrounds the intermediate filaments (Figure 1–28).

Matrix

Various estimates of the relative quantities of matrix to intermediate filament protein (amorphous to crystalline proteins) have been made for

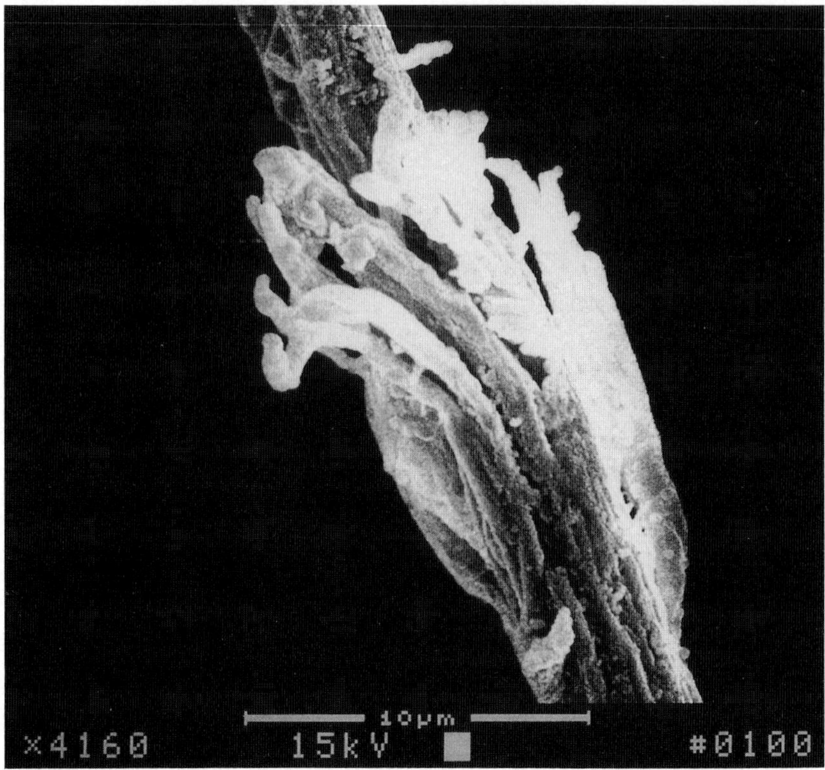

FIGURE 1–27. Scanning electron micrograph (SEM) of a cluster of macrofibrils in a cortical cell of a human hair fiber from a split hair.

both wool fiber and human hair [71,72]. Although the relative quantities vary [73], the matrix/intermediate filament ratio in human hair is generally greater than 1.0. Protein derived primarily from matrix (gamma keratose) can be isolated from keratin fibers by the method of Alexander and Earland [74]. This method involves oxidation of hair using peracetic acid. Analysis of the gamma keratose from human hair indicates a higher proportion of sulfur compared to the other keratose fractions or to whole fiber [31]. Corfield et al. [75] have isolated matrix material from merino wool by this procedure. Chemical analysis shows a relatively high proportion of sulfur and a correspondingly greater proportion of cystine compared to the other fractions or to whole fiber [31].

Electron microscopy takes advantage of the high cystine content of matrix to reveal the fine structure of hair in the following manner. Reduction of the fibers followed by treatment with osmium tetroxide before sectioning produces a heavily stained matrix that reveals the relatively unstained intermediate filaments [76].

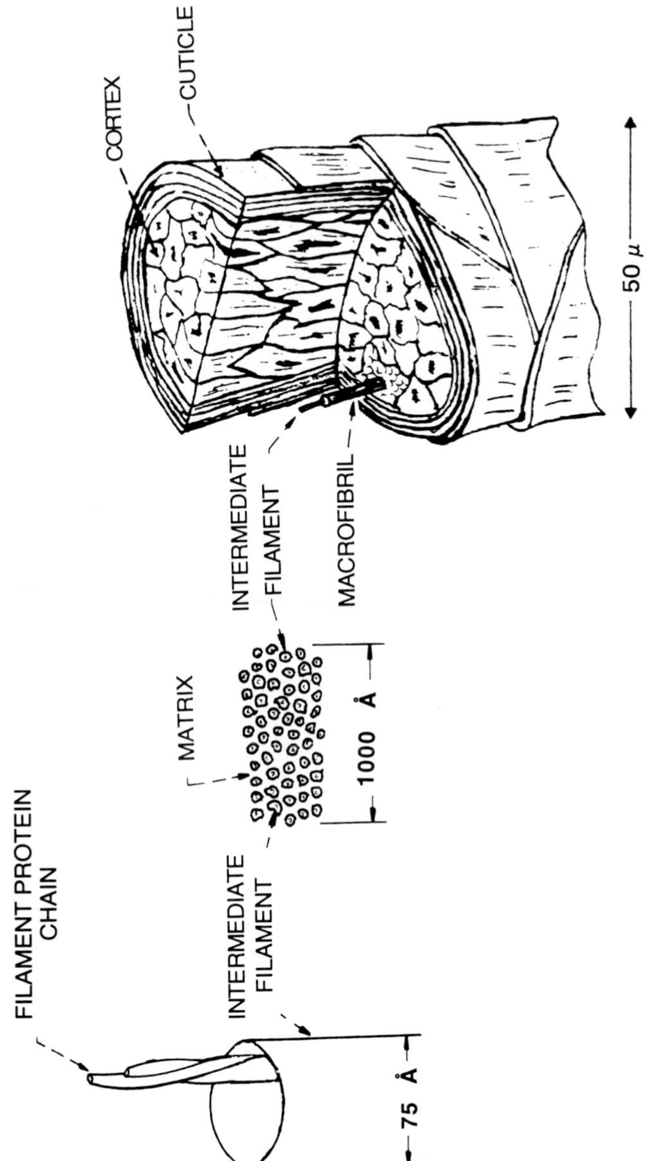

FIGURE 1–28. Stereogram of a human hair fiber includes intermediate filament matrix structures.

Matrix comprises the largest structural subunit of the cortex of human hair fibers. It contains the highest concentration of disulfide bonds, most of which are probably intrachain bonds rather than interchain bonds because the matrix swells considerably when wet with water and mechanically resembles a lightly cross-linked gel [80] rather than a highly cross-linked polymer. The matrix is often referred to as the amorphous region, although evidence suggests that it does contain some degree of structural organization [77,78]. A spacing of 28 Å has been demonstrated in mohair fiber, and Spei has attributed this spacing to structural repeat units of the matrix [79]. Some proteins of the matrix are sometimes referred to in sequence studies as the interfibrillar associated proteins (IFAP)

The Intermediate Filaments or the Microfibrils

As indicated, the macrofibrils in human hair contain subfilamentous structures called intermediate filaments (IF) or microfibrils (microfilaments), arranged in spiral formation in the cortical cells. The radius of each spiral, the macrofibril, is approximately 4,000 Å units [81], and the width or diameter of an intermediate filament is close to 75 Å (see Figure 1–16).

The exact organization within the intermediate filaments is still being determined, although several structures have been proposed for this important structural unit [82,83]. Crewther et al. [82] concluded that the intermediate filaments contain precise arrays of the low-sulfur proteins, containing short sections of α-helical proteins in coiled coil formation, showing a heptad repeat unit. The coiled coils are interrupted at three positions by nonhelical fragments and are terminated by nonhelical domains at both the nitrogen (N) and carbon (C) termini of the chain (Figure 1–29).

The amino acid sequences of the helical sections of the intermediate filaments are similar to those of related structures identified from skin and other tissues. The filamentous polypeptides, often classified as Type I and Type II, differ by their amino acid sequences resulting in acidic (Type I) and neutral to basic (Type II) proteins. The individual filament-like protein chains of Figure 1–29 are arranged into coiled coil dimers rather than the three-strand rope model originally proposed by Crick. These dimers then aggregate in an antiparallel arrangement to form structural units composed of four protein chains or tetramers [83]. Seven to 10 of these tetramer units are then believed to combine or aggregate into a larger helical structure, forming the intermediate filaments (the microfibril structures) of animal hairs.

The cystine content of the low-sulfur intermediate filament regions is about 6% and is not uniformly dispersed between domains of an intermediate filament chain. The rod domain contains about only 3% half-cystine or as little as 1 half-cystine residue, while the N-terminal domain contains about 11% half-cystine and the C-terminal unit about 17% half-cystine

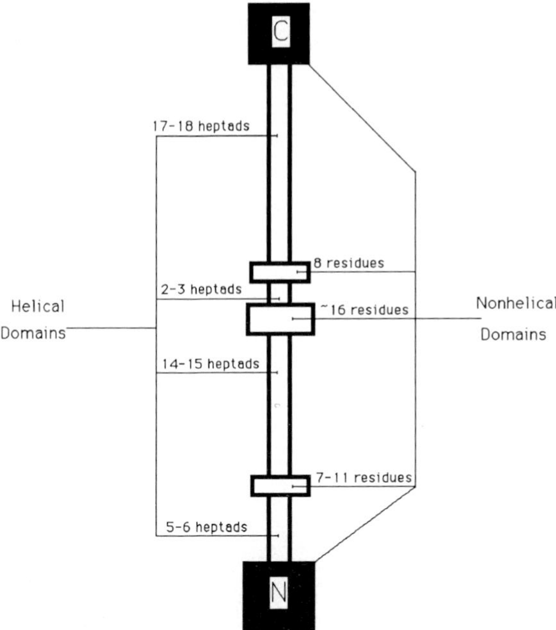

FIGURE 1–29. Schematic illustrates the structure of an intermediate filament protein chain.

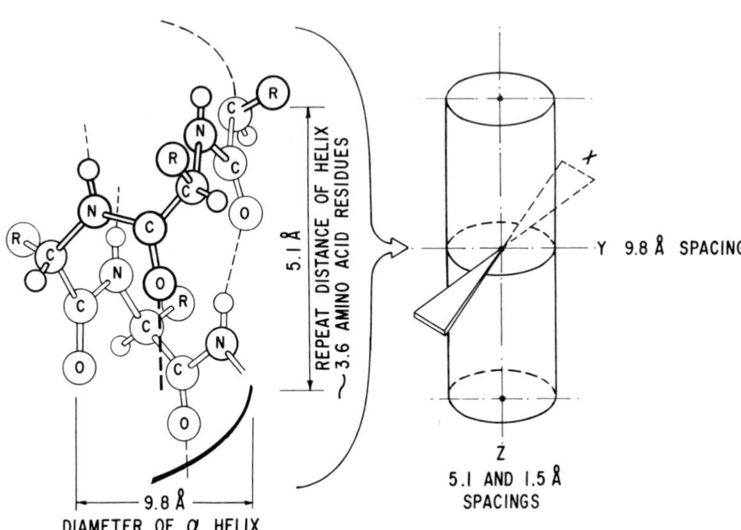

FIGURE 1–30. Structure of an α-helix proposed by Pauling and Corey.

[95]. Fraser also suggests that these half-cystine residues are involved in disulfide linkages and that most disulfide residues exist in the matrix.

The Type I intermediate filament proteins of human hair represent a class of proteins about 44 and 46 K in molecular weight, while the Type II proteins are about 50, 59, and 60 K [106]. Amino acid sequences for intermediate filament polypeptides from several proteins including wool fiber have been described by Crewther et al. [82].

Helical Structures of the Intermediate Filaments

The subunits that constitute the intermediate filaments of hair fibers are polypeptide chains of proteins (see Figure 1–29). The coiled sections or the helical domains of these protein chains are approximately 10 Å in

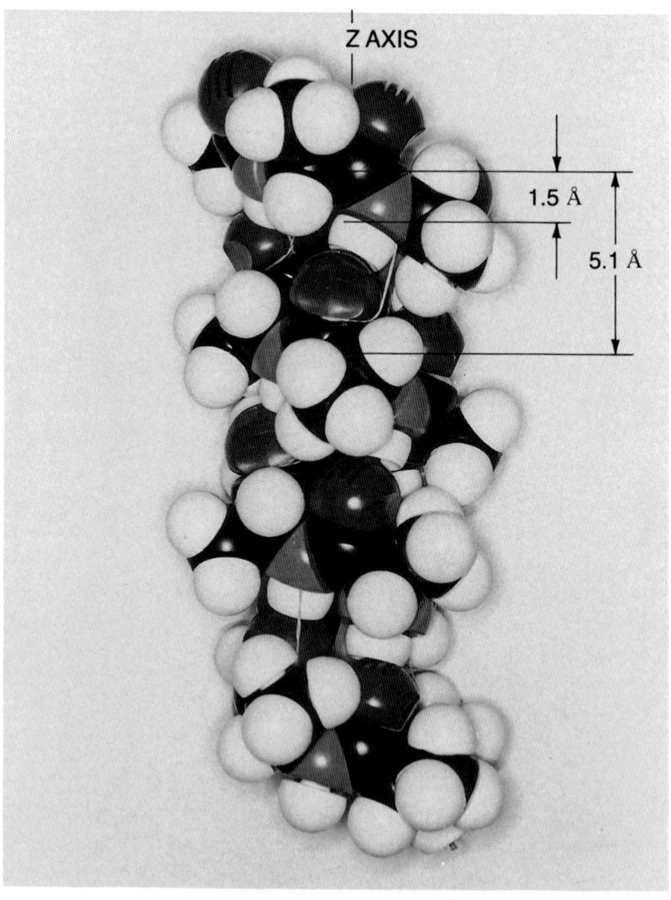

FIGURE 1–31. Molecular model of a left-hand helix of polyalanine. A right-hand helix (spiraling in the other direction) is the pattern found in most proteins, including animal hairs (see FIGURE 1–30).

diameter, including side chains, and are believed to approximate the form of an α-helix, as first proposed by Pauling and Corey [85–87] (Figures 1–30 and 1–31).

The α-helix of Pauling and Corey was proposed from the X-ray diffraction analysis of keratin fibers pioneered by Astbury et al. [88–90] and MacArthur [91,92]. Wide-angle X-ray diffractions (up to approximately 15 Å repeating units) of unstretched human hair and other keratin fibers (wool and porcupine quill) show several spacings, among which are an equatorial spacing (perpendicular to the fiber axis) of 9.8 Å and meridional spacings (parallel to the fiber axis) of 5.1 and 1.5 Å (see Figures 1–30 and 1–31).

Pauling and Corey interpreted the 1.5-Å spacing to represent the distance between each amino acid residue, the 5.1-Å spacing to represent the repeat distance for coiling, corresponding to 3.6 amino acid residues, and the 9.8-Å spacing to represent the center-to-center distance between each α-helix, approximating the thickness of the α-helix. A linear polypeptide alpha helix would have a repeat distance of 5.4 Å units, so coiling of each helix [93] was proposed to account for the shorter 5.1-Å meridional spacing. Further, it was originally suggested that two or three strands of polypeptides were coiled about each other analogous to a twisted rope [94–96]. This structure has been routinely referred to as the "coiled coil" model. The model that is now accepted for animal hairs is the two-strand rope polypeptide.

Stretching Hair

Stretching hair can produce splits or cracks in the endocuticle and transverse cracks in the cuticle layers as well as damage to the cortex; neverthe-

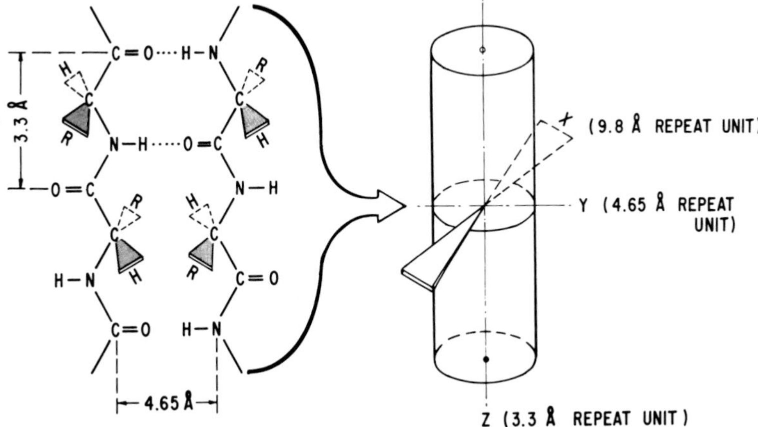

FIGURE 1–32. Portions of two polypeptide chains in the beta configuration. The cylinder represents a hair fiber and the axis identifies the orientation of the proteins

less, most of the scientific attention relating to stretching hair has been concerned with cortical effects, (see Chapter 8 for details). Water produces negligible effects to the wide-angle X-ray diagram of keratin fibers [89]. However, extension in water diminishes the intensities of the reflections corresponding to the α-helix and produces a pattern called β-keratin: a 3.3-Å reflection along the fiber axis (the Z axis in Figure 1–32), a 4.65-Å reflection at right angles to the Z axis (along the Y axis), and a 9.8-Å reflection at right angles to the Z axis (along the X axis) [86]. The interpretation of these reflections, in terms of molecular structure, is described by the molecular model of Figure 1–33.

Most explanations of this phenomenon invoke an alpha to beta transformation, the transformation of molecules of the α-helical structure into the pleated sheet arrangement of the beta structure.

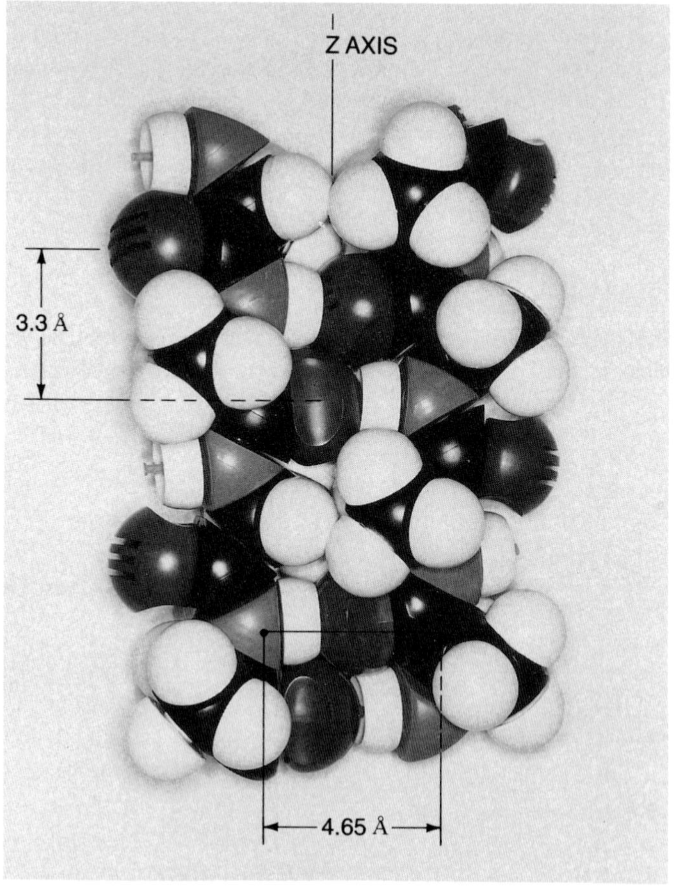

FIGURE 1–33. Molecular model of two polyalanine chains in the beta configuration.

To refine our understanding of the mechanical properties of keratin fibers, models involving the intermediate filament matrix level of organization have been employed. Feughelman [97] has suggested a two-phase model consisting of water-impenetrable rods (intermediate filaments) oriented parallel with the fiber axis embedded in a water-penetrable matrix. This two-phase model is useful for helping to explain the mechanical properties of keratin fibers including extension, bending, and torsional properties, and also the swelling behavior of keratins.

For the initial extension of hair, involving the Hookean region of Figure 8–1 (see Chapter 8), the α-helices of the intermediate filaments are strained, and the hydrogen bonds of the globular proteins of the matrix are also involved. On further extension into the yield and postyield regions, the alpha to beta transformation occurs in the intermediate filaments producing a loss of helical structure, which is recovered on relaxation. The globular proteins of the matrix act in parallel with the intermediate filaments. The matrix phase is weakened by the presence of water. On the other hand, the crystalline regions of the intermediate filaments are virtually inert to water over the entire load-extension curve. For additional details of the extension behavior of keratin fibers see Chapter 8, and for effects of extension on the cuticle see Chapters 5 and 8.

Swelling Behavior of Hair

The level of structural organization believed to control the swelling behavior of keratins is the secondary and tertiary structure of the intermediate filaments and the matrix [98,99]. As indicated previously, the intermediate filaments consist of proteins containing α-helical segments embedded in the less organized matrix of high cystine content.

In keratin fibers like human hair and wool fiber, the helical proteins of the intermediate filaments (microfibrils) are oriented parallel with the axis of the fiber (Figure 1–34), and the intermediate filaments help to maintain the structural integrity of the fibers while most of the volume swelling takes place in the matrix proteins [98–100]. This is consistent with Feughelman's two-phase model of water-impenetrable rods (intermediate filaments) in a water-penetrable matrix. As a result, maximum swelling occurs between the intermediate filaments and minimum swelling occurs along the axis of the intermediate filaments. Therefore, maximum swelling occurs in the diametral dimension of hair, and minimum swelling occurs in the longitudinal dimension or along the axis of the fibers (see Figure 1–34). For example, in 0% to 100% relative humidity hair increases nearly 14% in diameter but less than 2% in length [101]. Other reagents such as sodium lauryl sulfate, formic acid, and thioglycolic acid swell hair similarly; that is, they produce greater swelling in the diametral dimension than in the fiber length [99].

Swift [102] and others have provided evidence that the nonkeratin

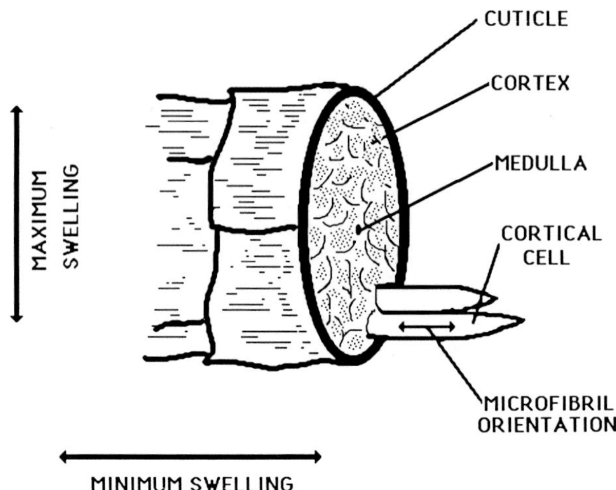

FIGURE 1–34. Directional swelling of human hair.

portions of hair are also important to fiber swelling. For example, Swift demonstrated by microscopic studies of the penetration of fluorescently labeled proteins in water that a large order swelling occurs in the nonkeratin regions of hair. The diametral swelling of hair by water from the dry state is about 14% to 16%. On the other hand, X-ray diffraction measurement of inter-intermediate filament separation distances indicates that swelling of only 5.5% occurs [103]. Swift, therefore, proposed that the difference can be explained by the large order of swelling that occurs in the nonkeratin portions of hair, primarily the cell membrane complex and the endocuticle of hair. For additional details on swelling of human hair, see Chapter 8.

Alpha Keratose

Protein from the intermediate filaments of human hair can be isolated by oxidation with peracetic acid according to the method of Alexander and Earland [74]. One fraction obtained by this procedure is called α-keratose. This material amounts to about 45% of the fiber mass, containing a substantially lower proportion of sulfur than the other two keratose fractions (see Table 2–8). The low sulfur content suggests a relatively low proportion of the amino acid cystine in the intermediate filaments and therefore a low proportion of cystine in the α-helical proteins. This conclusion is consistent with the amino acid analysis of α-keratose isolated from merino wool by Corfield et al. [75] that showed a relatively low percentage of cystine and a high percentage of the other bulky amino acids. This keratose fraction is in all probability not purely intermediate filament in origin and likewise not a pure α-helical protein. However, the fact that it can produce an X-ray pattern similar to that of α-keratin [74,104] and the other two keratose fractions cannot suggests that its origin is the intermediate filaments.

Curliness or Crimp

The manner in which natural curl or crimp is formed in animal hairs is not completely understood. Mercer [105], however, suggested that the shape of the follicle in the zone of keratinization (see Figure 1–4) determines the shape of the hair fiber. This suggests that the growing fiber takes the shape of the mold where hardening or keratinization occurs. Thus, if the follicle or sac in which the fiber is formed is highly curved in the zone of keratinization, the emerging hair fiber will be highly curled; if the follicle is relatively straight, however, the emerging hair will be straight. The fact that a sudden change in hair curliness can occur over the entire scalp after cessation of chemotherapy is difficult to explain by this mechanism.

An alternative and more likely explaination considers the bilateral structure of some keratin fibers like wool. A helical fiber will arise if opposite halves of the fiber grow at different rates or if opposite halves of the cortex differ chemically and thus will contract to different extents during moisture uptake or drying. This is analogous to the way a bilateral thermostat bends with changes in temperature. Nevertheless, in crimpy negroid hair or in coarse animal hairs the bilateral structure has not been demonstrated [106].

In the 1950s Mercer [107] and Rogers [108] independently identified two types of cortical cells in merino wool fibers; these two types of cells were named orthocortex and paracortex (see Figure 1–25). However, more recently, Kaplin and Whiteley [109] have been able to distinguish between three different types of cortical cells in high-crimp and low-crimp merino wool. Cells on one side of the cross section contain whorls of intermediate filaments, and these were called orthocortical cells. Those cells without whorls of microfibrils opposite to the orthocortical cells in fibers of high crimp and low crimp were named paracortical cells and mesocortical cells, respectively. Orwin and Woods [110] have demonstrated that the proportion of the cortex occupied by these three types of cortical cells varies with fiber diameter. In particular, the proportion of the cortex occupied by orthocortical cells increases with increasing fiber diameter.

Current evidence suggests that crimp frequency in wool is determined by the relative proportion of the three types of cortical cells, their location in the fiber cortex, and the protein composition of the matrix of the orthocortical cells [66]. Thus, it would appear that the dominant factor for determining crimp or curliness in human hair and other coarse animal hairs is the protein composition of the matrix of each cortical cell. The conclusion that the curliness or crimp of human hair is determined by differences in chemical composition of different cortical cells and their placement in the cortex could help to explain why a sudden change in hair curliness occurs in newly grown hair following chemotherapy, and it is the most plausible mechanism for explaining the natural curliness or crimp of human scalp hair.

Intercellular Matter

The cell membrane complex is the vital substance that consists of cell membranes and adhesive material that glues or binds the cuticle and cortical cells together. Figure 1–21 describes the cell membrane complex in schematic form. The cell membrane complex contains a low proportion of sulfur-containing amino acids including cystine compared with most of the intracellular proteins, and together with the endocuticle this complex is sometimes referred to as the "nonkeratinous regions."

The nonkeratinous regions are becoming more and more important to cosmetic and wool science because they are believed to be the primary pathway for entry or diffusion into hair and wool fibers (see Chapter 5 and Figure 5–4). In addition, during stretching or extension, signs of cuticle separation and damage occur in these regions, occurring sooner in tip ends (about 10% extension) of hair fibers than in root ends (about 20% extension) and much sooner than fiber breakage (about 36% extension) [111]. Thus, stretching or extending hair fibers as occurs in stretching or bending during combing or grooming operations can produce stress cracking in the nonkeratinous regions and the endocuticle and intercellular regions, and scale lifting can also occur in these damaged regions. If cleavage occurs within the endocuticle and the lifted portion of the scale is eroded, then a rough or granular surface will be exposed (see Figure 1–17 and Chapter 5 for additional discussion on fiber degradation during hair grooming).

Together these structures of cell membranes and adhesive material measure approximately 300 to 600 Å thick [112,113]. A number of sublayers of the cell membrane complex have been identified. The most important of these is the central delta layer, about 100 Å thick [49]. The delta layer is the intercellular cement; its proteins are low in cystine (<2%) and high in polar amino acids, containing about 12% basic amino acids and 17% acidic amino acids (see Chapter 2). This layer is sandwiched by other layers from each cell that are approximately 50 Å thick. These layers, sometimes called the inert beta layers [112], consist of lipids such as squalene and fatty acids that are rich in palmitic, stearic, and oleic acids. Marshall suggests that the composition of cell membrane complex material between cuticle cells may be different from the material between cortical cells [114].

Part of the intercellular complex may be isolated by oxidation of the fibers with peracetic acid according to the method of Alexander and Earland [74]. This isolated complex, called β-keratose, and amounts to about 15%

of the hair fiber mass [115] and consists of proteinaceous material with a low disulfide content. β-Keratose is not a pure substance, and its complete histological origin is unknown but is probably derived primarily from the cell membranes and intercellular matter from all three morphological regions of hair. The striking features of the intercellular complex are its long-range extensibility and its contrast in chemical reactivity to the intracellular matter.

The cell membrane complex is more reactive to proteolytic enzymes than are the intracellular proteins of hair. However, it is more resistant to alkali and to reducing agents, both of which readily attack the intracellular proteins of the cortex and the cuticle [113]. Intercellular cement is also attacked by acids, hot aqueous solutions, and reducing agents [116]. As expected, the lipid beta layers are readily attacked by lipid solvents and are more susceptible to surfactants than the intracellular proteins of hair (see Chapter 5).

Chemical analysis of β-keratose indicates a low proportion of sulfur compared to gamma keratose [74], a fraction containing some of the intracellular proteins. This helps to explain the differences in chemical reactivity between the intracellular and intercellular proteins of keratin fibers.

Medulla

Fine animal hairs, such as merino wool, consist of only cuticle and cortex [84], but as fiber diameter increases, a third type of cell, the medulla, can sometimes be found (see Figures 1–6 and 1–35). In thick animal hairs such as those from a horse tail or mane, or a porcupine quill, the medulla constitutes a relatively large percentage of the fiber mass. However, in human hair, the medulla, if present, generally makes up only a small percentage of this mass. The medulla may be either completely absent, continuous along the fiber axis, or discontinuous, and in some instances a double medulla may be observed (Figure 1–36).

Medullary cells are loosely packed, and during dehydration (formation) they shrivel up, leaving a series of vacuoles along the fiber axis. Because the medulla is believed to contribute negligibly to the chemical and mechanical properties of human hair fibers [115] and is difficult to isolate [104,117], it has received comparatively little scientific attention. The chemical composition of medullary protein derived from African porcupine quill has been reported by Rogers [100] and is described in Chapter 2.

Abnormalities

Fungal and nit infections produce gross hair fiber distortions. Figure 1–37 is a photomicrograph of a nit infection, illustrating the empty nits of the head louse *Pediculus capitis.*

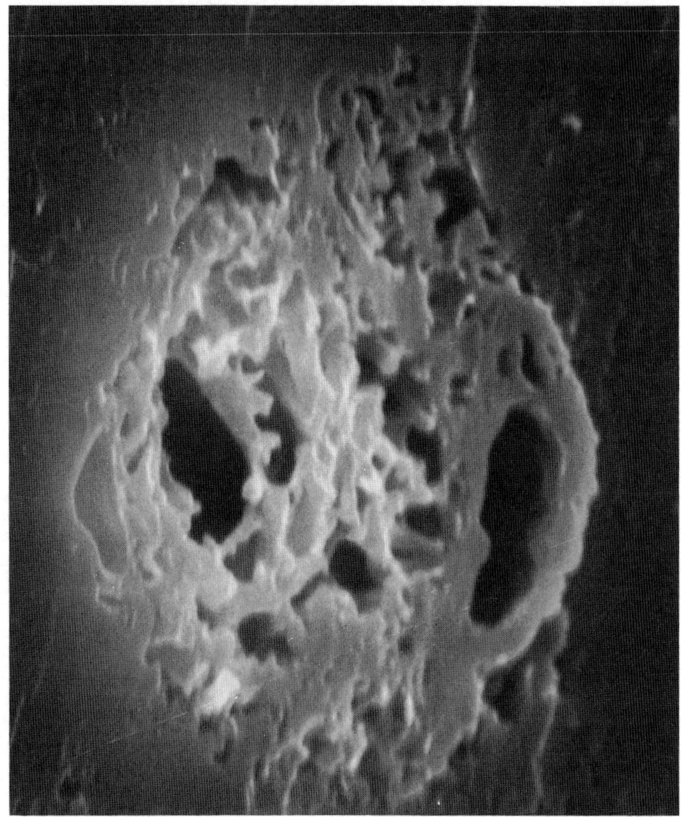

FIGURE 1–35. Scanning electron micrograph (SEM) illustrates the porous medulla of a hair fiber cross section.

Monilethrix, pili torti, pili annulati, and trichorrhexis nodosa are somewhat rare structural anomalies in human hair that are of congenital, genetic, or acquired origin. The structural changes occurring in these anomalies are so large that they may be readily observed microscopically.

Monilethrix is a congenital, hereditary disease resulting in abnormal human scalp hair. Monilethrix is also called moniliform hair or beaded hair, and it produces hair fibers with the appearance of a twisted ribbon, as illustrated by the light micrograph of Figure 1–38. With this condition, hair length generally does not exceed a few centimeters, particularly of hair with narrow internodes. This disease is also characterized by dry, fragile hair fibers.

Pili torti is a rare congenital deformity of the hair characterized by flattened fibers with multiple twists. In some cases, the hair grows to a normal length, although frequently this deformity produces short, twisted, broken hairs presenting the appearance of stubble. Pili torti provides a high fre-

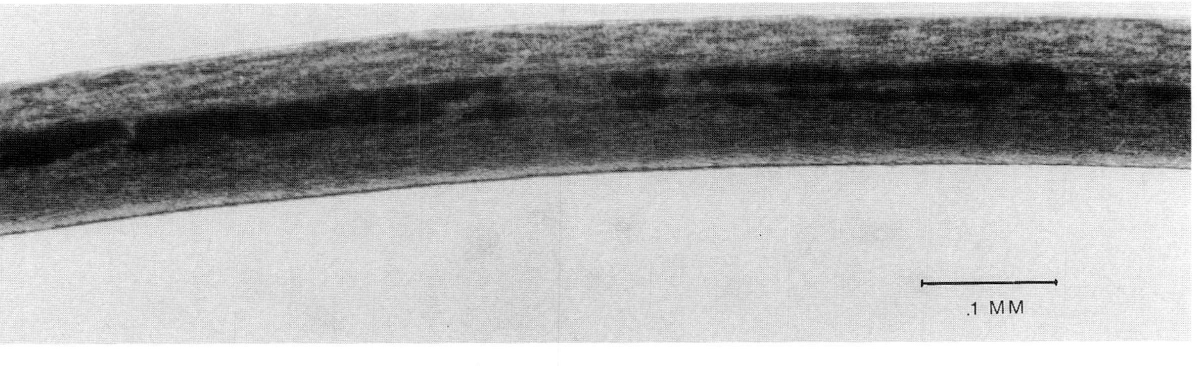

FIGURE 1–36. An optical section of a light micrograph illustrates a hair fiber with a divided or double medulla. Double medullas seem to be more common in facial than scalp hair. (Photograph courtesy of John T. Wilson.)

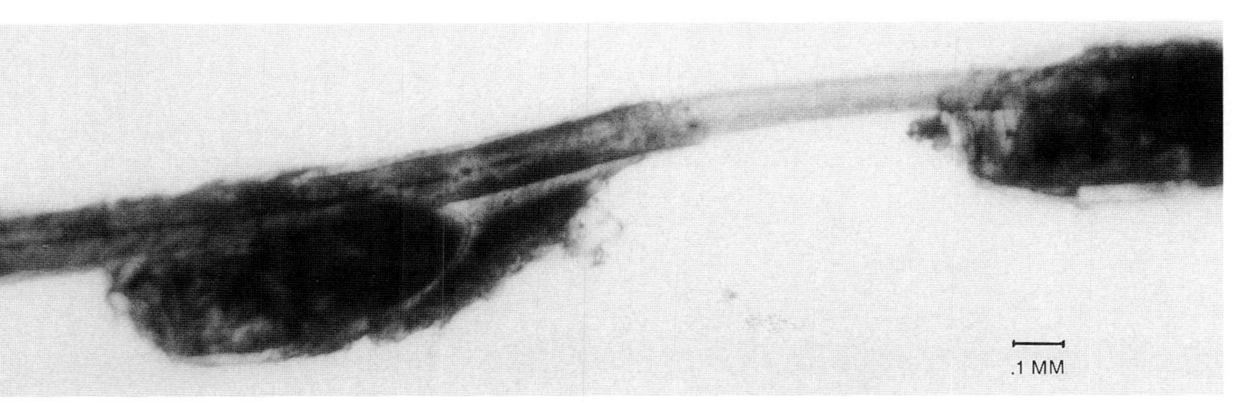

FIGURE 1–37. Light micrograph illustrates empty nits of the head louse *Pediculus capitis* on a human scalp hair. (Photograph courtesy of John T. Wilson.)

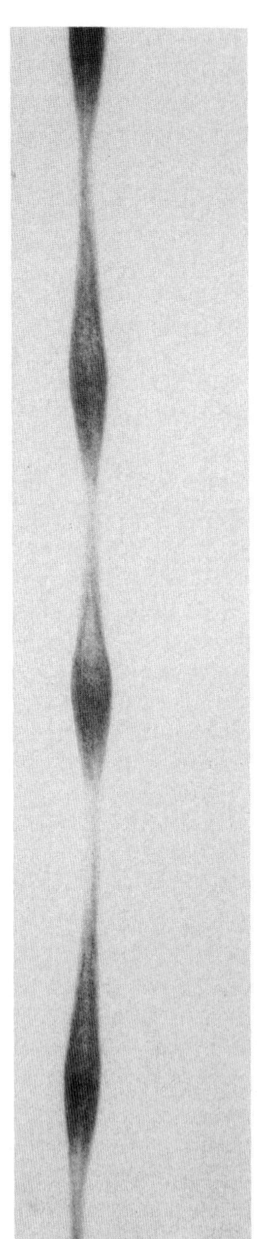

FIGURE 1–38. Monilethrix, a congenital and hereditary structural anomaly of human scalp hair. (Light micrograph courtesy of John T. Wilson.)

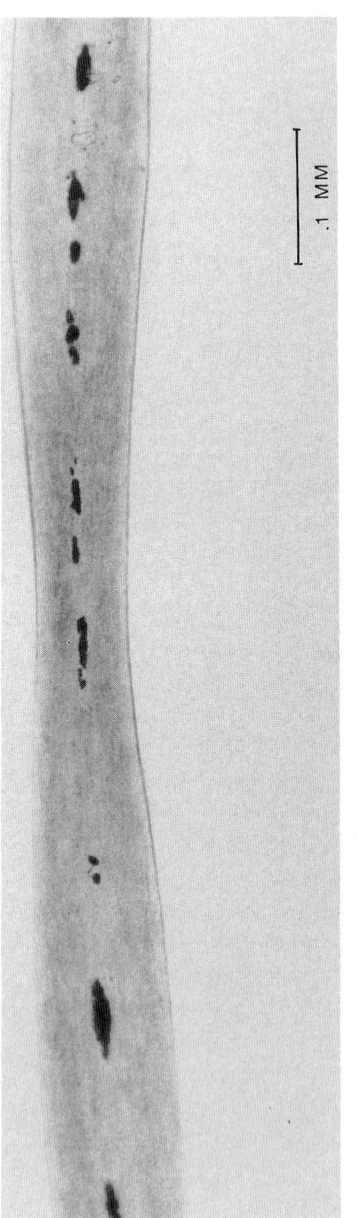

FIGURE 1–39. Pili torti, an uncommon hair shaft anomaly. (Light micrograph courtesy of John T. Wilson.)

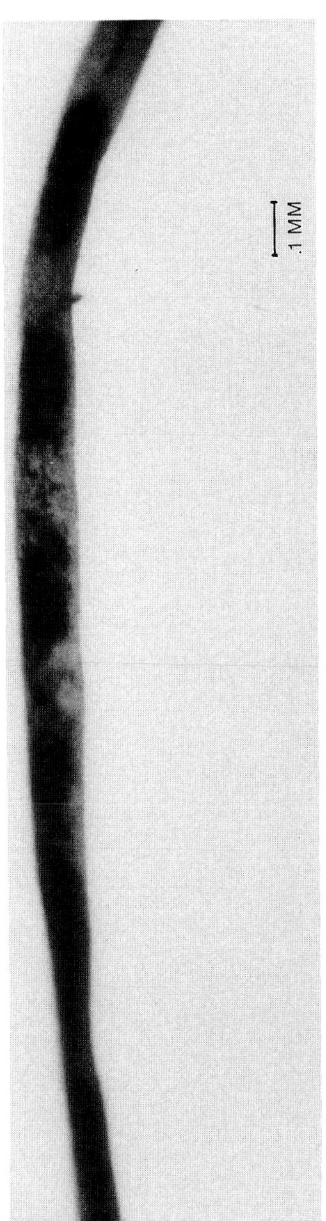

FIGURE 1–40. Pili annulati (ringed hair), an uncommon hereditary hair shaft anomaly. (Light micrograph courtesy of John T. Wilson.)

FIGURE 1–41. An intact hair fiber illustrates the condition of trichorrhexis nodosa. (Light micrograph courtesy of John T. Wilson.)

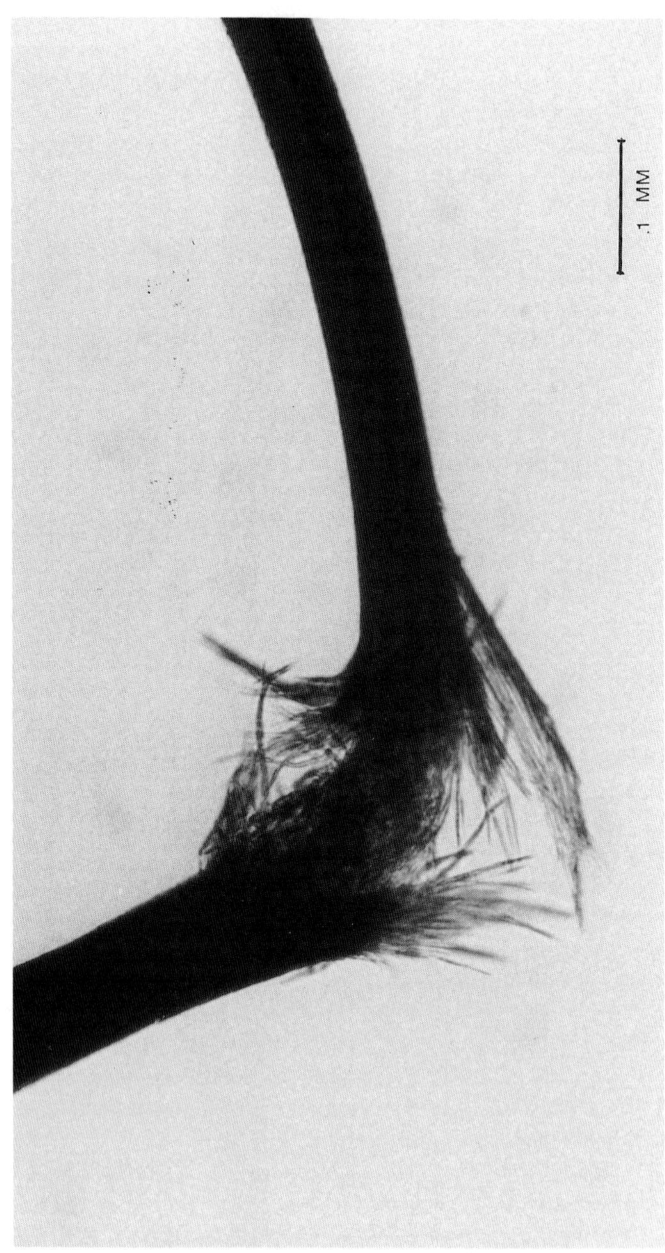

FIGURE 1–42. Trichorrhexis nodosa. "Broomlike" fractures at the nodes are symptoms of this hair shaft anomaly. (Light micrograph courtesy of John T. Wilson.)

quency of rotation and can resemble mildly affected monilethrix hair shafts (see Figure 1–39).

Pili annulati, sometimes called ringed hair (Figure 1–40), is a rare hereditary condition characterized by alternating light and dark bands produced by regions with and without medulla along the hair fiber axis. This condition produces hair fibers that display alternating bands of silvery gray and dark rings, although there is generally no abnormality in pigmentation.

Trichorrhexis nodosa (Figure 1–41) occurs more often in facial hair than in scalp hair, and produces bulbous-type nodes appearing as irregular thickenings along the hair shaft. These nodes are actually partial fractures, which under stress crack more completely forming broomlike breaks (Figure 1–42). For additional details relevant to these hair shaft anomalies, see the volume edited by Brown and Crounse [118] and articles written by John T. Wilson, who kindly provided the light micrographs of Figures 1–36 to 1–42.

References

1. Barnett, R.J.; Seligman, A.M. Science 116:323 (1952).
2. Kaswell, E.R. Textile Fibers Yarns and Fabrics, p. 52. Reinhold, New York (1953).
3. Randebrook, R.J. J. Soc. Cosmet. Chem. 15:691 (1964).
4. Bogaty, H.J. J. Soc. Cosmet. Chem. 20:159 (1969).
5. Piper, L.P.S. J. Textile Inst. 57T:185 (1966).
6. Venning, V.A.; Dawber, R.P.R. J. Am. Acad. Dermatol. 18(5):1073 (1988).
7. Barman, J.M.; Astore, I.; Pecoraro, V. J. Invest. Dermatol. 44:233 (1965).
8. Reeves, J.R.T.; Maibach, H.I. In Advances in Modern Toxicology, Marzulli, F.; Maibach, H.I., eds., Vol. 4, pp. 487–500. Hemisphere Publishing, Washington, D.C. (1977).
9. Van Scott, E.J. Clin. Obstet. Gynecol. 7:1062 (1964).
10. Stroud, J.P. Cutis 40:272 (1987).
11. Randall V.A.; Ebling, F.J.G. Br. J. Dermatol. 124:146 (1991).
12. Courtois, M.; Giland, S.; Fiquet, C. Int. Fed. SCC Paris, pp. 407–424 (1982).
13. Chidsey, C.A. III; Kahn, G. U.S. patent 4,596,812 (1986).
14. Montagna, W.; Carlisle, K.S. In Hair Research, Orfanos, C.E.; Montagna, W.; Stuttgen, G., eds., pp 3–12. Springer-Verlag, Berlin (1981).
15. Pinkus, H. In Hair Research, Orfanos, C.E.; Montagna, W.; Stuttgen, G. eds., pp. 237–243. Springer-Verlag, Berlin (1981).
16. Hamilton, J.B. Am. J. Anat. 71:451 (1942).
17. Gerstein, T. Cosmet. Toiletries 101:21 (1986).
18. Orentreich, N. In Hair Growth. Advances in Biology of the Skin, Montagna, W.; Dobson, R.L., eds., Vol. 9, pp. 99–108. Pergamon Press, Oxford (1967).
19. Schumacher-Stock, U. In Hair Research, Orfanos, C.E.; Montagna, W.; Stuttgen, G., eds., pp. 318–321. Springer-Verlag, Berlin (1981).
20. Barth, J.H. Drugs 35:83, (1988).

21. Sawaya, M.; Honig, L.S. Hsia, S.L. J. Invest. Dermatol. 91:101 (1988).
22. Griffin, J.E., et al. Am. J. Physiol. 243E:81, (1982).
23. Inaba, M.; Inaba, Y. Cosmet. Toiletries 105:77 (1990).
24. King, W.J. Nature 307:745 (1984).
25. Welshons, W.V.; Lieberman, M.E.; Gorski, J. Nature 307:747 (1984).
26. Sawaya, M.; Garland, L.D. Hsia, S.L. J. Invest. Dermatol. 92:91 (1989).
27. Orentreich, N. Ann. N.Y. Acad. Sci. 83:463 (1959).
28. Hamilton, J.B.; Terada, H.; Mestler, G.I. In Hair Growth, Advances in Biology of Skin, Montagna, W.; Dobson, R.L., eds., Vol. 9, pp. 143–145. Pergamon Press, Oxford (1967).
29. Hamilton, J.B. In The Biology of Hair Growth, Montagna, W.; Ellis, R.A., eds., pp. 418–419. Academic Press, New York (1958).
30. Montagna, W.; Ellis, R.A. Hair Growth. Pergamon Press, Oxford (1967).
31. Mercer, E.H. International Series of Monographs on Pure and Applied Biology, Vol. 12: Keratins and Keratinization, Alexander, P.; Bacq, Z.M., eds., Pergamon Press, New York (1961).
32. Shiell, R.C.; Norwood, O.T. In Hair Transplant Surgery, 2nd Ed., Shiell; Norwood, eds., pp. 328–333. Thomas, Springfield, IL (1984).
33. Bouhanna, P. J. Dermatol. Surg. Oncol. 10(7):551 (1984).
34. Geiger, W.J. Res. Natl. Bur. Stand. 32, 127 (1944).
35. Harris, M.; Smith, A. J. Res. Natl. Bur. Stand. 17:577 (1936).
36. Leeder, J.D.; Bradbury, J.H. Nature 218:694 (1968).
37. Bradbury, J.H.; Chapman, G.V.; King, N.L.R. Proc. 3rd Int. Wool Textile Res. Conf. (Paris) I:359 (1965).
38. Hock, C.W.; Ramsey, R.C.; Harris, M. J. Res. Natl. Bur. Stand. 27:181 (1941).
39. Holmes, A.W. Text. Res. J. 34:707 (1964).
40. Wortman, F.J. Text. Res. J. 52:479 (1982).
41. Rundall, K.M. Proc. Leeds Philos. Soc. 4(I):13 (1941).
42. Wolfram, L.J.; Lindemann, M. J. Soc. Cosmet. Chem. 22:839 (1971).
43. Stoves, J. J. Soc. Dyers Col. 63:65 (1947).
44. Garcia, M.L.; Epps, J.A.; Yare, R.S. J. Soc. Cosmet. Chem. 29:155 (1978) (and references therein).
45. Bradbury, J.H.; Chapman, G.V.; Hambly, A.N.; King, N.L.R. Nature 210:1333 (1966).
46. Blout, E.R.; de Loze, C.; Bloom, S.M.; Fasman, G.D. J. Am. Chem. Soc. 82:3787 (1960).
47. Astbury, W.T.; Street, A. Philos. Trans. A 230:75 (1931).
48. Langermalm, G.; Philip, B. Text. Res. J. 20:668 (1950).
49. Swift, J.A.; Holmes, A.W. Text. Res. J. 35:1014 (1965).
50. Fraser, R.D.B.; Rogers, G.E. Aust. J. Biol. Sci. 8:129 (1955).
51. Swift, J.; Bews, B. J. Soc. Cosmet. Chem. 25:355 (1974).
52. Swift, J.; Bews, B. J. Soc. Cosmet. Chem. 25:13 (1974).
53. Hunter, L., et al. Text. Res. J. 44:136 (1974).
54. Fraser, R.D.B.; MacRae, T.P.; Rogers, G.E. In Keratins, Their Composition, Structure, and Biosynthesis, Chap. 4. Thomas, Springfield, IL (1972).
55. Allworden, K.Z. Angew. Chem. 29:77 (1916).
56. Alexander, P.; Hudson, R.F.; Earland, C. Wool, Its Chemistry and Physics, pp. 7–8. Franklin Publishing, New Jersey (1963).

57. Lindberg, J.; Phillip, B.; Gralen, N. Nature 162:458 (1948).
58. Leeder, J.D.; Rippon, J.A.; Rivett, D.E. Tokyo Wool Research Conf. (1985).
59. Leeder, J.D.; Bradbury, J.H. Text. Res. J. 41:215 (1971).
60. Leeder, J.D.; Bishop, D.G.; Jones, L.N. Textile Res. J. 53:402 (1983).
61. Holmes, A.W. Nature 923 (1961).
62. Holmes, A.W. Textile Res. J. 34:777 (1964).
63. Kidd, F. Proc. 3rd Int. Wool Text. Res. Conf. (Paris) I:221 (1965).
64. Mercer, E.H. Text. Res. J. 23:388 (1953).
65. Kassenbeck, P. In Hair Research, Orfanos, C.E.; Montagna, W.; Stuttgen, G., eds., pp. 52–64. Springer-Verlag, Berlin (1981).
66. Mowat, I, et al. J. Text. Inst. 73:246 (1982).
67. Gjesdal, F. Acta Pathol. Microbiol. Scand. 133:112 (1959).
68. Tolgyesi, E.; Coble, D.W.; Fang, F.S.; Kairimen E.O. J. Soc. Cosmet. Chem. 34:361 (1983).
69. Birbeck, M.S.C.; Mercer, E.H. J. Biophys. Biochem. Cytol. 3:203 (1956).
70. Piper, L.P.S. J. Text. Inst. 57T:185 (1966).
71. Consdon, R.; Gordon, A.H. Biochem J. 46:8–20 (1950).
72. Hailwood, A.J.; Harrobein, S. Trans. Faraday Soc. 42B:84 (1946).
73. Gillespie, J.M.; Reis, P.J.; Schinekel, P.G. Aust. J. Biol. Sci. 17:548 (1964).
74. Alexander, P.; Earland, C. Nature 166:396 (1950).
75. Corfield, M.C.; Robson, A.; Skinner, B. Biochem. J. 68:348 (1958).
76. Filshie, B.K.; Rogers, G.E. J. Mol. Biol. 3:784 (1964).
77. Johnson, D.J.; Sikorski, J. Nature 194:31 (1962).
78. Bailey, C.J.; Tyson, C.N.; Wooos, H.J. Proc. 3rd Int. Wool Text. Res. Conf. (Paris) I:21 (1965).
79. Spei, M. 5th Int. Wool Textile Conf. Aachen II:90 (1975).
80. Bendit, E.G.; Feughelman, M. In Encyclopedia of Polymer Science & Technology, Vol. 8, p. 1. Wiley, New York (1968).
81. Johnson, D.J.; Sikorski, J. Proc. 3rd Int. Wool Text. Res. Conf. (Paris) I:53 (1965).
82. Crewther, W.G., et al. Int. J. Biol. Macromol. 5:267 (1983).
83. Fraser, R.B.D., et al. Int. J. Biol. Macromol. 10:106 (1988).
84. Fraser, R.D.B., et al. Nature 193:1052 (1962).
85. Pauling, L.; Corey, R.B. J. Am. Chem. Soc. 72:5349 (1950).
86. Pauling, L.; Corey, R.B. Proc. Natl. Acad. Sci. USA 37:261 (1951).
87. Pauling, L.; Corey, R.B. Sci. Am. 190:3 (1954).
88. Astbury, W.T.; Street, A. Philos. Trans. R. Soc. 230A:73 (1931).
89. Astbury, W.T. Trans. Faraday Soc. 29:193 (1933).
90. Astbury, W.T.; Woods, H.J. Philos. Trans. R. Soc. 232A:333 (1933).
91. MacArthur, I. Nature 152:38 (1943).
92. MacArthur, I. Symposium on Fibrous Proteins. Soc. Dyers Col., pp. 5–14, and references therein (1946).
93. Pauling, L.; Corey, R.B. Nature 171:59 (1953).
94. Fraser, R.D.B., et al. Proc. 3rd Int. Wool Tex. Res. Conf. (Paris) I:6 (1965).
95. Pauling, L.; Corey, R.B. Nature 171:59 (1953).
96. Crick, F.H.C. Nature 170:882 (1952).
97. Feughelman, M. J. Soc. Cosmet. Chem. 33:385 (1982).
98. Menefee, E. Text. Res. J. 38:1149 (1968).
99. Robbins, C.R.; Fernee, K.M. J. Soc. Cosmet. Chem. 34:21 (1983).

100. Rogers, G.E. In The Epidermis, Montagna, W.; Ellis, R.A., eds., p. 205. Academic Press, New York (1964).
101. Stam, P.; Katy, R.F.; White Jr. H.J. Text. Res. J. 22:448 (1952).
102. Swift, J.A. Proceedings of the 8th International Hair Science Symposium of the DWI, (Keil, Germany) September 9–11, 1992.
103. Spei, M.; Zahn, H. Melliand Textilber. 60(7):523 (1979).
104. Hoppy, F. Nature 166:397 (1950).
105. Mercer, E.H. International Series of Monographs on Pure and Applied Biology, Vol. 12, Keratin and Keratinization, Alexander, P.; Bacq, Z.M., eds., p. 156. Pergamon Press, New York (1961).
106. Mercer, E.H. Ibid., p 274.
107. Mercer, E.H. Text. Res. J. 23:387 (1953).
108. Rogers, G.E. J. Ultrastruct. Res. 2:309 (1959).
109. Kaplin, I.J.; Whiteley, K.J. Aust. J. Biol. Sci. 31:231 (1978).
110. Orwin, D.F.G.; Woods, J.L. J. Text. Inst. 71:315 (1980).
111. TRI Princeton Progress Report No. 3, pp. 20–40, October 1, 1991–March 31, 1992.
112. Fraser, R.D.B.; MacRae, T.P.; Rogers, G.E. In Keratins, Their Composition, Structure, and Biosynthesis, p. 70. Thomas, Springfield, IL (1972).
113. Mercer, E.H. J. Soc. Cosmet. Chem. 16:507 (1965).
114. Marshall, R.C. Proceedings of the 8th International Wool Textile Research Conference, (New Zealand) Vol. 1, pp 169–185 (1990).
115. Menkart, J.; Wolfram, L.J.; Mao, I. J. Soc. Cosmet. Chem. 17:769 (1966).
116. Dobb, M.G. Nature 203:48 (1964).
117. Fraser, R.D.B.; MacRae, T.P.; Rogers, G.E. Nature 183:592 (1959).
118. Brown, A.C.; Crounse, R.G., eds. Hair Trace Elements and Human Illness. Praeger, New York (1980).
119. Robbins, C.R. Skin. In Encyclopedia of Human Biology, Vol. 7, Dulbecco, R., ed. Academic Press, San Diego (1991).
120. Robbins, C.R. Hair. In Encyclopedia of Human Bology, Vol. 4, Dulbecco, R., ed. Academic Press, San Diego (1991).

2

Chemical Composition

Human hair is a complex tissue consisting of several morphological components (see Chapter 1), and each component consists of several different chemical species. It is an integrated system in terms of both its structure and its chemical and physical behavior wherein its components can act separately or as a unit. For example, the frictional behavior of hair is related primarily to the cuticle, but the softness of hair is determined by the cuticle, the cortex, and its intercellular components acting in concert. Although the tensile behavior of human hair is determined largely by the cortex, the physical integrity of the fiber to combing and grooming forces is determined by all three of these essential components—cuticle, cortex, and intercellular components—acting together.

For simplicity and ease of discussion, the different types of chemicals that comprise human hair are described separately in this chapter. However, for a clear understanding of its chemical and physical behavior, one should keep in mind that this fiber is an integrated system wherein several or all of its components can act simultaneously.

Depending on its moisture content (up to 32% by weight), human hair consists of approximately 65% to 95% proteins. Proteins are condensation polymers of amino acids, and the structures of those amino acids found in human hair are depicted in Table 2–1. The remaining constituents are water, lipids (structural and free), pigment, and trace elements (generally not free, but combined chemically with side chains of protein groups or with fatty acid groups of sorbed or bound lipid). These different components of hair—proteins, water, lipids, pigment, and trace elements—are described separately in this chapter.

TABLE 2–1. Structures of amino acids found in hydrolyzates from human hair.

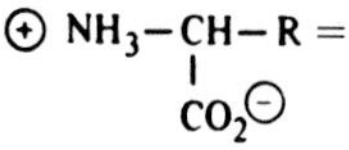

$\oplus\ NH_3-CH(CO_2^{\ominus})-R$ = Amino acid

Aliphatic hydrocarbon R group

Glycine	$\overset{\oplus}{N}H_3-CH_2-CO_2^{\ominus}$
Alanine	$\overset{\oplus}{N}H_3-CH(CO_2^{\ominus})-CH_3$
Valine	$\overset{\oplus}{N}H_3-CH(CO_2^{\ominus})-CH(CH_3)_2$
Isoleucine	$\overset{\oplus}{N}H_3-CH(CO_2^{\ominus})-CH(CH_3)-CH_2-CH_3$
Leucine	$\overset{\oplus}{N}H_3-CH(CO_2^{\ominus})-CH_2-CH(CH_3)_2$

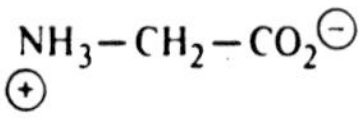

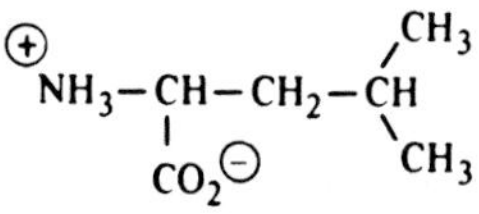

Aromatic hydrocarbon R group

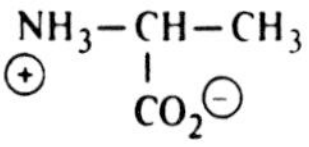

Phenylalanine	$\overset{\oplus}{N}H_3-CH(CO_2^{\ominus})-CH_2-C_6H_5$
Tyrosine	$\overset{\oplus}{N}H_3CH(CO_2^{\ominus})-CH_2-C_6H_4-OH$

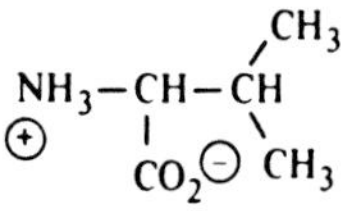

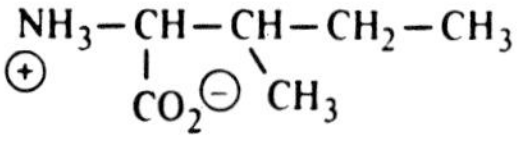

Dibasic

Lysine	$\overset{\oplus}{N}H_3-CH(CO_2^{\ominus})-CH_2-CH_2-CH_2-CH_2-NH_2$
Arginine	$\overset{\oplus}{N}H_3-CH(CO_2^{\ominus})-CH_2-CH_2-CH_2-NH-C(=NH)-NH_2$
Histidine	$\overset{\oplus}{N}H_3-CH(CO_2^{\ominus})-CH_2-C{=}CH-N(H)-CH{=}N$ (imidazole ring: C—N, $C{=}C(H)$, $N{=}CH$, $N-H$)

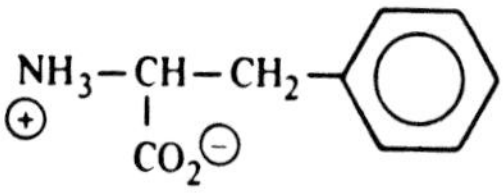

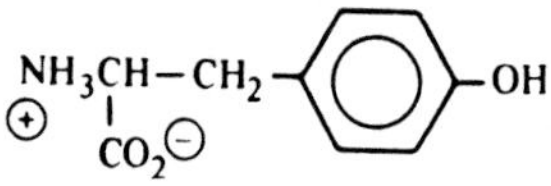

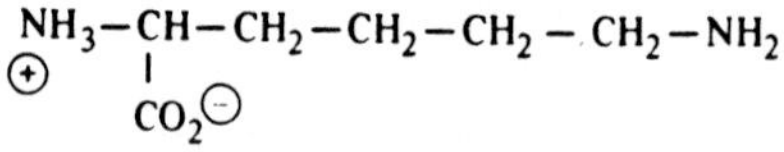

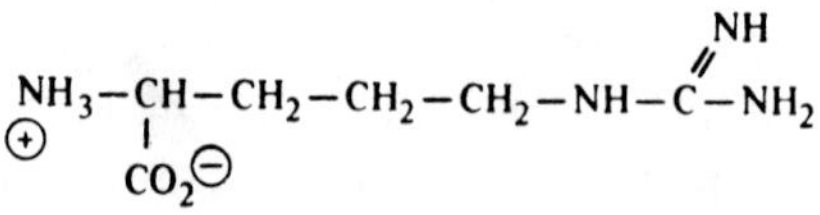

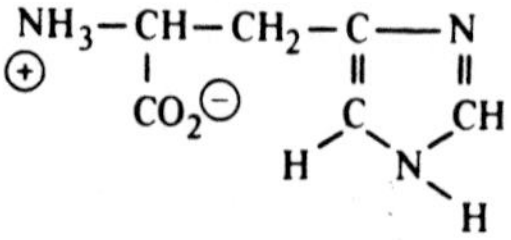

TABLE 2–1. (*Continued*)

Citrulline	$\overset{\oplus}{NH_3}-\underset{\underset{CO_2^{\ominus}}{\mid}}{CH}-CH_2-CH_2-CH_2-NH-\overset{\overset{O}{\Vert}}{C}-NH_2$
Diacidic	
Aspartic acid[a]	$\overset{\oplus}{NH_3}-\underset{\underset{CO_2^{\ominus}}{\mid}}{CH}-CH_2-CO_2H$
Glutamic acid[a]	$\overset{\oplus}{NH_3}-\underset{\underset{CO_2^{\ominus}}{\mid}}{CH}-CH_2-CH_2-CO_2H$
Hydroxyl R group	
Threonine	$\overset{\oplus}{NH_3}-\underset{\underset{CO_2^{\ominus}}{\mid}}{CH}-\overset{\overset{OH}{\mid}}{CH}-CH_3$
Serine	$\overset{\oplus}{NH_3}-\underset{\underset{CO_2^{\ominus}}{\mid}}{CH}-CH_2-OH$
Sulfur-containing R group	
Cystine	$\overset{\oplus}{NH_3}-\underset{\underset{CO_2^{\ominus}}{\mid}}{CH}-CH_2-S-S-CH_2-\underset{\underset{CO_2^{\ominus}}{\mid}}{CH}-\overset{\oplus}{NH_3}$
Methionine	$\overset{\oplus}{NH_3}-\underset{\underset{CO_2^{\ominus}}{\mid}}{CH}-CH_2-CH_2-S-CH_3$
Cysteine	$\overset{\oplus}{NH_3}-\underset{\underset{CO_2^{\ominus}}{\mid}}{CH}-CH_2-SH$
Cysteic acid	$\overset{\oplus}{NH_3}-\underset{\underset{CO_2^{\ominus}}{\mid}}{CH}-CH_2SO_3H$

[a] Aspartic and glutamic acids exist as the primary amides as well as the free acids in human hair.

TABLE 2–1. (*Continued*)

Heterocyclic R group

Proline

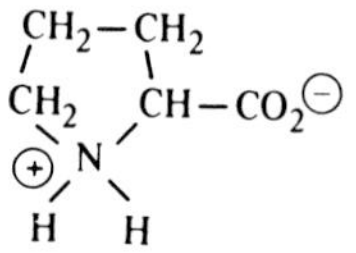

$$\begin{array}{l} CH_2-CH_2 \\ / \quad\quad \backslash \\ CH_2 \quad CH-CO_2^{\ominus} \\ \quad \backslash \quad / \\ \oplus N \\ / \quad \backslash \\ H \quad\quad H \end{array}$$

Tryptophan

$$\text{(indol-3-yl)}-CH_2-CH(NH_3^{\oplus})-CO_2^{\ominus}$$

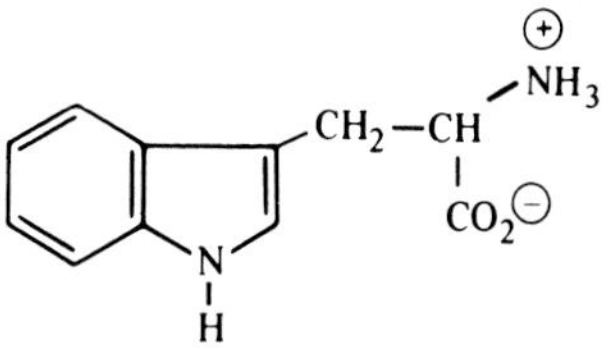

Studies of the proteinaceous matter of human hair may be classified according to the following types of investigation:

Studies of individual or several amino acids
Analysis of types of amino acids
Fractionation and peptide analysis.

Most studies of individual amino acids of keratin fibers involve the amino acids cystine or tryptophan. Quantitation of cystine can be accomplished via chemical analysis of mercaptan with [1,2] or without hydrolysis [3], or spectrophotometrically on intact hair [4,5]. With increasing sophistication in instrumental analysis, electron spectroscopy for chemical analysis (ESCA), secondary ion mass spectrometry (SIMS), and different absorbance, reflectance, and fluorescence techniques, spectrophotometric analysis of intact hair is becoming increasingly important. Chemical analyses for tryptophan have been described by Block and Bolling [6] and are all hydrolytic procedures. Quantitative determination of several amino acids in human hair has become increasingly widespread since the development of the ion exchange chromatographic systems of Moore et al. [7] and, more recently, protein sequencing techniques. Studies of types of amino acids, which identify specific functional groups that are characteristic of more than one amino acid, are also useful; examples are the titration of basic groups [8] and of acidic groups [8].

Fractionation and peptide analysis is concerned primarily with fractionation into similar peptide types or even fractionation into the different morphological components. This area of research currently receives more scientific attention for wool fiber than for human hair. In fact, the major research concerned with the chemical composition of hair and wool fibers, during the past decade, has involved:

Sequencing of the amino acids of the major protein types of keratin fibers, including the intermediate filaments and the matrix or the intermediate filament-associated proteins;
The structure of the cell membrane complex;
Forensic studies concerned with DNA analysis of hair, and
The analysis of drugs in human hair.

Each of these respective subjects is summarized here, and leading references to each area of research are also included in this chapter. Of course, the first two subjects are highly relevant to the structure and the cosmetic behavior of human hair, and therefore are emphasized in this chapter. On the other hand, the latter two subjects are important to the field of criminology and fiber identification, but are of less significance to the cosmetic behavior of human hair.

Composition of the Proteins of Human Hair and the Different Morphological Regions

Whole-Fiber Amino Acid Studies

A large number of investigations have been described regarding the analysis of the amino acids of whole human hair fibers. Whole-fiber amino acid analysis has several limitations, because it provides average values for the amino acid contents of the average proteinaceous substances of the fibers. Therefore, for whole-fiber results, cross-sectional and axial differences in the composition of the fibers are averaged.

A second complicating factor is hydrolytic decomposition of certain amino acids. The most commonly used medium for keratin fiber hydrolysis is 5,6-N hydrochloric acid. In studies involving acid hydrolysis of keratins, partial decomposition has been reported for cystine, threonine, tyrosine [9], phenylalanine, and arginine [10] with virtually complete destruction of tryptophan [11].

With the foregoing limitations in mind, the following discussion describes several important factors contributing to differences in the whole-fiber amino acid analysis results of human hair reported in the literature.

Unaltered or "Virgin" Human Hair

Unaltered human hair is hair that has not been chemically modified by treatment with bleaches, permanent waves, straighteners, or hair dyes. Numerous publications [6,12–27] describe results of the amino acid analysis of unaltered human hair. Table 2–1 depicts the structures for 21 amino

TABLE 2–2. Amino acids in whole unaltered[a] human hair (in micromoles per gram of dry hair).

Amino Acid	Robbins and Kelly [12]	Ward and Lundgren [13]	Other References
Aspartic acid	444–453[a]	292–578[c]	
Threonine	648–673[b]	588–714	
Serine	1013–1091[b]	705–1090	
Glutamic acid	995–1036[b]	930–970	
Proline	646–708[d]	374–694[c]	
Glycine	463–513[d]	548–560	
Alanine	362–384[d]	314	
Half-cystine	1407–1512[d]	1380–1500	784–1534 [14][d]
Valine	477–513[d]	470	
Methionine	50–56[b]	47–67	
Isoleucine	244–255[d]	366	
Leucine	502–529[d]	489[c]	
Tyrosine	177–195[d]	121–171[c]	
Phenylalanine	132–149[d]	151–226	
Cysteic acid	22–40[d]	—	
Lysine	206–222[b]	130–212[c]	
Histidine	64–86[d]	40–77	
Arginine	499–550[d]	511–620	
Cysteine	—	41–66	17–70 [14][c]
Tryptophan	—	20–64	
Citrulline	—	—	11 [16]
Percent of nitrogen as ammonia		15.5%–16.9%	16.5% [15][e]

[a] Hair is assumed to be cosmetically unaltered for references 13, 14, and 16.
[b] No significant differences among samples analyzed.
[c] These values are results of a microbiological assay by Long and Lucas [17].
[d] Significant differences indicated among samples analyzed.
[e] These results are a compilation of results from several laboratories and therefore contain no basis for statistical comparison.

acids that have been identified in human hair. Cysteic acid and other amino acids, derived from those amino acids of Table 2–1, are also present in either weathered or cosmetically altered hair. Table 2–2 summarizes results from several sources describing quantitative whole fiber analyses of these 21 amino acids. These same amino acids, classified according to functional group, are summarized in Table 2–3.

Note the high frequencies of hydrocarbon, hydroxyl, primary amide, and basic amino acid functions in addition to the relatively large disulfide content. The high frequency of hydrocarbon-containing amino acids confirms that hydrophobic interactions will play a strong role in the reactivity of hair toward cosmetic ingredients. Hydroxyl and amide groups interact through hydrogen-bonding interactions, and the basic groups interact through hydrogen-bonding and ion-exchange type interactions.

TABLE 2–3. Approximate composition of unaltered human hair by amino acid side-chain type.

Amino acid side-chain type[a]	Approximate micromoles per gram hair
Hydrocarbon (except phenylalanine):	2800
glycine, alanine, valine, leucine, isoleucine, proline	
Hydroxyl:	1750
serine and threonine	
Primary amide + carboxylic acid:	1450
primary amide (ammonia estimation);	1125
carboxylic acid (by difference)	325
Basic amino acids:	800
arginine, lysine, histidine	
Disulfide	750
Cystine	
Phenolic	180
Tyrosine	

[a] See Table 2–2.

Of particular note is that all these functional groups occur at higher frequencies than the disulfide bond in hair. However, these frequencies are whole-fiber frequencies, and considering, the whole fiber assumes that hair is a homogeneous substrate. This is certainly not the case, as subsequent sections of this chapter demonstrate.

Table 2–2 shows substantial variation in the quantities of some of the amino acids, notably aspartic acid, proline, cystine, and serine, while considerably less dispersion is indicated for other amino acids, primarily valine, glutamic acid, glycine, alanine, leucine, and arginine. The following factors can produce differences in whole-fiber amino acid analysis results: genetics, weathering (primarily sunlight exposure), cosmetic treatment, experimental procedures, and diet (not the normal diets of healthy individuals, but protein-deficient diets).

Marshall and Gillespie [28] have proposed special mathematical relationships between cystine and leucine: first, that leucine (residue %) = −0.31 × half-cystine (residue %) + 11.3, and second, between cystine and proline, to determine abnormal variations, that proline (residue %) = 0.26 × half-cystine (residue %) + 3.8. These relationships are based on the fact that leucine and cystine are common components of the low-sulfur proteins and that proline and cystine are primary components of the high-sulfur proteins. Marshall and Gillespie further suggest that the cystine content should be about 17% to 18% and that large variations beyond the calculated values for these three amino acids indicate some cause of variation such as genetic, environmental (sunlight exposure), cosmetic treatment, or diet. Variation from these factors is described next.

Factors Relating to Genetics

The variation of cystine and cysteine in human hair has been studied extensively. Clay et al. [14] quantitatively analyzed hair from 120 different persons for cystine and cysteine (see Table 2–2). The hair in this study was selected from both males and females of varying age and pigmentation, and analysis was by the hydrolytic method of Shinohara [29].

These results show a wide spread in disulfide content varying from 784 to 1,534 μmol half-cystine/g of hair (8.7%–17%), substantially different from the cystine level concluded by Marshall and Gillespie for "normal" hair. Significantly more cystine was found in hair from males than from females. Also, dark hair generally contained more cystine than light hair. A similar relationship between cystine content and hair color has been reported by Ogura et al. [30].

No consistent relationship was found between age and cystine content. Although factors such as diet (malnutrition), cosmetic treatment, and environmental effects (sunlight degradation) may have contributed to variation among these samples, these were not considered in this study.

With regard to racial variation, nothing has been definitely established. Hawk's data [22] appear to show subtle differences in the relative percentages of various amino acids found in the hydrolysates of hair from Blacks hair as compared to Caucasian hair. Wolfram [31] has compiled a more complete set of data from the literature of whole-fiber amino acid analysis of the three races that show overlap in the amounts of all the amino acids from scalp hair for the three major racial groups. Polypeptide isolation and amino acid sequencing rather than whole-fiber amino acid analysis will provide the best means for determining with certainty if any differences exist in the proteins of hair of different races.

Weathering

The photochemical degradation of cystine (see Chapter 4) provides a major cause for variation in this amino acid found among different hair samples. Weathering effects [32] in human hair may be explored by comparing tip ends (which have been exposed longer) to root ends. In a study of this type, the cystine and cysteine contents of tip ends were shown to be lower than in root ends [33]. Complementary to these results, larger amounts of cysteic acid have been reported in hydrolysates of tip ends of human hair than in root ends [34]. Evidence for cysteic acid in weathered wool has also been provided by Strasheim and Buijs by means of infrared spectroscopy [35].

These results suggest conversion of the cystinyl groups in human hair to higher oxidation states by the elements. This conclusion is supported by the work of Harris and Smith [36], who have shown that ultraviolet light disrupts the disulfide bond of dry wool. In another study, Robbins and

Bahl [5] have examined the effects of ultraviolet light on both hair and hair from root and tip sections from several persons using ESCA to examine sulfur in hair. Their data suggest that weathering of cystine in hair is primarily a photochemical reaction proceeding through the C–S fission route producing cystine S-sulfonate residues as a primary end product. This reaction also occurs to a greater extent near the fiber surface. Tryptophan has also been shown to be sensitive to photochemical degradation. For additional details on the photochemical degradation of hair, see Chapter 4.

Significantly lower quantities of lysine and histidine have been reported in tip ends of human hair compared to root ends [34]. Oxidation at the amide carbon has also been shown to occur, producing carbonyl groups [6]. In addition, weathering degradation of several other amino acids has been reported in wool fiber [37,38] (see Chapter 4).

Dietary Insufficiencies

The hair of persons suffering from severe protein malnutrition is weaker mechanically, with fibers that are finer in diameter, sparser, and less pigmented. The cystine, arginine, and methionine contents of human hair have been reported to be influenced by diet that is insufficient in protein content. Koyanagi and Takanohashi [39] conducted a study among 8- to 9-year-old Japanese children who had been fed millet and very little animal protein. Analysis of the hair from these children revealed cystine contents as low as 8.1% (675 μmol half-cystine/g of hair). Diet supplementation with shark liver oil produced a significant increase in the cystine content of the hair, and diet supplementation with skim milk for 6 months produced an even greater increase in cystine.

Cystine, methionine, and sulfur contents of the hair of children suffering from kwashiorkor have been reported to be lower than that of normal children [40]. The arginine content of hair has also been reported to decrease as a result of kwashiorkor [41]. In fact, Noer and Garrigues [41] have reported arginine contents of human hair in severe cases of kwashiorkor as low as one-half the normal value.

By analogy with the effects of malnutrition and sulfur enrichment on the high-sulfur proteins in wool fiber [42,43], the lower cystine content in hair is probably caused by decreased synthesis of the sulfur-rich proteins because of malnutrition. Studies of the effects of diet in persons suffering malnutrition, specifically protein deficiencies, show that diet supplementation can influence the protein composition of human hair. However, such effects have only been demonstrated among persons suffering from severe malnutrition and never among healthy persons [see the section on the intermediate filament-associated proteins (IFAP) of human hair later in this chapter].

Experimental Procedures

The inconsistent use of correction factors to compensate for hydrolytic decomposition of certain of the amino acids has already been described. In addition, methods of analysis described in the literature have ranged from wet chemical [19] to chromatographic [12] to microbiological [17]. Reexamination of Table 2–2 with this latter condition in mind shows values for aspartic acid, proline, tyrosine, and lysine as determined by microbiological assay to be in relatively poor agreement with the other values for these same amino acids as determined by wet chemical and chromatographic procedures. The values for the microbiological and chromatographic procedures for valine are in close agreement, suggesting that for certain amino acids (e.g., valine) the microbiological assay is satisfactory, whereas for other amino acids (aspartic acid, proline, tyrosine, and lysine), the microbiological method is questionable.

Stability of Hair Keratin

A few years ago, a well-preserved cadaver was discovered by archaeologists in the Han Tomb No. 1 near Changsha, China [44]. The occupant of the casket wore a well-preserved hairpiece that was more than 2,000 years old. Although this hair was not analyzed for amino acid content, it was analyzed by X-ray diffraction, revealing that the α-helical content had been well preserved. Nevertheless, some minor disruption of the low-ordered matrix had occurred from reaction with a mercurial preservative in the casket. This suggests that the basic structure of the microfibrils of human hair remains unchanged over centuries and that its essential structural features are extraordinarily stable.

Cosmetically Altered Hair

Bleached Hair

The whole-fiber amino acid composition of human hair when bleached on the head with commercial hair-bleaching agents such as alkaline hydrogen peroxide or, more generally, alkaline peroxide/persulfate [45], has been described in the literature [9]. This investigation [9] defines the amino acids found in hydrolysates of hair bleached to varying extents on the head. Data describing frosted (extensively bleached) hair bleached on the head about 1 month before sampling versus nonbleached hair from the same person are summarized in Table 2–4. These data suggest that the primary chemical differences between extensively bleached hair and unaltered hair are a lower cystine content, a higher cysteic acid content, and lower amounts of tyrosine and methionine in the bleached hair. Mildly to mod-

TABLE 2–4. Amino acids from frosted compared to nonfrosted hair.

Amino acid	Micromoles per gram hair: Nonfrosted fibers	Micromoles per gram hair: Frosted fibers	Significant difference for frequencies at $\alpha = .01$
Aspartic acid	437	432	—
Threonine	616	588	—
Serine	1085	973	—
Glutamic acid	1030	999	—
Proline	639	582	—
Glycine	450	415	—
Alanine	370	357	—
Half cystine	1509	731	Yes
Valine	487	464	—
Methionine	50	38	Yes
Isoleucine	227	220	—
Leucine	509	485	—
Tyrosine	183	146	Yes
Phenylalanine	139	129	—
Cysteic acid	27	655	Yes
Lysine	198	180	—
Histidine	65	55	—
Arginine	511	486	—

erately bleached hair shows only significantly lower cystine and correspondingly more cysteic acid than unaltered hair. These results support Zahn's original conclusion that the reaction of bleaching agents with human hair protein occurs primarily at the disulfide bonds [46]. Fewer total micromoles of amino acids per gram of hair are found in bleached than in unaltered hair (see Table 2–4), which may be attributed to the addition of oxygen to the sulfur-containing amino acids and to other effects, such as solubilization of protein or protein-derived species into the bleach bath [47].

Products of disulfide oxidation, intermediate in oxidation state between cystine and cysteic acid (Table 2–5), have been shown to be present in wool oxidized by aqueous peracetic acid [48–50]. These same cystine oxides have been demonstrated at low levels in bleached hair [51]; however, disulfide oxidation intermediates have not been shown to exist in more than trace amounts in hair oxidized by currently used bleaching products [52].

The actual presence of large amounts of cysteic acid in bleached hair had at one time been in doubt [53,54]. It had been theorized that the cysteic acid found in bleached hair hydrolysates was formed by decomposition of intermediate oxidation products of cystine during hydrolysis before the analytical procedure [53]. However, differential infrared spectroscopy [4]

TABLE 2–5. Some possible oxidation products of the disulfide bond.

Formula	Name
$R-SO-S-R$	Disulfide monoxide
$R-SO_2-S-R$	Disulfide dioxide
$R-SO_2-SOR$	Disulfide trioxide
$R-SO_2-SO_2-R$	Disulfide tetroxide
$R-SO_3H$	Sulfonic acid

and electron spectroscopy for chemical analysis [5] on intact unhydrolyzed hair have conclusively demonstrated the existence of relatively large quantities of cysteic acid residues in chemically bleached hair. Evidence for other sulfur acids such as sulfinic or sulfenic acids in bleached hair has not been provided, and it is unlikely that these compounds exist in high concentrations in hair because these chemical species are relatively unstable. For details concerning the mechanism of oxidation of sulfur in hair, see Chapter 4.

Permanent-Waved Hair

Nineteen amino acids in human hair have been studied for possible modification during permanent waving; that is, all the amino acids of Table 2–1 except tryptophan and citrulline. Significant decreases in cystine (2%–14%) and corresponding increases in cysteic acid [1,10] and in cysteine [1] have been reported for human hair that has been treated either on the head by home permanent-waving products or as isolated samples in the laboratory by thioglycolic acid and hydrogen peroxide in a simulated permanent-waving process.

Trace quantities (<10 micromol/gm) of thioacetylated lysine and sorbed thioglycolic acid have also been reported in human hair treated by cold-waving reagents [1]. Small quantities of mixed disulfide [5], sorbed dithiodiglycolic acid [55], and methionine sulfone [10] have also been found in hydrolysates of hair treated by the thioglycolate cold-waving process.

$$NH_2-\underset{\displaystyle CO_2H}{\underset{|}{CH}}-(CH_2)_4-NH-CO-CH_2SH$$

Thioacetylated lysine

$$CH_3SO_2-CH_2-CH_2-\underset{\displaystyle CO_2H}{\underset{|}{CH}}-NH_2$$

Methionine sulfone

$$NH_2-\underset{\displaystyle CO_2H}{\underset{|}{CH}}-CH_2-S-S-CH_2-CO_2H$$

Mixed disulfide

$$HOOC-CH_2-S-S-CH_2-COOH$$

Dithiodiglycolic acid

Methionine sulfone is presumably formed by reaction of the neutralizer with methionine residues; thioacetylated lyine is probably formed by reaction of lysine with thioglycolide impurity in the thioglycolic acid [1]. The mixed disulfide is presumably formed by displacement of thioglycolate on the cystine residues in hair (see Chapter 3 for mechanistic details).

Zahn et al. [56,57] have reported that thioglycolate can accelerate the rate of formation of thioether residues (lanthionyl) in wool fiber (see Chapter 3). Therefore, one might expect to find trace quantities of this amino acid in hair permanent-waved in an alkaline medium. Chao et al. [58] have demonstrated small quantities of lanthionine and carboxymethyl thiocysteine (see Chapter 3) in hair reduced by thioglycolic acid.

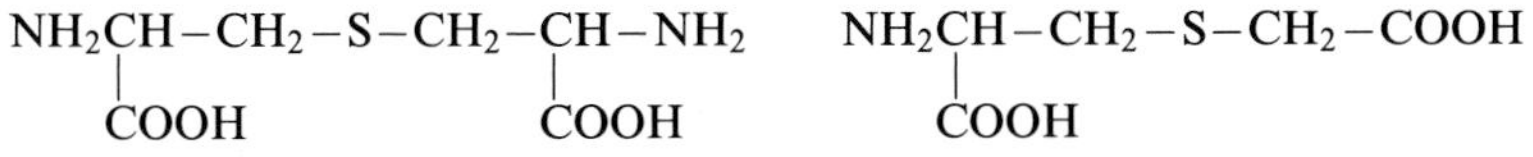

$$NH_2\underset{\substack{|\\ COOH}}{CH}-CH_2-S-CH_2-\underset{\substack{|\\ COOH}}{CH}-NH_2 \qquad NH_2\underset{\substack{|\\ COOH}}{CH}-CH_2-S-CH_2-COOH$$

Lanthionine *Carboxymethyl thiocysteine*

Analytical procedures involving reduction and determination of mercaptan are not accurate determinations of either cystine or cysteine in permanent-waved hair or in hair treated with mercaptan, because mixed disulfide is reduced to mercaptan during analysis, and adsorbed mercaptan can also interfere in the determination. Procedures that do not involve reduction of hair such as ninhydrin detection (α-amino group) or dinitrofluorobenzene (DNFB) reaction followed by chromatographic separation [1,56] discriminate between mercaptans and therefore should be better analytical procedures for detecting the different types of mercaptans and disulfides actually present in permanent waved hair.

Analysis of Acidic and Basic Groups in Whole Human Hair

Both the acid-combining capacity [59,60] and the acid dye-combining capacity [34,61] of unaltered keratin fibers have been used to estimate the frequency of basic groups. Similarly, the base-combining capacity can be used to estimate the frequency of acidic groups in hair [59]. The acid-combining capacity of unaltered human hair fibers is approximately 820 μmol/g [62] and provides an estimate of the frequency of basic amino acid residues, including N-terminal groups (approximately 15 μmol/g) [63] and sorbed alkaline matter, whereas the base-combining capacity provides an estimation of the titratable acidic groups in the fibers, including C-terminal amino acid residues and any sorbed acidic matter.

Alterations to the fibers that affect the apparent frequency of acidic or basic groups, such as hydrolysis, susceptibility to hydrolysis, or the introduction of sulfonic acid groups [24,64], can affect the acid- or base-combining capacity of hair. Therefore, permanent waving, and especially bleaching, can affect these titration parameters [8]. The effects of cosmetic treatments and environment on these parameter are described in detail in Chapter 5.

Chemical Composition of the Different Morphological Components

Cuticle

Bradbury et al. [16] have suggested that the cuticle of human hair contains more cystine, cysteic acid, proline, serine, threonine, isoleucine, methionine, leucine, tyrosine, phenylalanine, and arginine than does the whole fiber. Data calculated from the results of Bradbury and those of Robbins [12] on whole human hair fibers are summarized in Table 2–6. Swift and

TABLE 2–6. Amino acid composition of the different morphological components of hair.[a]

Amino acid	Cuticle[b]	Whole fiber[c]	Medulla[d]
Aspartic acid	287	449	470
Threonine	524	664	140
Serine	1400	1077	270
Glutamic acid	819	1011	2700
Proline	994	667	160
Glycine	611	485	300
Alanine	—	374	400
Half-cystine	2102	1461	Trace
Valine	634	499	320
Methionine	38	53	40
Isoleucine	184	249	130
Leucine	418	516	700
Tyrosine	132	184	320
Phenylalanine	91	142	—
Cysteic acid	68	29	—
Lysine	—	217	740
Histidine	—	71	100
Arginine	360	529	180
Ammonia	—	—	(700)
Citrulline	45	11	—

[a] Data are expressed in micromoles of amino acid per gram of dry hair.
[b] The data for cuticle analysis are based on the work of Bradbury et al. [16], who analyzed cuticle and whole fiber from several keratin sources, including human hair, merino wool, mohair, and alpaca. These scientists concluded that there is very nearly the same difference between the amino acid composition of the cuticle and each of these fibers from which it was derived. They listed the average percentage differences used in these calculations. More recent analyses of cuticle and whole fiber of human hair [65,66] are in general agreement with these data.
[c] Whole-fiber results approximated by cortex analysis [12].
[d] These data are results of analysis of medulla derived from porcupine quill from Rogers [64].

Bews [66] have described comparative cuticle and cuticle-free hair analyses of certain amino acids in human hair, and their data are qualitatively similar to those of Bradbury. In addition, these authors suggest less tryptophan and histidine occur in cuticle than in whole fiber.

In general, these results show that cuticular cells contain a higher percentage than whole fiber of the amino acids that are not usually found in α-helical polypeptides [67]. Small amounts of citrulline (11 μmol/g) have been reported in whole human hair fibers, whereas cuticle is found to be somewhat richer in citrulline (45 μmol/g) with only trace quantities of ornithine (5 μmol/g) [16].

The two main fractions of the hair cuticle, the exocuticle and the endocuticle, have been separated after enzymatic digestion and analyzed [68]. Their chemical compositions are quite different. The proteins of the exocuticle and its A layer are highly cross linked by cystine (more than 30%) and are therefore extremely tough and resilient. In contrast, the proteins of the endocuticle contain very little cystine (3%–6%) and relatively large amounts of the dibasic and diacidic amino acids.

As a result of these large compositional differences, these two layers of the cuticle can be expected to react differently to permanent waves, bleaches, and even to water and surfactants. Raper et al. [69] have described a method to determine the cuticle composition from endocuticle of chemically treated wools. Such a procedure would be valuable to evaluate changes in the endocuticle of cosmetically modified human hair.

Cortex

Because the cortex comprises the major part of the fiber mass, results of whole-fiber analysis may be considered to be a good approximation of the composition of the cortex (see Table 2–6). Obviously, the largest errors resulting from this approximation will be in those amino acids occurring in smaller quantities in the cortex.

Average cortex is rich in cystine (although there is less cystine in cortex than in cuticle), and the cortex is richer in diacidic amino acids and lysine and histidine than is cuticle. However, the two main components of cortex, the intermediate filaments and the matrix, are very different in chemical composition. The intermediate filaments are rich in leucine, glutamic acid, and those amino acids that are generally found in α-helical proteins, although small quantities of cystine (~6%), lysine, and tyrosine are also regularly arranged in the microfibrils [70] (see the section on the intermediate filament proteins in this chapter). On the other hand, the matrix is rich in cystine (~21%, calculated from the sulfur content of γ-keratose of human hair), proline, and those amino acids that resist helix formation [70]. For additional information on the composition of the intermediate filaments and the matrix, see the section on the fractionation and peptide analysis of hair in this chapter.

Medulla

Complete chemical analysis of the medulla of human hair fibers has not been reported. Studies of the medulla of human hair are complicated, because it has poor solubility and is difficult to isolate. In fact, most of the experimental work on medulla has been on African porcupine quill, horse hair, or goat hair medulla rather than medulla of human hair fiber. Rogers [64] has described the amino acid composition of medullary protein isolated from porcupine quill (see Table 2–6).

Blackburn [71] has also determined some of the amino acids from medulla of wool fibers. Although most wool fibers do not contain a medulla, some coarse wools like kemp do contain this porous component. Although Blackburn's results are more qualitative, they agree in general with the data of Rogers, suggesting a low cystine content as compared to whole fiber and relatively large amounts of acidic and basic amino acids. The low cystine content of medullary protein is consistent with its low sulfur content [72].

If one assumes that medullary protein of porcupine quill is representative of medullary protein of human hair, some interesting comparisons can be made of the three morphological regions of human hair. Among the gross differences is the fact that cuticle has an even higher cystine content than does whole fiber and that medulla has only trace quantities of cystine. Medulla also appears to have relatively small amounts of hydroxy-amino acids and relatively large amounts of basic and acidic amino acids, as compared to the other two morphological components of animal hairs. This suggests that the medulla will be more susceptible to reactions with acids and alkalis and to ion exchange reactions; for example, reactions with anionic and cationic surfactants and ionic dyes. However, medulla will also be less sensitive to reaction with reducing agents. One must also consider that because medulla is located at the core of the fiber it is protected both by the cuticle and the cortex and by the slow rate of diffusion through these two morphological regions.

Epicuticle

A proteinaceous substance called epicuticle (a portion of the cell membrane complex; see Chapter 1) has been isolated from wool fiber by Golden et al. [73]. The protein portion of this cell membrane complex material was found to be rich in the dicarboxylic amino acids, aspartic acid, and glutamic acid.

Holmes [74] has isolated a fatty acid protein complex (cell membrane complex) from human hair that appears to protect the hair during papain digestion. Analysis of this complex indicates 20% to 30% fatty acid (lipid material) and 60% to 70% protein, rich in the amino acid lysine. Holmes suggests that this substance is either the epicuticle or a fraction of the epicuticle. To avoid confusion, this material should be called either the cell membrane complex or a portion of the cell membrane complex, which consists of protein and lipid components.

Cell Membrane Complex

Figure 1–12 describes the cell membrane complex in schematic form. The intercellular cement or the delta layer is primarily nonkeratinous protein [75–77], that is, low in cystine content (about 2%); the partial amino acid content is given in Table 2–7.

The cell membranes are usually isolated from keratins after digestion of the hair with an enzyme and a reducing agent such as papain in the presence of dithiothreitol and then extracted with a lipid solvent system such as chloroform: methanol, as in the procedure by Swift and Bews [80,81].

The inert beta layers of the cell membrane complex (Figure 1–12) are lipid-type structures [75] and in human scalp hair there is not as yet a concensus about its composition. Sakamoto et al. [82] claim that fatty acids and wax esters are the main components of the internal lipids of human hair. Zahn [79] also finds fatty acids, cholesteryl esters, and wax esters as main components; however, he also found polar lipids as major components. The fatty acids of this important component of human hair are predominately palmitic, stearic and oleic acids.

Leeder et al. [75] analyzed the lipid composition of wool fibers after removing surface grease. Continued extraction with solvent removed the beta layers as evidenced by electron microscopy; however, the extract contained free cholesterol, free fatty acid and triglycerides, but negligible quantities of the phospholipid normally associated with biological membrane lipids. Koch [83], in his work with internal lipid of human hair, did not report significant quantities of phospholipid. Thus, more work needs to be done to more fully identify both the lipid components and the protein composition of the cell membrane complex of human hair.

TABLE 2–7. Amino acids in intercellular cement of hair and wool fibers.

Amino acid[a]	Wool, mol % [78]	Hair, mol % [79]
Arginine	6.2	8.2
Lysine	4.7	6.0
Histidine	1.7	1.7
Aspartic acid	7.3	5.8
Glutamic acid	10.5	8.0
Half cystine	—	0.8

[a] Protein content is 35.4% of sample for hair.

TABLE 2–8. N-Terminal amino acids in human hair (relative ratios).

Amino acid	μ moles per gram of hair	Reference 63	Reference 7
Valine	4.0	8	4
Threonine	4.0	8	6
Glycine	3.9	8	8
Alanine	1.0	2	2
Serine	1.0	2	2
Glutamic acid	1.0	2	2
Aspartic acid	0.5	1	1
Total	15.4		

Fractionation and Peptide Analysis of Hair

N-Terminal Amino Acids

Kerr and Godin [84], using the dinitrophenylation method of Sanger [85], have identified valine, threonine, glycine, alanine, serine, glutamic acid, and aspartic acid as N-terminal amino acids in human hair. Quantitative data by Speakman [63] and Hahnel [86] for N-terminal amino acids of human hair, using this same procedure, are summarized in Table 2–8. All three of these references identify the same seven amino acids as N-terminal residues in human hair. In addition, there is agreement on the relative quantities of glycine, alanine, serine, glutamic acid, and aspartic acid as N-terminal groups. However, the quantitative data for valine and threonine are in discord. The apparent disagreement of these data may result from differences in the relative ratios of the different proteins in the different samples caused by either sampling or experimental procedures. Speakman [63] has reported these same seven amino acids as N-terminal residues in three different types of wool fiber.

C-Terminal Amino Acids

The C-terminal amino acids in human hair have been identified by Kerr and Godin [84] using the hydrazinolysis method of Niu and Fraenkel-Conrat [87]. These amino acids are threonine, glycine, alanine, serine, glutamic acid, and aspartic acid. Interestingly, all six of these amino acids also serve as N-terminal residues. These same six C-terminal amino acids have been identified as C-terminal residues in wool fiber [88].

Fractionation Procedures

More extensive peptide investigations of keratin fibers generally consist of solubilization of the keratin; separation of the resultant mixture by means of solubility, chromatography, or electrophoresis; and analysis of the resultant fractions.

The most commonly used method for preparing keratins for sequencing or peptide analysis consists of solubilizing the keratins with strong reducing solutions, usually salts of dithiothreitol or thioglycolic acid [89]. In this procedure, the reduced keratin is generally reacted with iodoacetic acid, forming the *S*-carboxy methyl keratin (SCMK) derivatives [90] to enhance the solubility of the proteinaceous matter and to prevent reoxidation of the thiol groups. Radiolabeled iodoacetic acid is often used to tag the fractions, and gel electrophoresis is used to separate the different protein fractions.

$$\mathrm{K{-}SH + I{-}CH_2{-}COOH \longrightarrow \underset{(SCMK)}{K{-}S{-}CH_2{-}COOH} + HI}$$

Considerable effort has gone into the fractionation of wool fiber into its major protein components and the characterization of the resultant fractions, and the mysteries underlying the detailed structures of the major polypeptides of wool and human hair fibers are beginning to be revealed. The papers by Crewther et al. [91,92], Gillespie and Marshall [43], Corfield et al. [93], Cole et al. [94], Chapters 2 and 3 in the book by Fraser et al. [70], Fraser's paper [95], the book by Rogers et al., [96], and the paper by Zahn [97] are leading entries into this work.

Major Protein Fractions of Hair

During the past decade, a considerable amount of work has been done on the fractionation and amino acid sequencing of the major proteins of human hair (the term major is used here in the sense of more, or the highest concentration of proteins in the fibers). The following abbreviations are commonly used to describe the more important protein types under investigation:

IF, intermediate filament proteins
HS, high-sulfur proteins
UHS, ultrahigh-sulfur proteins
IFAP, intermediate filament-associated proteins
HT, high-tyrosine proteins
HGT, high-glycine tyrosine proteins.

Rogers [96] suggests extracting the hair with dithiothreitol in alkali and 8 M urea, labeling with C-14 iodoacetic acid at pH 8, and separating by polyacrylamide gel electrophoresis (PAGE) in sodium dodecyl sulfate (SDS) solution. This procedure provides a separation for human hair into two major fractions, consisting of high-sulfur proteins that are from the matrix and classified as IFAP proteins and a second fraction of low-sulfur proteins which are IF material. A third fraction from wool fibers that is not present in human hair consists of HGT proteins, which are also matrix or IFAP proteins.

The IFAP Proteins of Human Hair

The HS proteins generally contain about 20% of the residues as half-cystine and the UHS proteins usually contain between 30% and 35% residues of half-cystine. These latter proteins in wool have been shown to be affected by the cystine/cysteine level in the wool follicle [98], which is determined by the cystine/cysteine level in the plasma. Proline generally occurs in the HS proteins at a relatively constant level (~7%–9%) and has been suggested as an indicator for HS proteins. Marshall [28] suggested that the half-cystine content of normal hair should be in the range of 17% to 18% and should not vary from age but should vary only from sunlight, cosmetic treatment, and biochemical abnormalities.

Although the HS and UHS proteins are rich in half-cystine, they contain virtually no methionine. Thus, methionine in the diet is important to these proteins because it can be converted into cysteine [99]. The important role of cystine/cysteine in protein synthesis and to hair growth in the follicle is summarized well by Reis [99].

Cosmetic advertisments abound with the suggested or implied nutrient or health benefit claims provided by proteins or vitamins or even provitamins in cosmetic products. Marshall and Gillespie [100] offer the following conclusion with regard to nutrition and hair. "In healthy humans, it is unlikely that any significant variation in the proteins of hair will result from normal changes in nutrition." If it is not likely that changes in the proteins (which determine the structure) of hair can be induced by ingestion or injection of nutrients into humans who are healthy, that is not malnourished, then it is much less likely that such changes could ever be induced from these same ingredients or their precursors when applied topically in a shampoo or a hair conditioner.

Indeed, there have been no systematic studies of the effects of nutrients such as vitamins on the rate of wool or hair growth or structure. However, there are some indications that in cases of dietary insufficiencies supplements of folic acid (a B-complex vitamin) or pyridoxine (a B-complex vitamin, B6) could be helpful to hair growth because these vitamins play a role in cysteine metabolism [99]. On the other hand, panthenol, the precursor to pantothenic acid (another B-complex vitamin) has never been demonstrated by any published scientific study to affect the nutrition or growth of hair. In a review on nutrition and hair, Flesch [101] reported, "There is no objective evidence available to support the assumption that pantothenic acid has a biochemical role in the production of hair." The more recent review by Reis [99] refers to this review by Flesch and confirms Flesch's conclusions about vitamins and hair. Thus, in spite of current advertising about provitamin B5, there is no current objective evidence to support a nutritional benefit to hair from this species.

Among sheep with dietary insufficiencies, the minerals copper and zinc, when supplemented to the diet, have been shown to be important to wool fiber growth. This effectiveness is attributed to the important roles these minerals play in sulfur amino acid metabolism: copper serves to catalyze the oxidation of cysteine to cystine during fiber synthesis [102], and zinc is required for cell division and also appears to play a role in protein metabolism [103].

The UHS proteins were discovered more than two decades ago, but their sequences are still not well characterized.

The Intermediate Filaments

The intermediate filaments in different tissues appear similar in form (see Figure 1–29) but they differ considerably in their exact composition. The most similar structural feature is the central helical rod [104]. On the other hand, the primary differences are in the amino and carboxyl domains, which vary in both amino acid sequences and size [105].

The intermediate filament molecules in keratins are composed of two different types of polypeptides designated as Type I (acidic side chains) and Type II (neutral to basic side chains). These chains are coiled about each other forming a two strand coiled-coil rope; each filament thus requires one acidic polypeptide that coils about a basic polypeptide partner or mate. In human hair, there are at least five of these low-sulfur proteins, two Type I and three Type II proteins [106].

The cystine content of the low-sulfur intermediate filament regions is about 6% and is not uniformly dispersed between domains of an intermediate filament chain. The rod domain contains about only 3% half-cystine or as little as 1 half-cystine residue, while the N-terminal domain contains about 11% half-cystine and the C-terminal unit about 17% half-cystine [95]. Fraser [95] also suggests that these half-cystine residues are involved in disulfide linkages and that most disulfide residues exist in the matrix.

The Type I intermediate filament proteins of human hair represent a class of proteins about 44K and 46K in molecular weight, while the Type II proteins are about 50K, 59K, and 60K [106]. Amino acid sequences for the intermediate filament polypeptides from several proteins including wool fiber have been described by Crewther et al. [92].

Tricohyalin Protein

Tricohyalin is a granular, proteinaceous material found in the cytoplasm of cells of the inner root sheath, which envelops the growing hair fiber (see Figure 1–4). It is a major protein synthesized during hair growth and can also be found in the medulla of fully formed hair fibers. However, its role in the growth of human hair fibers is not fully understood at this time. The amino acid composition of tricohyalin protein found in sheep and guinea pig follicles has been reported by Rogers et al. [107].

TABLE 2–9. Percent keratoses in human hair.

Fiber type	α-Keratose	β-Keratose	γ-Keratose	Total
Merino wool	56 (1.88)[a]	10 (2.13)	25 (5.84)	91
Caucasian hair	43 (2.38)	14 (4.00)	33 (6.60)	90

[a] Percent sulfur in parentheses.

Other Fractionation Methods

Another method, that of Alexander and Earland [108], consists of oxidation of the disulfide bonds of the keratin to sulfonic acid groups, using aqueous peracetic acid solution, and separation of the oxidized proteins, generally by means of differences in solubilities of the different components of the mixture. The first three fractions in this separation are called keratoses. The amino acid composition of these three fractions as isolated from merino wool has been reported by Corfield et al. [109].

Fractionation of human hair into keratoses by the method of Alexander and Earland [108], as modified by Corfield et al. [109], has been reported by Menkart et al. [24] (Table 2–9). This procedure consists of oxidation of the fibers with aqueous peracetic acid and solubilization in dilute alkali. The insoluble fraction is called β-keratose and is believed to consist of proteins derived primarily from cell membranes and similar matter. Acidification of the solution to pH 4.0 produces a precipitate called α-keratose, believed to originate primarily in the crystalline or fibrillar regions of the cortex. The material remaining in solution has been labeled γ-keratose, which is the fraction containing the largest percentage of sulfur (see Table 2–9) and is believed to consist of proteins derived primarily from the amorphous regions of the fibers (primarily from the matrix). Of special interest is the significantly larger γ-keratose fraction from human hair compared to merino wool (see Table 2–9); this is consistent with the higher cystine content in human hair.

Using a similar procedure, Crounse [110] examined a portion of the α-keratose fraction by quantitative amino acid analysis. He found simi lar quantities of amino acids, except cystine, cysteine, and glycine, when this fraction was obtained from human hair as compared to nails and epidermis.

A modified version of this procedure, described by Wolfram and Milligan [111], involves esterification of the carboxyl groups which are believed to reside primarily on the α-helical proteins and proteins of the hair surface. Esterification decreases the solubility of these proteins, allowing the nonesterified proteins (of the matrix) to be extracted more easily. The soluble fraction of this procedure is called γ*keratose; it resembles γ-keratose but provides a higher yield. The insoluble residue exhibits birefringence and is called the α-β*keratose fraction.

Other fractionations of human hair have been reported by Andrews [112] and by Lustig et al. [113]. The former paper describes a hydrolytic separation and the latter a fractionation by sulfonation followed by reduction [18]. These procedures have not been pursued to a great extent because of the inherent amino acid degradation in the initial solublization reaction.

Water: A Fundamental Component of Human Hair

Table 2–10 summarizes the effects of relative humidity (RH) on the water content of human hair [114]. (Additional details are given in Tables 8–14, 8–15, and 8–16.) Obviously, the determined moisture content of keratin fibers depends on the conditions selected as the state of dryness of the fibers [115] as well as on the RH. The amount of moisture in hair also plays a critical role in its physical and cosmetic properties, as described in Chapters 3 and 8. The data of Table 2–10 were obtained by dehydration of the fibers in a dry box over calcium chloride and determining the water regained at increasing humidities. Chamberlain and Speakman [116] have reported the moisture content of human hair by moisture regain from the dry state and by way of dehydration from 100% RH. Their data show a hysteresis in which the moisture contents at intermediate humidities are slightly lower by the hydration method than by dehydration. This phenomenon is described in more detail in Chapter 8.

Similarily, hair dried with heat will exhibit a lower moisture content than hair dried at room temperature [117]. After heat-drying, hair absorbs moisture but does not return to the room temperature-dried moisture level until it is either rewet with water or conditioned at a higher relative humidity. Thus, a hysteresis exists between heat-dried hair and room temperature dried-hair similar to that from absorption versus desorption of moisture.

TABLE 2–10. Water content of hair at different relative humidities.

RH[a]	Approximate moisture content[b]
29.2	6.0%
40.3	7.6%
50.0	9.8%
65.0	12.8%
70.3	13.6%

[a] Temperature, 74°F.
[b] Each value is an average of five determinations on dark-brown hair purchased from a hair supplier. The hair was not extracted with solvent.

Hysteresis phenomena in the water sorption by high polymers [118] and by other proteins such as wool fiber [118] and casein [119] have also been described. Smith [118] suggests that hysteresis is a result of differences in the ratio of "bound" to "free" water in the substrate, with a larger amount of bound water present on desorption than on absorption.

Undoubtedly, the several hydrophilic side chains (guanidino, amino, carboxyl, hydroxyl, phenolic, etc.) and peptide bonds of keratin fibers contribute to water sorption, although there is controversy over the primary water-binding groups. Leeder and Watt [120], in a very interesting study involving water sorption of unaltered and deaminated wool fibers, concluded that the binding of water by amino and guanidino groups is responsible for a large percentage of the water sorption capacity of keratin fibers, especially at low humidities. On the other hand, Breuer concluded that the peptide bonds are preferential sites for hydration [62].

The conclusions of Leeder and Watt are supported by Pauling [121], who describes the negligible attraction of water by the polypeptide nylon and the apparent agreement between the number of molecules of water initially sorbed by several proteins and the number of polar side-chain groups in those proteins.

Spectroscopic studies of the nuclear magnetic resonance (NMR) of both human hair [122] and wool fiber [123] indicate that the protons of water in keratin fibers are hydrogen bonded and are less mobile than in the bulk liquid. At RHs less than 25%, water molecules are principally bonded to hydrophilic sites of the fiber by hydrogen bonds and can be described by Langmuir's fundamental theory for the absorption of gases on solids [118]. As the humidity increases, additional water is sorbed, producing a decrease in the energy of binding of water already associated with the protein. At very high RHs, greater than 80%, multimolecular sorption (water on water) becomes increasingly important.

Feughelman and Haly [124] and Cassie [125] have suggested two different models for estimating the amounts of bound "unmobile" and mobile "free or liquid" water present in keratin fibers. Feughelman and Haly define bound water as water associated with the keratin structure and mobile water as water not associated with the keratin structure. This model considers the decrease in energy of binding of water molecules already associated with the keratin structure with increasing water content. King [126] discusses two- and three-phase adsorption theories to explain the adsorption of moisture by textile materials. His conclusions and cautions are pertinent to the same phenomenon in human hair. King suggests that it is relatively easy to derive a sorption isotherm that fits an empirical relation using two or three adjustable coefficients, and he cautions others in keratin science to make sure the theory they consider does not contradict accepted physical principles. The effects of water on swelling, friction, tensile, and other properties of human hair are described in Chapter 8.

Composition of Human Hair Lipid

Lipid extracted from human hair is similar in composition to scalp lipid [127]. The bulk of the extractable lipid in hair thus is free lipid; however, cell membrane complex lipid is also partially removed by extraction of hair with lipid solvents or surfactants. In a sense, the scalp serves as a lipid supply system for the hair, with sebum being produced continuously by the sebaceous glands [128]. Sebum production is controlled hormonally by androgens which increase cell proliferation in the sebaceous glands, and this in turn increases sebum production [128,129], although seasonal and even daily variations in the rate of sebum production do occur [130].

The aging of the sebaceous glands in man is primarily controlled by endocrine secretions [129]. For children, sebaceous secretion is low until puberty, when a large increase in sebaceous activity occurs (Figure 2–1). Note that the data did not permit a plot of the entire age curve for females; however, the same general effect of low sebaceous activity for males and females before puberty does exist. For all ages, sebaceous gland activity is lower for women than for men [129], and at menopause (generally between ages 40 and 50), there is a distinct decrease in sebum secretion to even lower levels, (see Figure 2–1). For men, there is more of a gradual reduction in sebum secretion with age after 30 years. Strauss and Pochi [129] concluded that androgenic secretions are directly responsible for sebaceous gland activity and development in both males and females.

Extraction of human hair with "fat solvents" removes approximately 1% to 9% matter. Ethanol, a solvent that swells hair, removes more lipid from

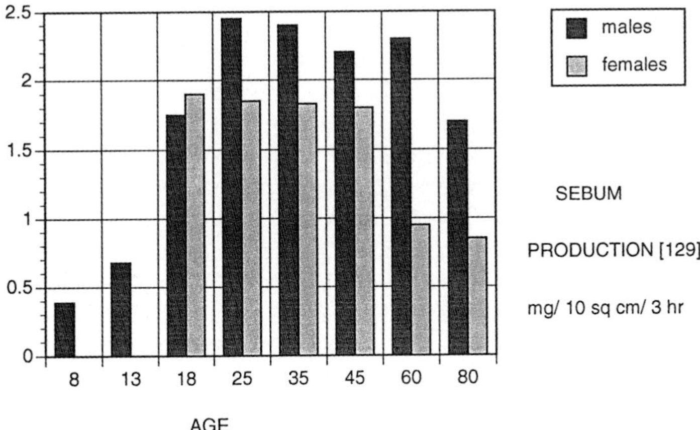

FIGURE 2–1. The variation of sebum production with age. (The data are from Pochi et al. [129].)

TABLE 2–11. Spangler synthetic sebum.

Lipid ingredient	Percentage
Olive oil (TG)	20
Coconut oil (TG)	15
Palmitic acid (FFA)	10
Stearic acid (FFA)	5
Oleic acid (FFA)	15
Paraffin wax (P)	10
Squalene (S)	5
Spermaceti (WE)	15
Cholesterol (C)	5

hair than nonswelling solvents such as benzene, ether, or chloroform. Hair consists of surface (external) and internal lipid. In addition, part of the internal lipid is free lipid and part is structural lipid of the cell membrane complex. The cell membrane complex is laminar in structure and is composed of both protein and lipid layers; however, this structural lipid is not phospholipid like those normally associated with bilayers of cell membranes (see the sections on epicuticle and intercellular matter in Chapters 1 and 2).

The data (1%–9% extracted hair lipid) represent total matter extracted from hair clippings of individual men and women. Although the conditions for extraction can influence the amount of matter extracted from hair [131], the values here represent "approximate" maxima and serve to indicate the variation in the amount of extractable material from hair among individuals. Presumably, the principal material in these extracts is derived from sebum and consists primarily of free fatty acids (FFA) and neutral fat (esters, waxes, hydrocarbons, and alcohols). Gloor [130] classifies the different components of sebum into six convenient groups: free fatty acids (FFA), triglycerides (TG), free cholesterol (C), cholesterol and wax esters (C and WE), paraffins (P), and squalene (S).

The Spangler synthetic sebum (Table 2–11) provides a working formula to represent an imitation of average sebum, and it contains components to represent each of the six components of Gloor's classification for sebum. Nicolaides and Foster [132], by examination of ether extracts of pooled hair clippings from adult males, found 56.1% as FFA and 41.6% as neutral fat. In contrast, daily soaking of the scalps of adults (males) in ether provided 30.7% FFA and 67.6% neutral fat. Nicolaides and Rothman [127] have suggested that this apparent discrepancy may be attributed to lipolytic hydrolysis of glycerides in the stored hair clippings.

Analysis of the FFA extracted from pooled hair clippings of adult males has been carried out by Weitkamp et al. [133]. Their study did not contain data concerning the effect of lipolysis on the structures of FFA found in hair fat. Saturated and unsaturated fatty acids ranging in chain length

TABLE 2–12. Composition of free fatty acids (FFA) in human hair lipid.

Chain length	Total FFA (%)	Unsaturated FFA of this chain length (%)
7	0.07	—
8	0.15	—
9	0.20	—
10	0.33	—
11	0.15	—
12	3.50	4
13	1.40	3
14	9.50	15
15	6.00	25
16	36.00	50
17	6.00	67
18	23.00	80
20	8.50	85
22	2.00	—
Residue	4.00	—
Total	100.80	

from 5 to 22 carbon atoms have been found in human hair fat [133, 134]. Location of the double bond in the unsaturated acids is suggested to occur at the 6,7 position, with some 8,9 and other isomers. Data from the study by Weitkamp et al. [133] are summarized in Table 2–12. In addition to the acids reported by Weitkamp et al. [133], Gershbein and Metcalf [134], by examination of the total fatty-acid content (following saponification) of human hair fat, found traces of C5 and C6 carboxylic acid and small quantities of C19 and C21 acids, as well as branched-chain isomers of many of the acids shown in Table 2–12 [134].

Comparison of the FFA content [133] with the total (hydrolyzed) fatty acid content [134] is summarized in Table 2–13. This comparison is made assuming that data from different laboratories are comparable.

With the exception of the C16 and C20 acids, the data in the second and third columns of Table 2–13 are very similar for each corresponding acid. Equivalence suggests that the relative amounts of each acid in ester form would be the same as the relative ratios of the free acids, and that hydrolysis may occur on standing (or other conditions) to increase the ratio of FFA to esters [127]. The noteworthy exceptions are the C16 saturated acid, which must exist in the ester form to a greater extent than is suggested by the relative ratios of free acids and the C20 unsaturated acid, which was found only in trace quantities by Gershbein and Metcalf [134]. A further conclusion from these studies is that the principal acyl groups present in human hair lipids are from the C16 fatty acids.

Analysis of some of the neutral material from human hair lipid, for example, triglycerides, cholesterol or wax esters, and paraffins, provides a mix-

TABLE 2–13. Comparison of FFA content of human hair with total fatty acid content.[a]

Chain length	Total FFA (%) [113]	Total Fatty Acids (%) [114]
12	3.36	2.19
13	1.36	—
14	8.10	8.40
15	5.50	6.70
16	18.00	24.90
17	2.00	2.30
18	5.00	4.60
20	1.30	—
	Relative ratio to C-14 FFA	Relative ratio to C-14 total fatty acids
12	0.4	0.3
13	0.2	—
14	1.0	1.0
15	0.7	0.8
16	2.2	3.0
17	0.3	0.3
18	0.6	0.6
20	0.2	—

[a] Only those acids with content greater than 1% are listed.

ture as complex as that of the fatty acids [127,132,133,135]. Although not all of the compounds of these different components of sebum have been fully analyzed, it is obvious from the discussion on fatty acids and the literature on wax alcohols in human hair lipid [135–138] that the variation in chain length and isomer distribution of all of these esters must be extremely complex.

It was indicated previously that the amount of sebaceous secretion changes with age near puberty. The composition of the sebaceous secretion also changes with age near puberty. Nicolaides and Rothman [139] have shown that the paraffinic hydrocarbon content of sebum is highest in children (boys), lower in men, and lowest in women, and these same two scientists have shown that the squalene content of the hair lipid of children, at approximately 1.35% of the total lipid content, is about one-fourth that of adults (sebum from boys aged 6–12 year was examined in this study and compared to that from both men and women). In addition, the cholesterol content of the hair lipid of children is less than that from adults: 3.7% versus 12.2% [139].

Nicolaides and Rothman [127] have shown with small sample sizes that hair from Blacks contains more lipid than hair from Caucasians. The distribution of fatty alcohols of human hair lipid was examined by Gershbein and O'Neill [135] to determine the relative amounts of fatty alcohols and sterols with regard to sex, race, and scalp condition. Samples originated from Caucasians and Blacks, both full-headed and balding, and from Caucasian women. The data indicated essentially no differences among these parameters between the two racial groups or between the sexes.

Several other factors relevant to differences in sebum composition on the scalp have been described in the literature. Anionic surfactants or ether extraction of the scalp does not stimulate the rate of refatting [140]; however, in a shampoo system selenium disulfide does increase sebum production [141,142] and also alters the ratio of triglycerides to free fatty acids found in sebum. Presumably, the latter effect involves reducing the microflora responsible for lipolytic enzymes on the scalp that hydrolyze triglycerides to free fatty acids. Zinc pyrithione appears to behave similarly and has been shown to increase hair greasiness [143], presumably in an analogous manner.

Several studies have shown significant differences in the composition of oily versus dry hair. Perhaps the most comprehensive study in this regard was by Koch et al. [144], who examined hair surface lipid from 20 dry- and oily-haired subjects, 3 days after shampooing, and found the following correlations with increasing hair oiliness:

an increasing percentage of wax esters in the lipid
an increasing ratio of unsaturated to saturated fatty acids
an increasing amount of monoglycerides
a decreasing percentage of cholesterol esters.

The quantity of total lipid was not found by Koch to correlate with hair oiliness. However, this is not surprising (in a several-day study), because the quantity of lipid on hair tends to level after a few days from the last shampoo owing to removal of excess lipid by rubbing against objects such as combs or brushes and even pillows and hats.

Koch et al. [144] explained oily versus dry hair by the rheological characteristics of the resultant scalp lipid. For example, increasing the ratio of unsaturated to saturated fatty acids should decrease the melting point of the sebum, making it more fluid or more oily. Monoglycerides are surface active and therefore should enhance the distribution of sebum over the hair [144]. Factors such as fiber cross-sectional area or hair curliness were kept constant in Koch's experiments and thus were not considered; however, one would expect the degree of oiliness to most affect straight, fine hair and to have the least cosmetic effect on curly coarse hair [145].

Bore et al. [146] found that the structures of the C18 fatty acids of oily and dry hair differ. For subjects with dry hair, the predominant isomer is octadecenoic acid (oleic acid), whereas 8-octadecenoic acid was the predominant isomer for subjects with oily hair. Thus oily hair is different from dry hair in its chemical composition and in its rheological character.

Hair lipid plays a critical role in shampoo evaluation [131] and in surface effects of hair, such as frictional effects [147]. (See Chapter 5 for discussion of the removal of hair lipid by shampoos.)

Trace Elements In Human Hair

There are a number of studies describing the quantitative determination of various elements of human hair other than carbon, hydrogen, nitrogen, oxygen, and sulfur. In particular, the inorganic constituents of human hair appear to be receiving some attention because of their potential in diagnostic medicine [148] and to a lesser degree in forensic science. The mineral content of human hair fibers is generally very low (<1%), and it is difficult to determine whether this inorganic matter is derived from an extraneous source (which much of it is) or whether it arises during fiber synthesis. Regardless of its origin, the principal metal content of human hair probably exists as an integral part of the fiber structure, that is, as salt linkages or coordination complexes with the side chains of the proteins or pigment, although the possiblity of mineral deposits or compound deposits as in soap deposition also exists.

Pautard [149] reports the total ash content of human hair to be as low as 0.26% of the dry weight of the fibers, but Dutcher and Rothman [150] report ash contents to vary from 0.55% to 0.94%. Among the trace elements reported in human hair are Ca, Mg, Sr, B, Al, Na, K, Zn, Cu, Mn, Fe, Ag, Au, Hg, As, Pb, Sb, Ti, W, V, Mo, I, P, and Se. The actual origin of most of these elements in human hair is from a variety of sources, which are described later. However, from a study involving quantitative analysis of 13 elements in human hair and in hair wash solutions, Bate et al. [151] concluded that a large portion of the trace elements in the hair they analyzed were from sweat deposits.

In the case of metals, the water supply generally provides calcium and magnesium to hair. Common transition metals such as iron and manganese can also deposit in hair from the water supply, and copper from swimming pools has been reported to turn blond hair green at very low concentrations [152]. Other sources of metals in hair are sweat deposits, diet, air pollution, and metabolic irregularities. Metal contamination can also arise from hair products that provide zinc or selenium (antidandruff products), potassium, sodium, or magnesium (soaps or shampoos), and even lead from lead acetate-containing hair dyes.

Although heavy metals occur at low concentrations in human hair, they sometimes accumulate at concentrations well above those levels present in blood or urine, and concentrations of metals such as cadmium, arsenic, mercury, and lead in hair tend to correlate with the amounts of these same metals in internal organs [148]; this is essentially why hair is being considered as a diagnostic tool. Wesenberg et al. [153] found a positive correlation between cadmium levels of hair and target organs (femur, kidney, liver, spleen, heart, muscle tissue, and adrenal glands of Wistar rats). Fowler [154] indicates that the highest levels of arsenic in humans are normally found in hair, nails, and skin, and it is well known that human hair serves as a tissue for the localization of arsenic during arsenic poisoning. Santoprete [155] reported high levels of mercury in hair, feces, urine, nails, skin, and perspiration of human volunteers given a mercury-containing diet. Heavy metals such as lead can also arise from air pollution; for example, significantly higher concentrations of lead were found in hair of 200 persons living within 5 km of a lead smelter than in a control group of 200 persons living at a distance more than 10 km from that same pollution source [156]. Dutcher and Rothman [150] have provided data showing that both the ash content and the iron content of red hair is higher than that of hair of other colors. Stoves [157] has provided evidence that black hair of dogs contains more iron than white hair. However, he has also shown that black hair has a greater capacity for sorption of iron as ferrous sulfate than does white hair. In addition, the iron content of hair can be used to detect iron deficiencies [148].

Analysis of hair can often serve as a lead to even more complicated disorders. For example, a study by Capel et al. [158] indicated significantly higher concentrations of cadmium in hair from dyslexic children than in a normal control group. These scientists suggest that cadmium analysis of hair may be used in early detection and that excessive cadmium may be involved in this type of learning disorder.

The hair of children with cystic fibrosis has been found to contain several times the normal level of sodium and considerably less than normal calcium [148], and persons suffering from phenylketonuria (phenyl ketones in the urine) have less-than-average concentrations of calcium and magnesium in their hair [148]. The hair of victims of kwashiorkor contains higher than normal levels of zinc [148] and low levels of sulfur and the cystine-rich proteins [42]. Hair analysis is also being considered as a screening tool for diabetes, because low levels of chromium in the hair have been demonstrated in victims of juvenile-onset diabetes [148]. Hair analysis offers possibilites for diagnosis of several other maladies or disabilities. For a more complete summary of this subject, see the review by Maugh [148] and the book edited by Brown and Crounse [159].

Hair Analysis for Drugs

Hair analysis for drugs of abuse has been described to detect cocaine [160,161], marijuana [162], nicotine [163], opiates [164], and amphetamines [163,165,166]. Originally, drugs were extracted from the hair followed by gas chromatograph/mass spectrometry (GC/MS) analysis for the drug. More recently, the hair is dissolved and antibodies used in radioimmunoassay (RIA) act as specific agents for extraction/analysis

[162,167,168]. The analysis is generally by GC/MS. Some distinct advantages exist over urinalysis, such as the detection of long-term drug usage. However, drug usage over the most recent few days is not detectable by hair analysis, because of its slow growth and the slow incorporation of drugs into the hair fiber. All analytical methods have limits. Although hair analysis does appear to offer real potential, the limits for hair analysis are still in the process of being defined [167].

A review of hair analysis for drugs of abuse in the paper by Baumgartner [168] addresses some of these limits. One concern expressed in the literature for hair analysis is the potential for false positives created by contamination by passive environmental exposure, for example, smoking of PCP, marijuana, etc. The review by Baumgartner speaks to this concern by prewashing the hair to remove the passive contaminants and not the material deposited in the cortex through the bloodstream. Environmental contamination via smoke should provide for only superficial rather than deeply penetrated drugs, and the route of penetration by passive environmental exposure should be the intercellular rather than the transcellular route, which might also provide for discrimination.

Forensic Studies and DNA Analysis

Hair fibers are frequently found at crime scenes, and they are usually evaluated first for a large number of macroscopic and microscopic comparisons for identification [169,170]. Characteristics such as color, pigment size, pigment distribution, pigment density, whether the fiber has been dyed, type of medulla, maximum and minimum diameter, type of cut at tip, length, scale count, and various cross-sectional characteristics are used in this evaluation. Such comparisons have been invaluable for either excluding or incriminating suspects in crimes. However, more recently, several newer techniques have been developed including blood grouping analysis [171], DNA analysis [171–176], and drug analysis (see the previous section), which together provide for even more conclusive evidence for either excluding or targeting a suspect.

For DNA analysis, the specimen must be either plucked or shed, because it must contain root or root sheath material for DNA to be extracted for further workup and identification. Extraction of DNA from the biological specimen has been described by Walsh et al. (of the Roche Molecular Systems, Emeryville, CA) [176]. After extraction, two methods are used for further analysis: restriction fragment length polymorphism (RFLP) and the technique, polymerase chain reaction (PCR). The RFLP technique [171] provides a very high discriminating power because it discriminates by size and the number of the fragment lengths of the DNA sample. However, it cannot be used with highly degraded DNA and requires much more DNA material than the PCR technique, generally more than is provided by a single hair fiber.

The PCR technique offers many advantages because it requires minimal amounts of DNA and even permits typing from degraded DNA. After extraction, the PCR technique is used to replicate specific sections of a strand of DNA to increase the amount of material for analysis. (For further information, see bulletins describing the Gene Amp Polymerase Chain Reaction Technology and the AmpliType HLA DQα Forensic Typing Kit available from the Cetus Corporation, Emeryville, CA).

The PCR DQα typing technique targets a specific segment of chromosome 6, constituting the strands of an approximately 240-base-pair (240-bp) sequence of the HLA DQa segment. This is accomplished by adding DNA polymerase (enzyme), nucleotides, and primers or short stretches of the known gene sequence of the HLA DQα region to the extracted DNA. Repetitive heating and cooling cycles help to control the replication process. The two strands of DNA that form the double helix of the gene are separated by heating. The primers then bind to the separated strands at the specific location to be replicated, and the enzyme coordinates with the primers to copy the region of interest on each strand and joins the nucleotides together in the exact sequence of the extracted DNA template. This process is repeated until sufficient material of the region of interest is constructed for analysis.

Typing or analysis consists of reaction and color development of the replicated DNA sequence. The main technique at this time involves reaction with eight different oligonucleotide probes followed by color development [174]. These probes distinguish six alleles (A1.1, A1.2, A1.3, A2, A3, A4) that define 21 genotypes with frequences of 0.005 to 0.15 [172,174,176]. Because PCR analysis permits identification of degraded DNA, it can be used to analyze ancient samples of human or even nonhuman DNA. Even newer techniques are under development today to permit quantitation and even faster and more convenient qualitative identification of DNA for forensic, archaeological, and clinical research [177].

References

1. Zahn, H.; Gerthsen, T.; Kehren, M. J. Soc. Cosmet. Chem. 14:529 (1963).
2. Stein, H.; Guarnaccio, J. Anal. Chem. Acta 23:89 (1960).
3. Leach, S.J. Aust. J. Chem. 13:520 (1960).
4. Robbins, C.R. Text. Res. J. 37:811 (1967).
5. Robbins, C.R.; Bahl, M.J. Soc. Cosmet. Chem. 35:379 (1984).
6. Block, R.J.; Bolling, D. The Amino Acid Composition of Proteins and Foods. Thomas, Springfield, IL (1952).
7. Moore, S.; Spackman, D.H.; Stein, W.H. Anal. Chem. 30:1185 (1958).
8. Sagal, J., Jr. Text. Res. J. 35:672 (1965).
9. Robbins, C.R.; Kelly, C.H. J. Soc. Cosmet. Chem. 20:555 (1969).
10. Corfield, M.C.; Robson, A. Biochem. J. 59:62 (1955).
11. Block, R.J.; Weiss, J. In Amino Acid Handbook, p. 6. Thomas, Springfield, IL (1956).

12. Robbins, C.R.; Kelly, C.H. Text. Res. J. 40:891 (1970).
13. Ward, W.H.; Lundgren, H.P. In Advances in Protein Chemistry, Vol. 9 (and references therein). Academic Press, New York (1955).
14. Clay, R.C.; Cook, K.; Routh, J.L. J. Am. Chem. Soc. 62:2709 (1940).
15. Simmonds, D.H. Text Res. J. 28:314 (1958).
16. Bradbury, J.H.; Chapman, G.V.; Hambly. A.N.; King, N.L.R. Nature 210: 1333 (1966).
17. Lang, J.; Lucas, C. Biochem. J. 52:84 (1952).
18. Lustig, B., Kondritzer, A. Arch. Biochem. 8:51 (1945).
19. Block, R.J.; Bolling, D.J. Biol. Chem. 128:181 (1939).
20. Cohn E.J.; Edsall J.T. Proteins, Amino Acids and Peptides. Monograph series. American Chemical Society, New York (1943).
21. Schmidt, C. The Chemistry of Amino Acids and Proteins. Thomas, Springfield, IL (1943).
22. Hawk, P.; et al. Hawk's Physiological Chemistry, Oser, B.L., ed., Chapters 4–6. McGraw-Hill, New York (1965).
23. Shigeyashi, S. Hiroshima Igaku 5:29 (1957).
24. Menkart, J.; Wolfram, L.J.; Mao, I. J. Soc. Cosmet. Chem. 17:769 (1966).
25. Mendel, L.; Vickery, H. Carnegie Inst. Washington Yearbook 28:376 (1929).
26. Beveridge, T.; Lucas, C. Biochem. J. 38:88 (1944).
27. Traham, C.; et al. J. Biol. Chem. 177:529 (1949).
28. Marshall, R.C.; Gillespie, J.M. Proc. 8th Int. Wool Text Res. Conf., Wool Res. Org. N.Z. Publ. I:256 (1990).
29. Shinohara, K.J. Biol. Chem. 109:665 (1935).
30. Ogura, R.; Knox, J.M.; Griffin, A.C.; Kusuhara, M. J. Invest. Dermatol. 38:69 (1962).
31. Wolfram, L.J. In Hair Research, Orfanos, C.E.; Montagna, W.; Stuttgen, G. eds., p. 491. Springer-Verlag, Berlin (1981).
32. Veldsman, D.P. Wool Sci. Rev. 29:33 (1966).
33. Tadokuro, T.; Ugami, K. J. Biochem. (Tokyo) 12:187 (1930).
34. Robbins, C.R.; Scott, G.V.; Barnhurst, J.D. Text. Res. J. 38:1130 (1968).
35. Strasheim, A.; Buijs, K. Biochim. Biophys. Acta 47:538 (1961).
36. Harris, M.; Smith, A. J. Res. N.B.S. 20:563 (1938).
37. Launer, H.F. Text. Res. J. 35:395 (1965).
38. Inglis, A.S.; Lennox, F.G. Text. Res. J. 33:431 (1963).
39. Koyanagi, T.; Takanohashi, T. Nature 192:457 (1961).
40. Bigwood, E.J.; Robazza, R. Volding 16:251 (1955).
41. Noer, A.; Garrigues, J.C. Arch. Mal. Appar. Dig. Mal. Nutr. 45:557 (1956).
42. Gillespie, J.M. In Biology of the Skin and Hair Growth, Lynne, A.G.; Short, B.F., eds. Augus and Robertson, Sydney, Australia (1965).
43. Gillespie, J.M.; Marshall, R.C. Cosmet. Toiletries 95:29 (1980).
44. Kenney, D. Cosmet. Toiletries 96:121 (1981).
45. Cook, M.K. Drug Cosmet. Ind. 99:47 (1966).
46. Zahn, H. J. Soc. Cosmet. Chem. 17:687 (1966).
47. Inglis, A. Text. Res. J. 36:995 (1967).
48. Maclaren, J.A.; Leach, S.J.; Swan J.M. J. Text. Inst. T51:665 (1960).
49. Sweetman, B.J.; Eager, J.; Maclaren, J.A.; Savige, W.E. Proc. 3rd Int. Text. Res. Conf. (Paris) II:62 (1965).
50. Maclaren, J.A.; Savige, W.E.; Sweetman, B.J. Aust. J. Chem. 18:1655 (1965).

51. Zahn, H.; et al. 4th Int. Hair Sci. Symp., Syburg, W. Germany (1984).
52. Nachtigal, J.; Robbins, C. Text. Res. J. 40:454 (1970).
53. Stein, H.; Guarnaccio, J. Text. Res. J. 29:492 (1959).
54. Harris, M.; Smith, A. J. Res. N.B.S. 18:623D (1937).
55. Gerthsen, T.; Gohlke, C. Parfeum Kosmet. 45:277 (1964).
56. Zuber, H.; et al. Proc. Int. Wool Text. Res. Conf. (Australia) C:127 (1955).
57. Zahn, H.; et al. J. Text. Inst. T51:740 (1960).
58. Chao, J.; Newsom, E.; Wainright, I.; Matheus, R.A. J. Soc. Cosmet. Chem. 30:401 (1979).
59. Steinhardt, J.; Harris, M. J. Res. N.B.S. 24:335 (1940).
60. Speakman, J.B.; Elliot, G.H. Symp. Fibrous Proteins Soc. Dyers Col. Leeds 116 (1946).
61. Maclaren, J.A. Arch. Biochem. Biophys. 86:175 (1960).
62. Breuer, M. J. Phys. Chem. 68:2067 (1964).
63. Speakman, J.B. Meilland Text. Ber. 33:823 (1952).
64. Rogers, G.E. In The Epidermis, Montagna, W.; Lobitz, W., eds., p. 202. Academic Press, New York (1964).
65. Wolfram, L.; Lindemann, M. J. Soc. Cosmet. Chem. 22:839 (1971).
66. Swift, J.; Bews, B. J. Soc. Cosmet. Chem. 25:13 (1974).
67. Blout, E.R.; de Loze, C.; Bloom, S.M.; Fasman, G.D. J. Am. Chem. Soc. 82:3787 (1960).
68. Swift, J.; Bews, B. J. Soc. Cosmet. Chem. 27:289 (1976).
69. Raper, K., et al. Text. Res. J. 55:140 (1984).
70. Fraser, R.D.B.; MacRae, T.P.; Rogers, G.E. In Keratins, Their Composition, Structure and Biosynthesis, Chapters 2 and 3. Thomas, Springfield, IL (1972).
71. Blackburn, S. Biochem. J. 43:114 (1948).
72. Barritt, J.; King, A.T. Biochem. J. 25:1075 (1931).
73. Golden, R.L.; et al. Text. Res. J. 25:334 (1955).
74. Holmes, A.W. Nature 189: 923 (1961).
75. Leeder, J.D.; Bishop, D.G.; Jones, L.N. Text. Res. J. 53:402 (1983).
76. Bradbury, J.H., et al. Proc. 3rd Int. Wool Textile Res. Conf. (Paris) 1:359 (1965).
77. Swift, J.A.; Holmes, A.W. Text. Res. J. 35:1014 (1965).
78. Leeder J.D.; Rippon, J.A. J. Text. Inst. 73:149 (1982).
79. Hilterhaus-Bong, S.; Zahn, H. Int. J. Cosmet. Sci. 11:167 (1989).
80. Swift, J.A.; Bews, B. J. Soc. Cosmet. Chem. 25:355 (1974).
81. Swift, J.A.; Bews, B. J. Text. Inst. 65:222 (1974).
82. Sakamoto, O.; et al. J. Am. Oil. Chem. Sci. 54:143A (1977).
83. Koch, J. J. Soc. Cosmet. Chem. 33:317 (1982).
84. Kerr, M.F.; Godin, C. Can. J. Chem. 37:11 (1959).
85. Sanger, F. Biochem. J. 39:507 (1945).
86. Hahnel, R. Arch. Klin. Exp. Dermatol. 209:97 (1959).
87. Niu, C.; Fraenkel-Conrat, H. J. Am. Chem. Soc. 11:5882 (1955).
88. Bradbury, J.H. Biochem. J. 68:482 (1958).
89. Gillespie, J.M.; Lennox, F.G. Biochim. Biophys. Acta 12:481 (1953).
90. Crewther, W.G.; et al. Proc. 3rd Int. Wool Text. Res. Conf. (Paris) I:303 (1965).
91. Crewther, W.G.; Fraser, R.D.B.; Lennox, F.G.; Ludley, H. Adv. Protein Chem. 20:191 (and references therein) (1965).

92. Crewther, W.G.; et al. Int. J. Biol. Macromol. 5:267 (1983).
93. Corfield, M.C.; et al. Proc. 3rd Int. Text. Res. Conf. (Paris) I:205 (and references therein) (1965).
94. Cole, M.; et al. Proc. 3rd Int. Text. Res. Conf. (Paris) I:196 (and references therein) (1965).
95. Fraser, R.B.D.; et al. Int. J. Biol. Macromol. 10:106 (1988).
96. Rogers, G.E.; Reis, P.J.; Ward, K.A.; Marshall, R.C. The Biology of Wool and Hair. Chapman & Hall, London (1989).
97. Zahn, H. Chimia 42:289, (1988).
98. Reis, P.J. In Physiological and Environmental Limitations to Wool Growth, Black H.; Reis P.J., eds., pp. 223–242. University of New England Armidale (1979).
99. Reis, P.J. In The Biology of Wool and Hair, Rogers, G.E.; Reis, P.J.; Ward, K.A.; Marshall, R.L., eds., pp. 185–201. Chapman & Hall, London (1989).
100. Marshall, R.C.; Gillespie, J.M. In The Biology of Wool and Hair, Rogers, G.E.; Reis, P.J.; Ward, K.A.; Marshall, R.C., eds., p. 117. Chapman & Hall, London (1989).
101. Flesch, P. In Physiology and Biochemistry of the Skin, Rothman, S. ed., p. 624, University of Chicago Press, Chicago (1955).
102. Gillespie, J.M. In Biochemistry and Physiology of the Skin, Goldsmith, H., ed., pp. 475–510. Oxford University Press, New York (1983).
103. Hsu, J.M.; et al. Proc. Aust. Soc. Anim. Prod. 15:120, (1984).
104. Steinert, P.; Jones, J.; Goldman, R.D. J. Cell. Biol. 99:225, (1984).
105. Goldman, R.D.; Dessev, G.N. In The Biology of Wool and Hair, Rogers, G.E.; Reis, P.J.; Ward, K.A.; Marshall, R.C., eds., pp. 87–95. Chapman & Hall, London (1989).
106. O'Guin, W.M.; et al. In The Biology of Wool and Hair", Rogers G.E.; Reis, P.J.; Ward, K.A.; Marshall R.C., eds., pp. 37–49. Chapman & Hall, London (1989).
107. Rogers, G.E.; Kuczek, E.S.; MacKinnon, P.J.; Feitz, M.J. et al. In The Biology of Wool and Hair, Rogers G.E.; Reis, P.J.; Ward, K.A.; Marshall, R.C., eds., pp. 69–85. Chapman & Hall, London (1989).
108. Alexander, P.; Earland, C. Nature 166:396 (1950).
109. Corfield, M.C.; et al. Biochem. J. 68:348 (1958).
110. Crounse, R.G. In Biology of the Skin and Hair, Lynn A.G.; Short B.F., eds., p. 307. Angus and Robertson, Sydney (1965).
111. Wolfram, L.J.; Milligan, B. Proc. 5th Int. Wool Text. Res. Conf. (Aachen) 3:242 (1975).
112. Andrews, J.C. Arch. Biochem. 17:115 (1948).
113. Lustig, B.; et al. Arch. Biochem. 8:57 (1945).
114. Anzuino, G. Personal communication.
115. Downes, J.G. Text. Res. J. 31:66 (1961).
116. Chamberlain, N.; Speakman, J.B. J. Electrochem. 37:374 (1931).
117. Crawford, R.J.; Robbins, C.R. J. Soc. Cosmet. Chem. 32:27 (1981).
118. Smith, S. J. Am. Chem. Soc. 69:646 (1947).
119. Mellon, E.F.; et al. J. Am. Chem. Soc. 70:1144 (1948).
120. Leeder, J.D.; Watt, I.C. J. Phys. Chem. 69:3280 (1965).
121. Pauling, L. J. Am. Chem. Soc. 67:555 (1945).
122. Clifford, J.; Sheard, B. Biopolymers 4:1057 (1966).
123. West, A.; et al. Text. Res. J. 31:899 (1961).

124. Feughelman, M.; Haly, A.R. Text. Res. J. 32:966 (1962).
125. Cassie, A.B. Trans. Faraday Soc. 41:458 (1962).
126. King, G. In Moisture in Textiles, Chapter 6, Hearle J.W.S.; Peters R.H., eds. Interscience, New York (1960).
127. Nicolaides, N.; Rothman, S. J. Invest. Dermatol. 21:9 (1953).
128. Kligman, A.M.; Shelly, W.D. J. Invest. Dermatol. 30:99 (1958).
129. Strauss, J.; Pochi, P. In Advances in Biology of Skin, Vol. 4, The Sebaceous Glands, pp. 220–254. Pergamon Press, New York (1963).
130. Gloor, M. In Cosmetic Sciences, Vol. 1, Breuer, M., ed., p. 218. Academic Press, New York (1978).
131. Ester, W.C.; Henkin, H.; Longfellow, J.M. Proc. Sci. Sect. Toilet Goods Association. 20:8 (1953).
132. Nicolaides, N.; Foster, R.C., Jr. J. Am. OilChem. Soc. 33:404 (1956).
133. Weitkamp, A.W.; Smiljanic, A.M.; Rothman, S. J. Am. Chem. Soc. 69:1936 (1938).
134. Gershbein, L.L.; Metcalf, L.D. J. Invest. Dermatol. 46:477 (1966).
135. Gershbein, L.; O'Neill, H.J. J. Invest. Dermatol. 47:16 (1966).
136. Brown, R.A., Young, Y.S.; Nicolaides, N. Anal. Chem. 26:1653 (1954).
137. Haugen, F.W. Biochem. J. 59:302 (1955).
138. Singh, E.; Gershbein, L.L.; O'Neill, H.J. J. Invest. Dermatol. 48:96 (1967).
139. Nicolaides, N.; Rothman, S. J. Invest. Dermatol. 19:389 (1952).
140. Burton, J.L. J. Soc. Cosmet. Chem. 23:241 (1972).
141. Goldschmidt, D.; Kligman, A. Acta Dermatol. Neurol. 48:489 (1968).
142. Bereston, E.S. JAMA 156:1246 (1954).
143. Knott, C.A.; Daykin, K.; Ryan, J. Int. J. Cosmet. Sci. 5:77 (1983).
144. Koch, J.; Aitzemuller, K.; Bittorf, G.; Waibel, J. J. Soc. Cosmet. Chem. 33:317 (1982).
145. Robbins, C.R.; Reich, C. 4th Int. Hair Sci. Symp., Syburg, W. Germany (1984).
146. Bore, P.; Goetz, N.; Cann, L. Int. J. Cosmet. Sci. 2:177 (1980).
147. Scott, G.V.; Robbins, C. J. Soc. Cosmet. Chem. 31:179 (1980).
148. Maugh, T.N. Science 202:1271 (1978).
149. Pautard, F.G.E. Nature 199:531 (1963).
150. Dutcher, T.F.; Rothman, S. J. Invest. Dermatol. 17:65 (1951).
151. Bate, L.C.; Healy, W.B.; Ludwig, T.G. N.Z. J. Sci. 9(3):559 (1966).
152. Bhat, G.R.; Lukenbach, E.R.; Kennedy, R.R. J. Soc. Cosmet. Chem. 30:1 (1979).
153. Wesenberg, G.; et al. Int. J. Environ. Stud. 16(3–4);147 (1981).
154. Fowler, B.A. Handb. Toxicol. Met. p. 293 (1979).
155. Santoprete, G.G. Ital. Med. Lav. 2(3–4):199 (1980).
156. Milosevic, M., et al. Arkh. Hig. Radiat Toksikol. 31(3):209 (1980).
157. Stoves, J.L. Research (Lond.) 6:295 (1953).
158. Capel, I.D.; et al. Soc. Clin. Chem. 27(6):879 (1981).
159. Brown, A.C.; Crounse, R.G., eds. Hair Trace Elements and Human Illness. Praeger, New York (1980).
160. Baumgartner, W.A.; Jones, P.F.; Black, C.T.; Bland, W.H. J. Nuclear Med. (Suppl.) 23(9):790 (1982).
161. Graham, K.; Koren, G.; Klein, J.; Greenwald, M. JAMA 262(23):3328 (1989).
162. Baumgartner, W.A.; Jones, P.F.; Bland, W.H. J. Nuclear Med.

(Suppl.)29(5):980 (1988).
163. Ishiyama, I.; Nagai, T.; Toshida, S. J. Forensic Sci. 28:380 (1983).
164. Marigo, M.; Tagliaro, F.; Polesi, C.; Neri, C. J. Anal. Toxicol. 10:158 (1986).
165. Suzuki, O.; Hahori, H.; Asano, M. J. Forensic Sci. 29:611 (1984).
166. Nagai, T.; et al. Z. Rechtsmed. 101:151 (1988).
167. Bailey, D. JAMA 262(23):3331 (1989).
168. Baumgartner, W.; Hill, V.A.; Bland, W.H. J. Forensic Sci. 34(6):1433 (1989).
169. Gaudette, B.D.; Keeping, E.S. J. Forensic Sci. 19:599 (1974).
170. Gaudette, B.D. J. Forensic Sci. 27:279 (1982).
171. Reynolds, R.; Sensabaugh, G.; Blake, E. Anal. Chem. 63(1):3 (1991).
172. VonBeroldingen, C.H.; Blake, E.T.; Higuchi, R.; Erlich, H.A. Applications of PCR to the analysis of biological evidence. In PCR Technology, Chap. 17, p. 209. Stockton Press, NY (1989).
173. Higuchi, R. Nature 332(6164):543 (1988).
174. Blake, E.; Crim, D.; Mihalovich, J. et al. J. Forensic Sci. 37:700 (1992).
175. Schneider, P.M.; Veit, A.; Rittner, C. PCR-typing of the Human HLA-DQa Locus: population genetics and application in forensic casework. In DNA Technology and Its Forensic Application, Berghaus; Brinkmann; Rittner; Staak, eds. Springer-Verlag, Berlin-Heidelberg (1991).
176. Walsh, P.S.; Erlich, H.A.; Higuchi, R. Research, PCR Methods and Applications, p. 241. Cold Spring Harbor Laboratory, Cold Spring Harbor, NY (1992).
177. Comey, C.T. J. Forensic Sci. 36:1633 (1991).
178. Walsh, P.S. Nucleic Acids Res. 20(19):5061 (1992).

3

Reducing Human Hair

The primary reactions involved in permanent-waving, straightening (relaxing), and depilation of human hair begin with reduction of the disulfide bond. In permanent-waving and hair straightening, reduced hair is stressed, that is, curled or combed straight, while molecular reorientation takes place primarily through a disulfide–mercaptan interchange process. Neutralization is then achieved either through mild oxidation or treatment with alkali (for some sulfite treatments).

Because reduction of the disulfide bond and its subsequent reactions are vital to several important cosmetic products, a large amount of research has been conducted that is relevant to these chemical processes. This chapter is concerned with reduction of the disulfide bond in hair by mercaptans, sulfites, and other reducing agents. Reactions of reduced hair are also considered, followed by a discussion of water setting, set and supercontraction, and swelling of hair, processes especially relevant to permanent-waving. In spite of the fact that research on permanent waving has decreased over the past several decades, a few significant publications have been made recently. Wortmann and Kure [1] have developed a new model that explains and clarifies that the bending stiffness of reduced and oxidized fibers controls the permanent waving behavior of human hair, and a new promising "neutralizer" has been described that involves treating reduced hair with dithioglycolate ester derivatives of polyoxyethylene (mol Wt. 550–750) [2].

The permanent-waving process is considered in detail in this chapter, including the model of Wortmann and Kure and a second useful model by

Feughelman [3], followed by several cold wave and depilatory compositions and procedures for making these same products in the laboratory. The last section of this chapter describes literature relevant to the safety of reducing agents and permanent-wave products.

Reduction of the Disulfide Bond

Equilibrium Constants, Redox Potentials, and pH

Experiments relating to equilibrium reactions of disulfides with mercaptans commonly use reaction times up to 24 h or longer. Although this may seem unrealistic, extremely valuable information with practical implications has been gained from these studies.

The cleavage of the disulfide bond in keratin fibers (I) by mercaptans (II) is a reversible equilibrium reaction summarized by equation A, where the K substituent signifies keratin:

$$\underset{\text{(I)}}{\text{K–S–S–K}} + \underset{\text{(II)}}{2\,\text{R–SH}} \overset{K_A}{\rightleftharpoons} \underset{\text{(III)}}{\text{R–S–S–R}} + \underset{\text{(IV)}}{2\,\text{K–SH}} \tag{A}$$

This reaction actually proceeds through two steps, each a nucleophilic displacement reaction by mercaptide ion on the symmetrical disulfide (I in equation B), and then on the mixed disulfide (V in equation C).

$$\underset{\text{(I)}}{\text{K–S–S–K}} + \text{R–SH} \overset{K_B}{\rightleftharpoons} \text{K–S–S–R} + \text{K–SH} \tag{B}$$

$$\underset{\text{(V)}}{\text{K–S–S–R}} + \text{R–SH} \overset{K_C}{\rightleftharpoons} \text{R–S–S–R} + \text{K–SH} \tag{C}$$

In considering these disulfide scission reactions, the equilibrium constant of the reaction shown in equation A tells to what extent the total process will go to completion.

$$\text{Equilibrium constant} = K_A = \frac{(\text{R–S–S–R})(\text{K–SH})^2}{(\text{K–S–S–K})(\text{R–SH})^2}$$

Among the ways to determine or approximate the equilibrium constant of this type of reaction are (1) analysis of ingredient concentrations at equilibrium, and (2) from redox potentials [4–6]. In either case, one may use cystine as a model for hair, because reports in the literature [4–6] show that the redox potential of "cystine-type" disulfides is virtually independent of the charge group about the disulfide bond. However, reduction potentials of mercaptans do vary with pH [5]. Therefore, equilibrium constants for these reactions will also vary with pH. Patterson et al. [7] have shown that when wool fiber is reacted with 0.2 M thioglycolic acid solution for 20 h, the extent of reduction increases with increasing pH above

6. Assuming equilibrium, this suggests that the difference in redox potential between thioglycolic acid and cysteine in keratin fibers increases with increasing pH above 6, and the equilibrium constant for this reaction increases similarly.

One may approximate the free energies and equilibrium constants of these reactions from these expressions:

$$\Delta F^o = -nfE^o \quad \text{and} \quad \Delta F^o = -RT \ln K_{eq}$$

The number of electrons transferred during the reaction (2) is designated by n, f is Faraday's constant (23,061 calories per volt equivalent), E_o is the difference in standard redox potentials of the two mercaptans in volts, F^o is the standard free energy, R is the gas constant (1.987 calories per degree mole), and T the absolute temperature (°K). These calculations assume standard conditions, that is, and reactants and products are at unit activity.

Equilibrium Constants and Chemical Structure

Equilibrium constants at pH 7 or lower, for the reduction of cystine by simple mercaptans such as cysteine (VI), thioglycolic acid (VII), or even more complex mercaptans such as glutathione (VIII), are all approximately 1 [4,5].

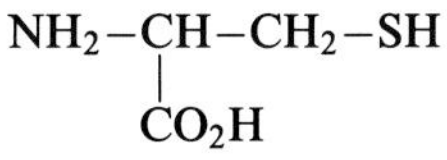

```
NH2–CH–CH2–SH
    |
    CO2H
Cysteine (VI)

HO2C–CH2–SH
Thioglycolic acid
(VII)

   NH–CH2–CO2H
   |
   C=O
   |
 H–C–CH2–SH
   |
   N–H
   |
 O=C–CH2–CH2–CH–CO2H
              |
              NH2
Glutathione (VIII)
```

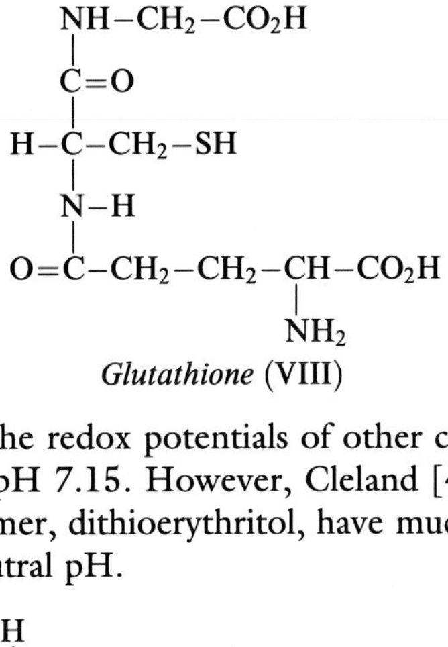

Fruton and Clark [5] have shown that the redox potentials of other cysteine-type mercaptans are very similar at pH 7.15. However, Cleland [4] showed that dithiothreitol (IX) and its isomer, dithioerythritol, have much lower redox potentials than cysteine at neutral pH.

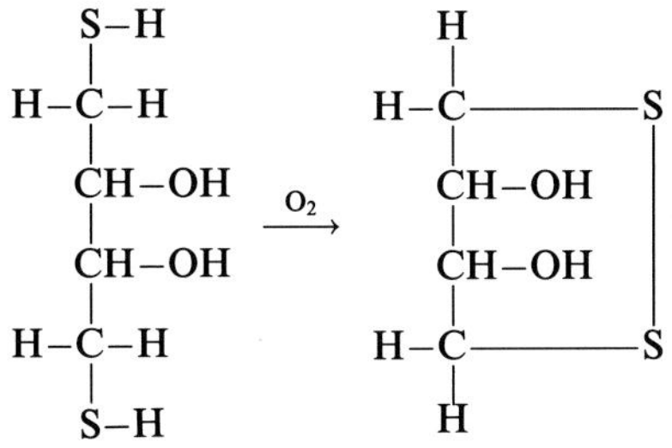

```
   S–H                  H
   |                    |
 H–C–H                H–C––––––S
   |                    |      |
   CH–OH     O2         CH–OH  |
   |        ––→         |      |
   CH–OH                CH–OH  |
   |                    |      |
 H–C–H                H–C––––––S
   |                    |
   S–H                  H
Dithiothreitol (IX)
```

Weigmann and Rebenfeld [8] have reacted IX with wool fiber, showing that complete reduction of cystinyl residues can be approached at pH 6 to 6.5 using only a fourfold excess of IX to keratin disulfide. Cleland suggested that the equilibrium constant K_B in Equation B (of dithiothreitol and cystine) should be close to 1. However, the cyclization of IX to a stable six-membered ring disulfide (X) during the reaction described in equation C provides an equilibrium constant of the order of $10^4 = K_C$, and therefore $K_B \times K_C = K_A$ is of the order of 10^4.

Wickett and Barman [9–11] have expanded this area of research through a series of studies that involve reduction of hair fibers under stress using analogs of dithiothreitol, dihydrolipoic acid (XI), and 1,3-dithiopropanol (XII). They have demonstrated that monothio analogs of dihydrolipoic acid reduce hair at a slower rate than the corresponding dithio compounds, which correlates with the higher equilibrium constant of reaction of dihydrolipoic acid and cystine. The dithio compounds can cyclize to form stable five-membered ring disulfide structures during reduction (analogous to dithiothreitol), but the monothio compounds cannot. This confirms that cyclization to stable ring structures during the reduction step can be an important driving force in this reaction.

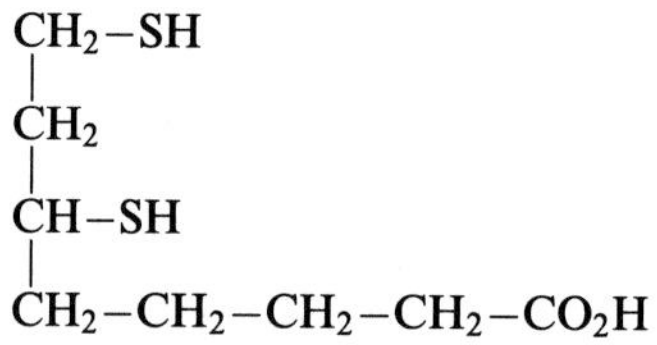

```
CH2–SH                                   CH2–SH
|                                        |
CH2                                      CH2
|                                        |
CH–SH                                    CH2–SH
|
CH2–CH2–CH2–CH2–CO2H                     1,3-Dithiopropanol (XII)
     Dihydrolipoic acid (XI)
```

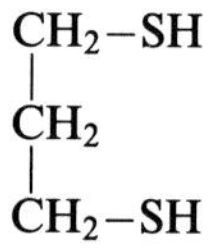

Wickett and Barman have further demonstrated that these five- and six-membered ring-forming reducing agents penetrate into hair via a moving boundary, which suggests nearly complete reduction as the thiol penetrates into the hair. Wickett and Barman have also demonstrated that thioglycolic acid below pH 9 does not exhibit moving boundary kinetics, but above pH 10 it does (see the section on kinetics in this chapter). They also studied structure–activity relationships of a variety of analogs of these three cyclizing dithiols, illustrating the effects of hydroxyl groups and alkyl chain groupings on the rate of this reaction.

One purpose of these studies was to try to achieve essentially complete reduction of a smaller cross section of the fiber and to determine if effective permanent-waving could still be achieved. A potential advantage of this type of process is to lessen cortical reduction and thereby to lessen cortical damage to the hair (the region primarily responsible for tensile properties) during the permanent-wave process. Complete reduction in the annulus or outer regions of the hair is not caused by thioglycolic acid in home permanent-wave products. Thioglycolic acid provides more diffuse reduction over a greater area of the fiber cross section [10]. This concept and its execution provide some very interesting implications for the mechanism of permanent waving, suggesting that permanent set retention is not governed solely by the cortex and cannot be explained by considering only matrix reduction and consequent matrix microfibril interactions. Moreover, strong cuticle interactions involving reduction and reshaping of the exocuticle and its A layer are probably relevant to permanent waving, and these cuticle changes should be considered in any explanation of the permanent-wave process.

These ring-forming reducing agents have never been successfully introduced into the marketplace, primarily because they are all sensitizing agents. Another possible concern must be greater cuticle damage by this type of action. It is conceivable that the effects of these extensive cuticle changes (essentially complete reduction of disulfide bonds in the exocuticle and A layer) on other hair properties and on long-term damaging effects from normal grooming operations (rubbing actions) could be prohibitive.

Further consideration of this two-step equilibrium process (equations B and C) suggests the possibility for approaching complete fission of keratin cystinyl residues while producing only about 50% of the possible cysteinyl residues through formation of an extremely stable mixed disulfide (V). This type of reaction could be described as one with an extremely high K_B and a K_C of much less than 1. Haefele and Broge [12] have suggested that thioglycolamide (XIII) is such a mercaptan, on the basis of its ability to produce excellent waving characteristics in addition to excellent wet strength. No further supporting evidence has been offered to confirm this conclusion.

$$NH_2{-}CO{-}CH_2{-}SH$$

Thioglycolamide (XIII)

Equilibrium and Removal of One of the Reaction Products

O'Donnell [13] has shown that wool fiber, when reacted with thoglycolic acid at pH 5.6, approaches complete reduction of keratin disulfide by removing cysteinyl residues (IV) by means of alkylation followed by retreatment with thioglycolic acid.

Equilibrium and Use of Excess Reactant

Leach and O'Donnell [14] have shown that the complete reduction of wool fiber with thioglycolic acid can be approached at pH 6.9 by employing extremely large concentrations of mercaptan (II) relative to keratin cystine (I). Similar results have been reported for the reduction of wool fiber with mercaptoethanol [15].

Cystinyl Residues of Different Reactivities in Keratin Fibers

Because human hair is a complex substrate consisting of different morphological regions composed of different proteins (see Chapter 2), finding different reactivities for similar functional groups is not surprising. Evidence for disulfide bonds of differing "reactivities" has been described [16,17]. Different reactivities could be caused by varying accessibilities or differences in the electronic nature of certain disulfide bonds in the fibers resulting from differing adjacent amino acids [18]. This latter suggestion, based on work with pure disulfides and not with fibers, is contrary to the findings of Fruton and Clark [5]. This idea is also contrary to the opinion of this author, who would expect differences in reaction rate from accessibility differences but not differences in the true equilibrium nature of the keratin disulfide reduction reaction from inductive effects.

The variation of equilibrium constant with structure, as a function of pH, has not been thoroughly explored. However, discussion of the behavior of keratin cystine in the presence of thioglycolic acid at different pHs is described in the first section of this chapter.

Because this reduction process is a reversible equilibrium reaction, removal of one of the products of reaction (either III or IV), or use of larger concentrations of mercaptan (II) than disulfide (I), should drive the reaction to completion. Both of these principles have been confirmed.

Kinetics of the Reduction

All cleavages of simple disulfides by mercaptans that have been studied kinetically are bimolecular ionic reactions of the SN 2 type, involving direct displacement by mercaptide ion on disulfide [19]. Because the active species in this disulfide scission process is the mercaptide ion [20] rather than the unionized mercaptan, pH is a critical factor. As a consequence, pH can determine the rate controlling step in the reductive cleavage of cystinyl residues in keratin fibers by mercaptans. For example, in the reaction of wool fiber with dithiothreitol, Weigmann [21] has shown that the rate-controlling step at pH 7.0 and above is diffusion of the reducing species into the fibers. However, at acidic pH (3.5), the chemical reaction itself appears to be rate limiting. A similar change in mechanism with pH has been suggested for the reduction of wool fiber by cysteine [22] and also for the reduction of human hair by several thiols, including thioglycolic acid [23].

Wickett [9] has shown that for the reaction of sodium thioglycolate with one lot of hair, at pH 9 or below, the rate of the reaction followed pseudo-first-order kinetics and therefore was reaction controlled. However, at pH 10 and above, moving boundary kinetics or diffusion of mercaptan into the hair controlled the reaction rate. Wickett further demonstrated under conditions closer to those of actual permanent waving (pH 9.5 and 0.6 M sodium thioglycolate) that for hair from one individual exhibiting high reactivity, the reaction rate followed pseudo-first-order kinetics. For hair from another individual that was more difficult to wave, diffusion of the reducing agent into the hair was the rate-determining step. For difficult-to-wave hair, therefore, the rate of the reaction of thioglycolate waves is governed by diffusion of the reducing agent into the hair.

In other words, for difficult-to-wave hair or at high pH, the concentration of mercaptide ion is so high that cleavage of the disulfide bond can occur faster than mercaptide can diffuse into the fibers. As the pH is decreased to the acid side, or for easy-to-wave hair, the rate of chemical reaction decreases faster than diffusion to the point at which the chemical reaction itself becomes rate limiting. With many mercaptans [20], further lowering of the pH to about 2 freezes or stops the reduction reaction.

In addition to pH, other important variables that influence the rate of reduction of keratin fibers by mercaptans are temperature, hair swelling, prior history of the hair, and structure of the mercaptan.

Factors Affecting Rate of Reduction Reaction

Because the rate-controlling step in this reaction can be diffusion of the reducing agent into the fibers or the chemical reaction itself, it is important to consider the rate in terms of these two potentially rate limiting factors.

The pH region most commonly employed for the reduction of hair fibers by mercaptans is above neutral (generally 9–9.5). In the professional field, glycerylmonothioglycolate (GMT) was introduced in Europe in the

$$\begin{array}{l} CH_2{-}OH \\ | \\ CH{-}OH \\ | \\ CH_2{-}O{-}CO{-}CH_2{-}SH \end{array}$$

Glycerylmonothioglycolate (GMT)

1960s and into the United States in the 1970s [24]. This thiol is the active ingredient used in commercial acid waves, that is, those in which the waving solution has a pH less than 7 (generally just under 7). It would appear that the reaction of GMT with hair is a reaction-controlled rate process, because the pH of the system is generally just under 7. The processing time for a GMT permanent is about twice as long as for a conventional thioglycolate wave, and it requires a covering cap and the heat of a dryer to enhance the rate of reduction. Wickett [9] has shown that for sodium thioglycolate under conditions where reaction rate control exists, the activation energy is lower than for diffusion rate control. Therefore, under these conditions an increase in temperature will have less effect on the reaction rate than if the reaction were a diffusion-controlled process. The acid wave supporters claim superiority because of reduced swelling and less damage, but no data could be found to support these claims. To date, GMT acid waves have been sold only to professionals and are not marketed in the retail field [24].

Effect of Temperature on Reaction Rate

The activation energy for the reduction of either human hair or wool fiber at alkaline pH is of the order of 12 to 28 kcal per degree mole [9,21,23]. Wickett [9] explains that when the mechanism is diffusion-rate controlled, the activation energy is higher (28.0 kcal per degree mole) [9], because the boundary movement depends on both reaction and diffusion. However, when the rate depends only on the chemical reaction, the activation energy is lower (19.7 kcal per degree mole).

Reaction rates for both these systems are therefore only moderately affected by increases in temperature. The activation energy for the chemical reaction at acid pH is slightly lower [21]. Therefore, the rate of reaction under acid conditions should be affected less by changes in temperature.

Effect of Hair Swelling and Hair Condition on Reaction Rate

Above the isoelectric point, the swelling of hair increases substantially with increasing pH [25] (see Chapter 8). Herrmann [23] has shown a corresponding increase in the rate of diffusion of mercaptans into hair fibers with increasing pH. Hydrogen bond-breaking agents (hair-swelling agents), namely urea and other amides, have been added to depilatory formulations for the purpose of enhancing the rate of reduction [23,26]. Heilingotter [27] has shown that the addition of urea to thioglycolic acid solution increases the rate of swelling of the fibers. Depilatory systems are generally high-pH mercaptan systems (pH 11–12) in which moving boundary kinetics exist under all conditions [9]; the most common depilatory ingredient is calcium thioglycolate (see Figure 3–1).

Undoubtedly, the condition of the hair also plays a role in the rate of reduction, especially under conditions where diffusion is rate limiting. Permanent waving [28] and bleaching [29] produce alterations to hair that result in increased swelling in solvents. One might also anticipate more rapid rates of reduction for fibers that have been previously bleached or permanent waved than for chemically unaltered fibers. As a consequence, weaker reducing systems are offered in the marketplace to permanent-wave hair that has been previously damaged by bleaches and other chemical treatments.

An initiation time for the reduction reaction was found by Weigmann [21] in his kinetic study of the reduction of wool fiber. Weigmann attributed the initiation time to the epicuticle, the initial barrier to reduction,

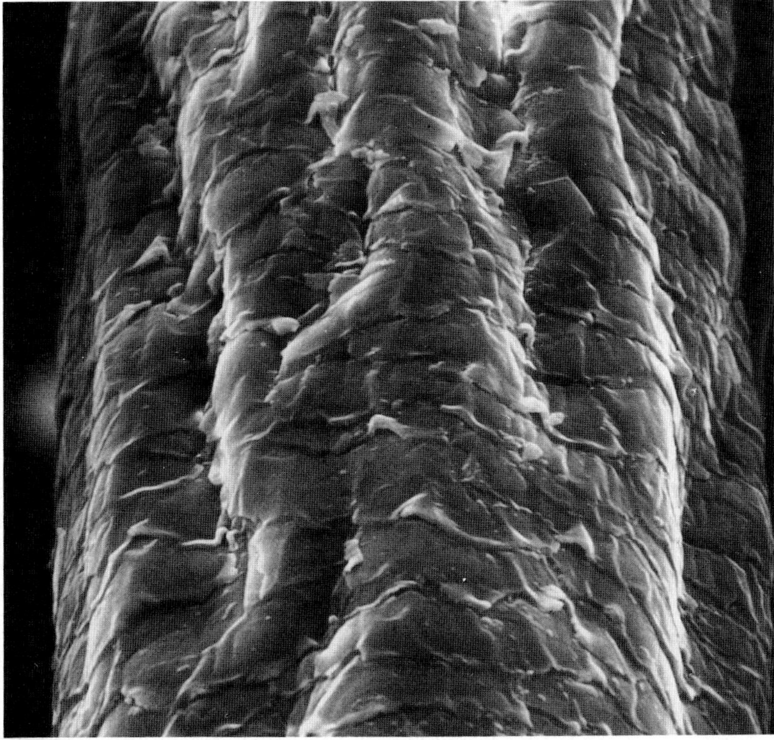

FIGURE 3–1. Scanning electron micrograph (SEM) of hair fiber after treatment with calcium thioglycolate (depilatory).

which is eliminated after a short reduction time. Weigmann suggested that the epicuticle is substantially altered during permanent-waving. If this is indeed the case, hair that has been permanent-waved or has undergone alterations to the epicuticle should provide no initiation time in subsequent reductions or reactions.

Diffusion rates are significantly greater in wool fiber than in human hair [30], so one might anticipate a more rapid rate of reduction for wool fiber than for human hair under conditions of diffusion-controlled reduction.

Effect of Mercaptan Structure on Reaction Rate

Electrostatic Effects

Herrmann [23] has described a minimum at acid pH for the rate of diffusion of a cationic containing thiol (thioglycolhydrazide) into human hair. He has also examined the influence of pH on the rate of diffusion of thio acids

$$HS{-}CH_2{-}CO{-}NH{-}NH_2$$

Thioglycolhydrazide

(thioglycolic and thiolactic acids) into human hair. For this latter type of mercaptan, the minimum occurs near neutral pH. These thio acids are anionic in character in alkaline media, and they diffuse faster in alkaline than in acidic media. Therefore, hair swelling must play a more important role than electrostatics for the diffusion of these simple mercaptans into human hair.

Nucleophilicity of the Mercapto Group

The nucleophilicity of the mercaptan grouping depends on the nature of the groups directly attached or in close proximity to the mercaptan functional group. In general, nucleophilicity increases with increasing basicity of the mercaptan function [31]. Over the range of conditions in which diffusion is rate limiting, changes to the nucleophilicity of the mercapto group will have little effect on the rate of reduction. Where the chemical reaction is rate controlling, however, the nucleophilicity of the mercapto group will be of considerable importance. Theoretically, in a diffusion-controlled reduction one could increase the rate of reduction by sacrificing nucleophilicity (decrease the basicity of the mercaptide ion) to increase diffusibility.

Haefele and Broge [32] have reported the mercapto acidities (pK RSH-acidity at which mercaptan is half ionized) for a large number of mercaptans (pK RSH 4.3–10.2). Hydrogen sulfide, the simplest mercaptan, has a pK RSH of 7.0 [32]. As one might predict, the substitution of electron-withdrawing groups (carbonyl, alkyl ester, alkyl amide) for a hydrogen atom increases the mercapto acidity; electron-donating groups (carboxy, alkyl) decrease mercaptan acidity.

Under conditions of lower pH, where this reduction process is reaction controlled rather than diffusion controlled, equation B or C can be rate limiting. If equation B is rate limiting, the reaction is simply second order, that is, first order with respect to mercaptan and first order with respect to keratin disulfide, and analysis is not as complicated as when equation C is rate limiting. In kinetic studies for a complex material such as human hair or wool fiber, an excess of thiol is most commonly employed, and one generally assumes the reaction in equation B to be rate controlling. The reaction is then described by pseudo-first-order kinetics (first-order with respect to keratin disulfide).

Steric Effects

The rate of diffusion of mercaptans into human hair is undoubtedly influenced by steric considerations. For example, molecular size (effective minimum molecular diameter) of the mercaptan molecule should affect the rate of diffusion into hair. Therefore, the rate of reduction of human hair by ethyl mercaptan in neutral to alkaline media, where diffusion is rate determining, should be faster than that of higher homologs. (The possible effects of variation in the structure of cystinyl residues in hair on the rate of reduction was considered in the previous section on cystinyl residues of differing reactivities.)

Counterion Effects

Ammonia or alkanolamines such as monoethanol amine are the primary neutralizing bases for reducing solutions of thioglycolate permanent waves. Ammonia is said to facilitate diffusion of thioglycolate through hair as compared to sodium hydroxide [33].

Heilingotter [34,35] has compared a large number of neutralizing bases including ammonia, monoethanol amine, sodium hydroxide, isopropanol amine, ethylene diamine, diethanol amine, and triethanol amine with regard to the ability of the corresponding salts of thioglycolic acid to decrease the 20% index (at a pH close to 9.2) . This criterion was used to assess the ability of these different thioglycolates to function as permanent-wave reducing agents. He found that ammonia and monoethanol amine provide the maximum effects. Furthermore, the reducing power of triethanolamine thioglycolate is so weak as to render it ineffective as a permanent-waving agent.

Heilingotter suggested that of the two most effective reducing systems, ammonium thioglycolate provides the more satisfactory waving characteristics. Rieger [36] suggests that this "catalytic activity" of nitrogen-containing bases derives from their ability to swell the hair, thus allowing faster diffusion of mercaptan into the interior of the hair.

Other salts of thioglycolic acid, including potassium [36], lithium [37], and magnesium [38], have been described as potential permanent-waving agents. Magnesium thioglycolate has been described as an odorless permanent wave, although this system has never achieved commercial success.

Side Reactions During the Reduction of Keratin Fibers with Mercaptans

The reaction of mercaptans with keratin fibers is a relatively specific reaction in mild acid. However, in alkaline media peptide bond hydrolysis and the formation of lanthionyl residues can also occur [39]. Zahn et al. [40] have suggested that mercaptides such as thioglycolate or cysteinate can accelerate the rate of formation of lanthionyl residues in wool fiber. (A more detailed discussion of the formation of lanthionyl residues in keratin fibers is presented later in this chapter.)

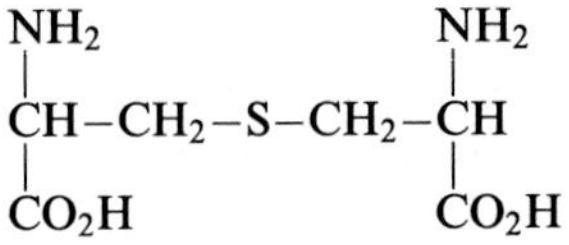

$$\begin{array}{c} NH_2 \\ | \\ CH-CH_2-S-CH_2-CH \\ | \\ CO_2H \end{array} \quad \begin{array}{c} NH_2 \\ | \\ \\ | \\ CO_2H \end{array}$$

Lanthionine

Hydrolysis of peptide and amide linkages is also a possible complication in an alkaline medium. Hydrolysis of the amide groups of the residues of aspartic and glutamic acids will increase the ratio of acidic to basic groups in the fibers, conceivably altering the isoelectric and/or isoionic points of the hair.

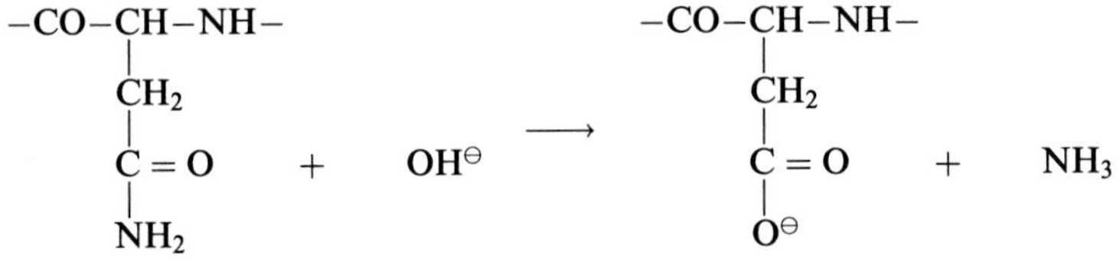

$$\begin{array}{c} -CO-CH-NH- \\ | \\ CH_2 \\ | \\ C=O \\ | \\ NH_2 \end{array} + OH^{\ominus} \longrightarrow \begin{array}{c} -CO-CH-NH- \\ | \\ CH_2 \\ | \\ C=O \\ | \\ O^{\ominus} \end{array} + NH_3$$

Amide of aspartic acid residue

Peptide bonds are the major repeating structural unit of polypeptides and proteins, and they form the structural backbone of human hair. Hydrolysis of peptide bonds can also occur at high pH, and both these reactions (hydrolysis of amide and peptide bonds) are far more prevalent in the action of depilatories, formulated near pH 12, than in permanent waves. Permanent-waving lotions are usually formulated at a pH of approximately 9.2 to 9.5.

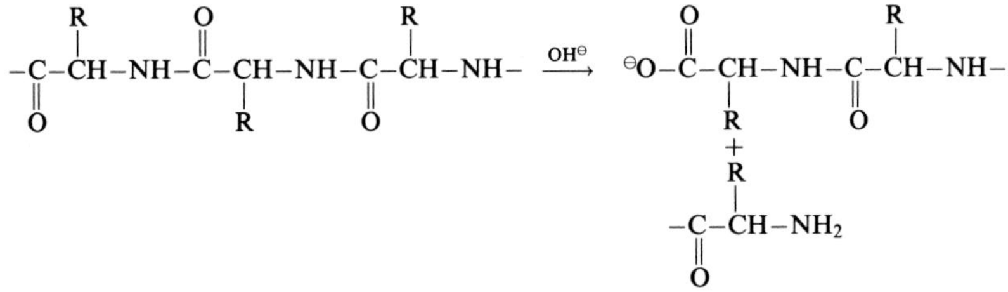

$$-\underset{\substack{\|\\O}}{C}-\overset{\substack{R\\|}}{CH}-NH-\overset{\substack{O\\\|}}{C}-\underset{\substack{|\\R}}{CH}-NH-\underset{\substack{\|\\O}}{C}-\overset{\substack{R\\|}}{CH}-NH- \xrightarrow{OH^{\ominus}} {}^{\ominus}O-\overset{\substack{O\\\|}}{C}-\underset{\substack{|\\R}}{CH}-NH-\underset{\substack{\|\\O}}{C}-\overset{\substack{R\\|}}{CH}-NH-$$

$$+ \; -\underset{\substack{\|\\O}}{C}-\overset{\substack{R\\|}}{CH}-NH_2$$

Alkaline hydrolysis of the peptide bond

Reduction of Hair with Sulfite or Bisulfite

Sulfites or bisulfite (depending on pH) are also important reducing agents for the disulfide bonds in commercial permanent waves. The reaction of sulfite with hair involves nucleophilic attack of sulfite ion on disulfide. This reaction produces one equivalent of mercaptan and one equivalent of Bunte salt [41].

$$K-S-S-K + M_2SO_3 \rightleftharpoons K-S-SO_3^{\ominus}M^{\oplus} + K-S^{\ominus}M^{\oplus}$$

Bunte salt

Reese and Eyring [42] have demonstrated that the reaction of sulfite with hair is a pseudo-first-order reaction. In other words, the chemical reaction of sulfite with the disulfide bond of hair is slower than diffusion of sulfite into hair. Elsworth and Phillips [43,44] and Volk [45] examined the sulfitolysis of keratin, demonstrating that the rate of cystine cleavage is optimal at acid pH. Wolfram and Underwood [46] found a broad optimum for cystine cleavage by sulfite at pH 4 to 6. The decrease in cystine cleavage at acidic pH (below pH 4) results from a decrease in the concentration of the nucleophilic sulfite species. On the other hand, the decrease in cystine cleavage as pH is raised (alkaline pH) results from alkaline hydrolysis of the Bunte salt [47].

The patent literature indicates that rebuilding disulfide bonds in keratin after sulfitolysis may be accomplished through water rinsing. However, reversal of sulfitolysis by rinsing is normally slow and inefficient [24], and Bunte salt is resistant to oxidizing agents. Therefore, neutralizers such as bromate or hydrogen peroxide are not totally efficient in rebuilding disulfide bonds in sulfite waves.

Sneath [48] has shown that the bisulfite waving treatment decreases the barrier function of the cell membrane complex as evidenced by cationic dye absorption; as part of this reaction is reversible, it probably involves the Bunte salt groupings. In addition, lipids are removed from the cell membrane complex during bisulfite waving and this part of the reaction is not reversible.

To summarize and to compare the two processes of thioglycolate and sulfite reduction of hair, we find that the thiol reacts with hair, producing cysteine residues in the following manner:

$$\text{K–S–S–K} + 2\,\text{R–SH} \rightleftharpoons 2\,\text{K–SH} + \text{R–S–S–R}$$

and this reaction can be reversed by rinsing and oxidation in air. However, the most effective reversal is achieved through mild chemical oxidants.

On the other hand, sulfite reacts with the disulfide bonds in hair to produce mercaptan and Bunte salt:

$$\text{K–S–S–K} + SO_3^{2+} \rightleftharpoons \text{K–S}^{\ominus} + \underbrace{\text{K–S–}SO_3^{\ominus}}_{\textit{Bunte salt}}$$

Rinsing of sulfite-treated hair slowly reverses the reaction, rebuilding the cystine bonds. The rate of the cystine reformation increases with increasing pH, and good set stability is achieved at pH 8 or higher. Because of the efficiency of reversal of the sulfitolysis reaction with alkali, Albrecht and Wolfram [49] suggest that for low cleavage levels sulfite is a more effective setting agent than thioglycolate. At higher cleavage levels, however, thioglycolate is the superior active ingredient.

Thioglycolate at alkaline pH is also more effective at higher cleavage levels, because thioglycolate is a stronger reducing agent than sulfite. Its greater effectiveness is borne out by the fact that under optimum conditions, for difficult-to-wave hair, the rate-controlling step for the thioglycolate reaction is diffusion of the reducing species into the fibers [9]. On the other hand, for sulfite at its optimum (acid pH), the rate-determining step is chemical reaction with the disulfide bond.

Reduction with Reagents Other than Mercaptan or Sulfite

In addition to mercaptans and sulfites, compounds that have been used for nucleophilic cleavage of the disulfide bond in hair and/or wool fiber are sulfides, hydroxide, water (steam), a phosphine, borohydride, dithionite (hydrosulfite), and sulfoxylate. The interactions of some of these compounds with the disulfide bond in hair are described next.

Sulfides

Salts of hydrogen sulfide are extremely potent reducing agents for hair and have been used in depilatory compositions [50]. In a sense, salts of hydrogen sulfide are the simplest and among the most diffusible of all mercaptans. The initial reaction with the disulfide bond in keratin fibers is described by equation D. Obviously, compound XIV can also ionize and react with cystinyl residues, forming organic polysulfides. Compound XIV can even react with hydrogen sulfide (anion) to form inorganic polysulfide.

$$\mathrm{K{-}S{-}S{-}K} + \mathrm{M^{\oplus}\,{}^{\ominus}S{-}H} \rightleftharpoons \underset{\text{(XIV)}}{\mathrm{K{-}S{-}SH}} + \mathrm{K{-}S^{\ominus}\,{}^{\oplus}M} \qquad \text{(D)}$$

Steam or Alkali

Setting of wool and hair by either steam or hot alkaline solutions is a very old technique [51]. Steam is also very effective for producing a permanent set. Alkali and steam are known to cleave the disulfide bond in keratins [52–54]. The reaction with hydroxide is summarized in equation E. Because sulfenic acids are generally unstable species [55], they have been suggested as intermediates that can react with the nucleophilic side chains in the keratin macromolecules [53].

$$\mathrm{K{-}S{-}S{-}K} + \mathrm{M^{\oplus}\,{}^{\ominus}O{-}H} \rightleftharpoons \underset{\textit{Sulfenic acid}}{\mathrm{K{-}S{-}OH}} + \mathrm{K{-}S^{\ominus}\,{}^{\oplus}M} \qquad \text{(E)}$$

As mentioned earlier, hydrolytic cleavage of peptide bonds in keratins, as well as formation of lanthionyl residues, can also occur in alkali. In addition to lanthionine, lysinoalanine [56] and β-aminoalanine [57] residues can be formed in some keratins under alkaline conditions.

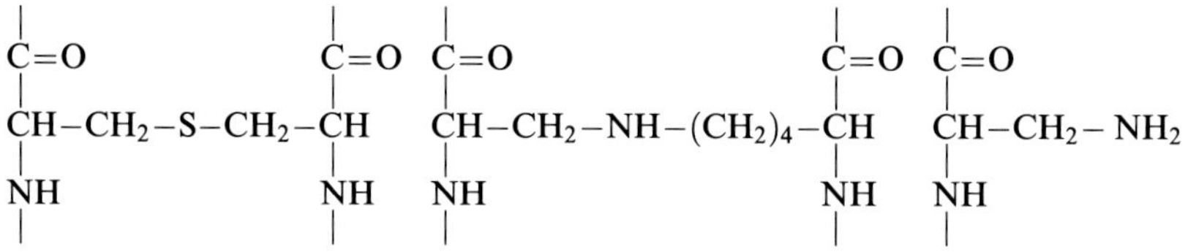

$$\begin{array}{ccc}
\begin{array}{c} | \\ C{=}O \\ | \\ CH{-}CH_2{-}S{-}CH_2{-}CH \\ | \\ NH \\ | \end{array} &
\begin{array}{c} | \\ C{=}O \\ | \\ CH{-}CH_2{-}NH{-}(CH_2)_4{-}CH \\ | \\ NH \\ | \end{array} &
\begin{array}{c} | \\ C{=}O \\ | \\ CH{-}CH_2{-}NH_2 \\ | \\ NH \\ | \end{array}
\end{array}$$

Lanthionyl residue *Lysinoalanine residue* *β-aminoalanine residue*

Formation of lanthionyl residues during alkaline treatment of keratin fibers was first suggested by Speakman [58] and later demonstrated by Horn et al. [59]. Lanthionyl residues may be formed from cystinyl residues in proteins under relatively mild alkaline conditions: 35°C and pH 9 to 14 [39]. However, lanthionine has not been identified from free cystine under these same reaction conditions. For that matter, thioethers have not been formed from organic disulfides other than cystine-containing proteins, using similar conditions [60]. At a higher reaction temperature (reflux), Swan [61] claims to have identified small quantities of lanthionine from reaction of alkali with cystine.

Earland and Raven [62] have examined the reaction of *N*-(mercapto-methyl)polyhexamethyleneadipamide disulfide (XV) with alkali. No thio-ether is formed from this polymeric disulfide under alkaline conditions that produce lanthionyl residues in wool; however, cyanide readily produces thioether from both (XV) and wool fiber. The mechanism for thioether formation thus must be different in these two reactions. Because this poly-meric disulfide (XV) contains no *β*-hydrogen atoms (beta to the disulfide group), a likely mechanism for formation of lanthionyl residues in keratins, under alkaline conditions, is the beta-elimination scheme [61] (the reac-tion depicted by equation F). Other mechanisms that have been suggested for this reaction are summarized by Danehy and Kreuz [63].

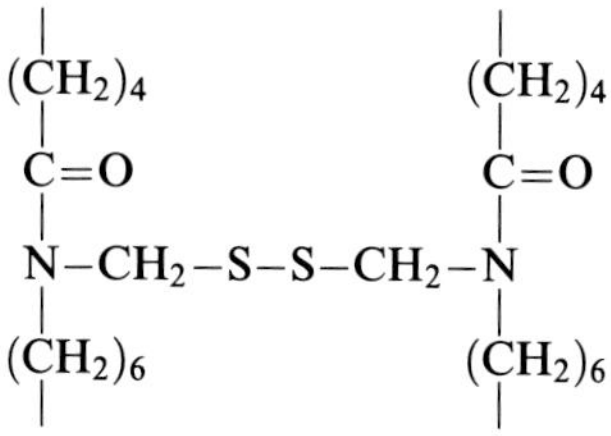

$$\begin{array}{ccc}
| & & | \\
(CH_2)_4 & & (CH_2)_4 \\
| & & | \\
C{=}O & & C{=}O \\
| & & | \\
N & {-}CH_2{-}S{-}S{-}CH_2{-} & N \\
| & & | \\
(CH_2)_6 & & (CH_2)_6 \\
| & & |
\end{array}$$

(XV)

The formation of lanthionine in keratin fibers is believed to involve two reaction sequences. The first sequence consists of beta elimination to form dehydroalanine residues in hair:

$$\begin{array}{c} | \\ C{=}O \\ | \\ CH{-}CH_2{-}S{-}S{-}CH_2{-}CH \\ | \\ NH \\ | \end{array} \;(\text{with } C{=}O,\ NH \text{ on right CH}) \;\underset{H_2O}{\overset{OH^{\ominus}}{\rightleftharpoons}}\; {}^{\ominus}C{-}CH_2{-}S{-}S{-}CH_2{-}CH$$

$$\begin{array}{c} | \\ C{=}O \\ | \\ C{=}CH_2 \\ | \\ NH \\ | \end{array} \quad + \quad {}^{\ominus}S{-}S{-}CH_2{-}CH \qquad \text{(F)}$$

Dehydroalanine intermediate

Reaction sequence 1: Beta elimination to form dehydroalanine residue

The disulfide anion (of reaction sequence 1) may then eliminate sulfur to form mercaptide ion. In addition, the dehydroalanine intermediate is a very reactive species. It may react with any nucleophilic species present, such as mercaptan or amine, including mercaptan or amine residues on the hair or such groups in solution, to form lanthionine (other thioethers), or lysinoalanine [60,64,65], or β-aminoalanine residues [64–67] as shown in the second sequence:

Cysteine or *mercaptan* → $CH{-}CH_2{-}S{-}CH_2{-}CH$ (each CH bearing C=O and NH) *Lanthionyl residue or thioether*

$\begin{array}{c} | \\ C{=}O \\ | \\ CH{-}CH_2 \\ | \\ NH \\ | \end{array}$ $\xrightarrow{\text{Lysine}}$ $CH{-}CH_2{-}NH{-}(CH_2)_4{-}CH$ (each CH bearing C=O and NH) *Lysinoalanine residue*

Dehydroalanine Intermediate $\xrightarrow{NH_3}$ $\begin{array}{c} | \\ C{=}O \\ | \\ CH{-}CH_2 - NH_2 \\ | \\ NH \\ | \end{array}$ *β-aminoalanine residue*

Second sequence of reactions: Nucleophilic addition to dehydroalanine

For wool fiber, all three residues, lanthionine, lysinoalanine, and β-aminoalanine, have been shown to form from reactions under alkaline conditions [64,67,68]. In the case of human hair, however, only lanthionine and lysinoalanine have been shown to form in alkali [65].

Amines

The foregoing discussion on the reaction of alkali with wool and hair shows that a very reactive intermediate, dehydroalanine, is formed in hair and wool in the presence of alkalinity at elevated temperature (30°–40°C or higher). Asquith [68] and Tolgyesi and Fang [65] have studied the reaction of wool and hair in the presence of alkaline amine solutions. Under these conditions, one might conclude that if amines are at a high enough concentration they might add to the dehydroalanine intermediate to form β-(*N*-alkylamino)alanine residues.

$$\begin{array}{c} | \\ C{=}O \\ | \\ C{=}CH_2 \\ | \\ NH \\ | \end{array} \xrightarrow{R-NH_2} H-\begin{array}{c} | \\ C{=}O \\ | \\ C \\ | \\ NH \\ | \end{array}-CH_2-NH-R$$

Dehydroalanine *β-(N-alkylamino)alanine residue*

Such is the case. However, the actual products formed depend on the substrate (hair versus wool), the structure of the amine, its concentration, and the reaction temperature.

Asquith [68] has demonstrated that with short-chain amines like ethyl or *n*-butyl amine in the presence of wool fiber in alkali, the amounts of lanthionine and lysinoalanine are less (compared to alkali alone), but these two species are still produced in detectable quantities. However, longer chain amines like pentyl amine react quantitatively with wool fiber, and virtually no lanthionine or lysinoalanine is formed.

Tolgyesi and Fang [65] have found that alkaline amine solutions react differently with human hair. With human hair, all amines examined, including pentyl amine, compete less effectively with the amino and mercaptan residues of the hair for the dehydroalanine intermediate. As a result, more lanthionine and lysinoalanine cross-links form than amine adduct, when human hair is the substrate. This is probably because diffusion rates into human hair are slower, decreasing the effective concentration of free amine in the fibers. As these species cannot compete as effectively for the dehydroalanine intermediate, therefore lanthionine and lysinoalanine are formed.

Cyanide

Salts of hydrogen cyanide have also been found to be capable of nucleophilic cleavage of the disulfide bond in keratin fibers [69]. In addition, nearly quantitative conversion of cystinyl residues to lanthionyl residues can be achieved in this reaction [70]. The most plausible mechanism is given in equations G and H [62]. This mechanism consists of two nucleophilic displacement reactions: the first is that of cyanide on sulfur, and the second is mercaptide ion on carbon. The mechanism is consistent with the observed formation of thioether from the reaction of *N*-(mercaptomethyl)polyhexamethyleneadipamide disulfide (XV) with cyanide, but not with alkali [62].

$$\text{K–S–S–K} + \text{M}^{\oplus}\,{}^{\ominus}\text{CN} \rightarrow \text{K–S–CN} + \text{M}^{\oplus}\,{}^{\ominus}\text{S–K} \tag{G}$$

$$\text{K–S–CN} + \text{M}^{\oplus}\,{}^{\ominus}\text{S–K} \rightarrow \text{K–S–K} + \text{M}^{\oplus}\,{}^{\ominus}\text{S–CN} \tag{H}$$

A Phosphine

Trihydroxymethyl phosphine (THP) or its precursor, tetrahydroxymethyl phosphonium chloride, has been used to reduce both human hair and wool fiber [71]. The mechanism of this reaction was studied by Jenkins and Wolfram [72], who discovered that this reaction proceeds by nucleophilic attack by the phosphine on sulfur, followed by hydrolysis of the intermediate addition compound to mercaptan and phosphine oxide.

$$\underset{}{\text{K–S–S–K}} + \underset{\text{THP}}{(\text{HO–CH}_2)_3\text{P}} \longrightarrow \text{K–S–}\overset{{}^{\oplus}\text{P–(CH}_2\text{OH)}_3}{\underset{\ominus}{\text{S}}}\text{–K} \;\overset{\text{H}_2\text{O}}{\rightleftharpoons}\; 2\,\text{K–SH} + \text{O=P–(CH}_2\text{–OH)}_3$$

Above pH 7, the rate of reaction of THP appears to be controlled by diffusion of the reagent into the fibers [73] and, like the reaction of mercaptans with hair, increases rapidly with increasing pH in the vicinity of pH 9 to 12. This reaction rate increase presumably results from increased swelling of the keratin substrate with increasing pH.

The equilibrium constant for the reaction of THP with cystyl residues in hair must be relatively large, since essentially complete reduction of human hair occurs with only a 10–fold excess of THP, at neutral pH [73].

Miscellaneous Reducing Agents

Borohydride (MBH_4) has also been used as a reducing agent for keratin fibers [74,75], as well as dithionite ($M_2S_2O_4$), sometimes called hydrosulfite [42], and sulfoxylate (M_2SO_2) [76] [or, more correctly, its ester salts, e.g., sodium formaldehyde sulfoxylate ($HO{-}CH_2{-}SO_2Na$)].

Reactions of the Mercaptan Group

The previous section described various reagents that have been used for the reduction of the disulfide bond in keratin fibers. Most of these reactions produce cysteinyl residues, or mercaptan groups, in the fibers.

The mercaptan group is one of the most reactive functional groups in all organic chemistry, and it readily undergoes oxidation, nucleophilic displacement, nucleophilic addition, and free radical addition and displacement reactions. This section discusses some of the chemical literature pertaining to these types of reaction in reduced keratin fibers, and illustrates the potential reactivity of the mercaptan group in hair.

Oxidation of Reduced Keratin Fibers

The oxidation of the mercaptan group can occur by two distinct pathways: the S–S fission route (pathway in the presence of most chemical oxidants), and the C–S fission route, the pathway for radiation-induced cleavage of the disulfide bond. Only the S–S fission route is discussed in this section, because it is the most relevant pathway in relation to permanent waves and reducing agents. For a more complete discussion of both these mechanistic schemes, see Chapter 4.

The oxidation of the mercaptan group can occur in several stages:

$$\underset{\textit{Mercaptan}}{2\,\text{K–SH}} \longrightarrow \underset{\textit{Disulfide}}{\text{K–S–S–K}} \longrightarrow \underset{\textit{Disulfide monoxide}}{\text{K–}\overset{\text{O}}{\overset{\|}{\text{S}}}\text{–S–K}} \longrightarrow$$

$$\underset{\textit{Disulfide dioxide}}{\text{K–}\overset{\text{O}}{\overset{\|}{\underset{\text{O}}{\underset{\|}{\text{S}}}}}\text{–S–K}} \longrightarrow \underset{\textit{Disulfide trioxide}}{\text{K–}\overset{\text{O}}{\overset{\|}{\underset{\text{O}}{\underset{\|}{\text{S}}}}}\text{–}\overset{\text{O}}{\overset{\|}{\text{S}}}\text{–K}} \longrightarrow \underset{\textit{Disulfide tetroxide}}{\text{K–}\overset{\text{O}}{\overset{\|}{\underset{\text{O}}{\underset{\|}{\text{S}}}}}\text{–}\overset{\text{O}}{\overset{\|}{\underset{\text{O}}{\underset{\|}{\text{S}}}}}\text{–K}} \longrightarrow \underset{\textit{Sulfonic acid}}{2\,\text{K–SO}_3\text{H}}$$

From this group of compounds, mercaptan, disulfide, and sulfonic acid have been isolated from the oxidation of reduced hair [77], the principal product being disulfide. Because the primary intent in the oxidation of reduced hair in permanent-waving is to stop at the disulfide stage, milder oxidizing conditions are used than for bleaching hair. Some of the reagents that have been used for oxidation of reduced hair are bromates [78,79], iodate [80], perborate [81], acidic hydrogen peroxide [77], monopersulfate [82], and even air oxidation or metal-catalyzed air oxidation [78].

Nucleophilic Displacement

The mercaptan group is an extremely powerful nucleophile and readily undergoes nucleophilic displacement reactions. This property is the basis of several quantitative tests for cysteine and/or cystine, including the Sullivan test, which involves nucleophilic displacement by mercaptide ion on iodoacetate [83].

$$K{-}SH + I{-}CH_2{-}CO_2H \longrightarrow K{-}S{-}CH_2{-}CO_2H + HI$$

Iodoacetic acid

Methyl iodide has been used as a mercaptan blocking group in studies on keratin fibers [84]. Other monofunctional alkyl halides, including benzyl chloride, heptyl bromide, and dodecyl bromide, have also been reacted with reduced keratin fibers [7]. Hall and Wolfram [85] have used this reaction (alkyl iodides with reduced hair) as a means to introduce alkyl groups or nonpolar residues into hair. These researchers found that methyl iodide was highly efficient in reacting with the mercaptan groups of reduced hair. Longer chain length alkyl iodides, however, were not nearly as efficient for introducing alkyl groups into reduced hair.

$$K{-}SH + R{-}X \longrightarrow K{-}S{-}R + HX$$

alkyl halide

The Bunte salt grouping has also been reacted with mercaptan in reduced keratin fibers [86] to form a mixed disulfide.

$$K{-}SH + R{-}S{-}SO_3Na \longrightarrow K{-}S{-}S{-}R + NaHSO_3$$

Bunte salt

Reaction of activated aryl halides such as 2,4–dinitrofluorobenzene with cysteine in unreduced and reduced hair has been described by Zahn [77] as a quantitative assay for mercaptan or disulfide in keratin fibers.

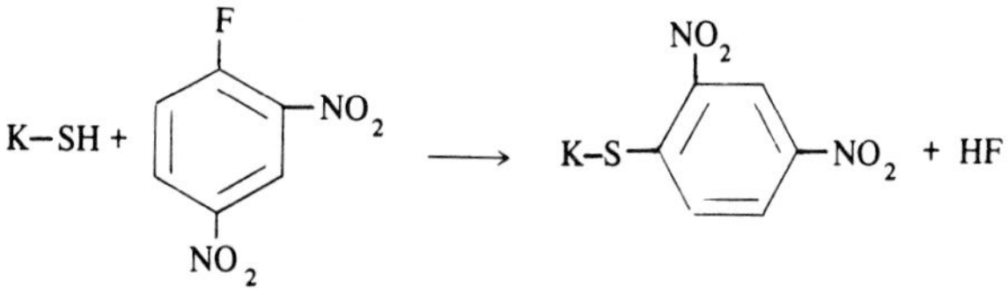

$$K{-}SH + C_6H_3F(NO_2)_2 \longrightarrow K{-}S{-}C_6H_3(NO_2)_2 + HF$$

Dinitrofluorobenzene

Halo mercury compounds such as methyl mercuric iodide also react readily with mercaptan in keratin fibers [87] and serve as the basis of Leach's method for cystine analysis.

$$K{-}SH + CH_3{-}Hg{-}I \longrightarrow K{-}S{-}Hg{-}CH_3 + HI$$

Methyl mercuric iodide

In fact, mercaptan in hair is capable of reacting with disulfide monoxide by nucleophilic displacement [88] or with most compounds that contain a group labile to nucleophilic displacement, if they are capable of diffusing into the hair.

Molecules containing two leaving groups similar to the previously described monofunctional compounds are capable of reacting with reduced keratin fibers and forming a new type of cross-link. Dihaloalkanes have been reacted with reduced wool fiber to provide a thioether cross-link [7]. This reaction is capable of promoting stability to moths in wool [89].

$$2\ K{-}SH + Br{-}(CH_2{-})_nBr \longrightarrow K{-}S{-}(CH_2)_n{-}S{-}K + 2\ HBr$$

Dihaloalkane

Di-Bunte salts have also been used to restore the cross-links in reduced keratin fibers through a bis-disulfide type of linkage [86,88].

$$2\ K{-}SH + NaO_3{-}S{-}S{-}(CH_2)_n{-}S{-}SO_3Na \longrightarrow$$
$$K{-}S{-}S{-}(CH_2)_n{-}S{-}S{-}K + 2\ NaHSO_3$$

Di-Bunte salt reaction

Other rather exotic difunctional reagents have been reacted with both reduced and unaltered wool fiber and are described in Section C of the *Proceedings of the International Wool Textile Research Conference* (1955).

Treatment of Reduced Hair with Dithioglycolate Ester Derivatives of Polyoxyethylene

A novel treatment in permanent waving involves treating reduced hair with polyoxyethylene esters of thioglycolic acid (mol wt, 550–750) described recently by Salce et al. [2]. These esters are reported to bind to the fibers by displacement of reduced disulfide in hair on the ester linkage of the additive resulting in the formation of mixed disulfides, producing more hydrophobic hair fibers and improvement in the curl relaxation from the increased hydrophobicity.

Nucleophilic Addition Reactions

Mercaptan groups in keratin fibers also undergo nucleophilic addition reactions with active olefins (olefins containing a strong electron-withdrawing group attached to the double bond). Schoberl [88] has shown that reduced wool fiber reacts with vinyl sulfones.

$$K{-}SH + CH_2{=}CH{-}SO_2{-}R \longrightarrow KS{-}CH_2{-}CH_2{-}SO_2{-}R$$

Vinyl sulfone

Maleimides are another example of activated olefins that react in this manner, for example, *N*-ethyl maleimide (NEMI) [88,90] reacts quantitatively with the mercaptan groups in reduced keratin fibers by a nucleophilic addition type of reaction. Hall and Wolfram [85] have used this reaction as a means of introducing *N*-substituted maleimide groups (*N*-ethyl, *N*-hexyl, and *N*-heptyl maleimides) into human hair to study the properties of hair modified by the introduction of nonpolar residues. They report enhanced settability and high set retention, at all humidities, for hair modified in this manner with greater than 50% disulfide cleavage. However, hair having less than 50% disulfide scission does not show improved set characteristics.

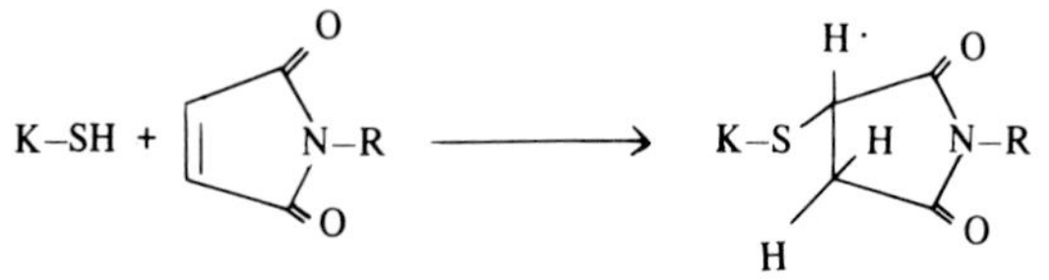

$$K{-}SH + \text{(maleimide ring: } O{=}C{-}CH{=}CH{-}C{=}O\text{, N}{-}R) \longrightarrow K{-}S{-}CH{-}CH_2 \text{ (succinimide ring, } N{-}R)$$

N-substituted maleimide

Acrylonitrile and phenyl acrylate have also been shown to react readily with the mercaptan groups of reduced hair [91].

$$K{-}SH + CH2{=}CH{-}CN \longrightarrow K{-}S{-}CH_2{-}CH_2{-}CN$$

Acrylonitrile

$$K{-}SH + CH_2{=}CH{-}CO{-}O{-}\emptyset \longrightarrow K{-}S{-}CH_2{-}CH_2{-}CO{-}O{-}\emptyset$$

Phenylacrylate

Difunctional reagents containing two active vinyl groups are capable of reacting with reduced keratin fibers and forming cross-links. Divinyl sulfone has been used for this purpose [88].

$$2\,K{-}SH + CH_2{=}CH{-}SO_2{-}CH{=}CH_2 \longrightarrow$$
$$K{-}S{-}CH_2{-}CH_2{-}SO_2{-}CH_2{-}CH_2{-}S{-}K$$

Divinyl sulfone reaction

Free Radical Addition and Polymerization Reactions

One form of polymerization that has been used in the chemistry of wool fiber involves reduction of the fibers followed by the addition of a vinyl monomer and an oxidizing agent [91,92]. These reactions have been carried out in an inert atmosphere and provide rather large polymer add-ons. Related procedures have also been described for polymerizing into human hair in an air atmosphere [93–95].

In this type of reaction, the mercaptan group of the reduced keratin may serve as the reducing agent in a redox system for generating free radicals. It may also serve as a site for grafting, and may serve as a chain transfer agent, limiting the degree of polymerization. Another advantage to this system is the increased swelling of the fibers accompanying reduction. This effect facilitates diffusion of all reagents necessary to polymerization into the fibers. (For additional details, see Chapter 7.)

Polymerization into wool fiber has also been accomplished using radiation grafting techniques [96,97], although no such procedures could be found with human hair as the substrate.

TABLE 3–1. Low-humidity effects on curl retention (all fibers set and dried at 60% RH).

	Percent curl retention (%)	
Time exposed (h)	60% RH	10% RH
2	73.5	61.3
24	59.6	58.3

Water-Setting Human Hair

If human hair is soaked in water and held in a given configuration while drying, it will tend to remain in that configuration. This is the basis of what is called a water-set in human hair. It is well known, however, that exposure of water-set hair to high humidity produces a loss of set.

Recently, Wolfram and Diaz [98] have demonstrated that exposure of water-set hair tresses to a lower humidity can also produce a loss of set. In addition, Robbins and Reich [99] have demonstrated this same phenomenon with single hair fibers. Thus, we conclude that exposure of water-set hair to "changes" in humidity results in the flow of additional moisture either into or out of hair. This moisture transfer, therefore, breaks hydrogen bonds critical to set stability, resulting in a decrease in water-set stability.

Table 3–1 summarizes one of the single-fiber experiments. In this experiment, single hairs were water-set in a curled configuration on glass rods and dried at 60% relative humidity (RH). After the fibers were removed from the rods, one group of hairs was exposed to a 60% RH atmosphere and another group to 10% RH; curl length was measured over time with a cathetometer. The data in Table 3–1 were then analyzed by a repeated measures analysis of variance (ANOVA). Highly significant time effects, significant humidity effects, and significant interactions were found. Therefore, one may conclude that changing the environment of single hairs water-set at a higher humidity (60% RH) to a lower humidity causes more rapid curl loss in short time intervals (2 h) than maintaining the hair at the higher humidity. This more rapid curl loss occurs in spite of the fact that hair equilibrated at a lower humidity will contain less water [100] and exhibit greater bending [101] and torsional stiffness [102] than hair equilibrated at a higher humidity.

The fibers when taken from 60% to 10% RH lose water until they re-equilibrate with the new environment (~16% moisture at 60% RH to 5% moisture at 10% RH [100]). During this transition stage, however, when water migrates from the fibers, hydrogen bonds are broken and reformed and more rapid curl loss (set loss) occurs.

At the longer time interval (24 h), the hair fibers that were maintained at the higher humidity are once again equal (in curl retention) to the fibers transferred to the lower humidity. Apparently, after equilibration of moisture at the lower humidity, the rate of curl loss becomes less than that for hair maintained at the higher humidity allowing the curl loss to equalize. Presumably, at even longer times the curl loss for the hair at the lower humidity would be less than for the hair at the higher humidity.

Wolfram's experiments were with hair tresses and in a sense were more pragmatic than the single-fiber experiments; however, they include interfiber complications excluded from the single-fiber data. Fiber friction increases with RH for keratin fibers [102], and effects from frictional contributions tend to enhance the set stability of the hair at the higher humidity. Nevertheless, Wolfram's results indicate the same general picture as with single hairs, and these two types of experiments thus complement each other with regard to providing a better understanding of the mechanism of water-setting human hair.

The following important insights are reflected in these results:
Exposure of water-set hair to changes in humidity results in moisture either entering or leaving the fibers. The flow of strongly bound water either into or out of hair produces cleavage of critical hydrogen bonds and a decrease in water-set stability. The behavior of hair equilibrated at different humidities may not reflect its behavior during the transition to different humidities, and changes in humidity are probably more likely to be encountered in the real world than constant humidity.

Water-setting hair provides a temporary reversible set to hair, because hydrogen bonds are involved. Therefore, a water-set can be removed by the transfer of moisture either into or out of hair. Permanent waves, in contrast, are more resistant to moisture transfer into and out of hair, because covalent bonding (the disulfide bond) is involved in permanent waves and covalent bonds are relatively inert to moisture changes in hair.

Set and Supercontracton

Set has been defined by Brown et al. [103] as a treatment that enables a keratin fiber to maintain a length greater than its original length. As a contrast, supercontraction is the condition in which a keratin fiber is fixed at a length less than its original length [103]. Set is usually determined by a procedure similar to that described by Speakman [52]. Fibers are stretched to 40%, treated, then rinsed and tested for their ability to retain the extended length when placed in water or buffer at elevated temperatures. The criterion for "permanent" set is the resistance to lengthwise shrinkage in boiling water. Supercontraction can also be followed by observing lengthwise changes in full-length fibers or by microscopic observation of fiber snippets [104].

The setting process is generally considered a three-stage process: stage 1 is the stretching stage, stage 2 is the period of structural rearrangement, consisting of the time period that the fibers are held in the stretched state, and stage 3 is the recovery period, the time after the external strain is removed.

Widely varying reaction conditions and gross alterations to the fibers have been made during the course of the study of the mechanism(s) of setting, although it appears that most of the literature on setting is concerned with establishing a single common mechanism for all treatments. Jenkins and Wolfram [105] have suggested and provided evidence indicating that more than one mechanism may exist for setting keratin and highly altered keratin fibers. The discussion in this section is primarily concerned with the conditions of wool setting that are related to the permanent-waving process. Therefore, a single mechanism is considered.

When keratin fibers are stretched in an aqueous medium in the presence of a reducing agent, several bonds are broken by internal stresses resulting from the imposed strain. When hydrogen bonds are broken, their resistance to the imposed strain decreases with increasing temperature [95]. The importance of hydrogen bonds to stage 3 (recovery) and to structural rearrangement (stage 2) has been demonstrated. Farnworth [106] showed that urea plus, a reducing agent, is capable of producing permanent set in keratin fibers under conditions in which neither reagent alone will produce permanent set. Therefore, the breaking of hydrogen bonds permits structural rearrangements to occur in the presence of reducing agent that reduction alone cannot achieve.

Weigmann et al. [107] and Milligan et al. [108] have clearly demonstrated the importance of the disulfide and mercaptan groups to all three stages of the setting process. Elimination of mercaptan before stretching prevents permanent set. On the other hand, permanent set is enhanced by elimination of mercaptan before releasing stretched fibers. Interestingly, mercaptan elimination in the latter circumstance may be accomplished either by reoxidation to the disulfide or higher oxidation products, or even by blocking mercaptan with active reagents such as iodoacetate [108]. In fact, Menkart et al. [109] have suggested that a larger amount of set results from blocking mercaptan than from reoxidation to disulfide.

These experiments collectively demonstrate that structural reorganization during the setting of keratin fibers in aqueous reducing agents occurs not only through disulfide bond breakdown but also, to a large extent, through disulfide–mercaptan interchange reactions.

Because keratin fibers undergo crystallographic changes on stretching, X-ray diffraction can be used as a tool for studying the setting mechanism. Setting of human hair by various means produces an alpha to beta transformation [79]. Therefore, the α-helices of the filamentous regions of the cortical cells are stretched, and a β-configuration arises during setting. A number of forces including covalent bonds, hydrogen bonds, salt linkages, van der Waals forces, and steric interferences oppose stretching and setting. The weak links in these "chains of forces" opposing strain probably exist primarily in the matrix of the cortex and in the A layer and exocuticle of the cuticle (the regions of high sulfur content). In aqueous reducing solutions, certainly, the weakest links in these "chains of forces" are the disulfide bonds and those hydrogen bonds that are broken and interchanged. These are the actions that permit the alpha to beta structural rearrangement in the microfibrils and other structural rearrangements to occur.

In addition to matrix changes, extensive changes occur in the cuticle during permanent waving. Wickett and Barman [10] have demonstrated that a satisfactory permanent wave can be achieved by greater cuticle/cortex reduction than with existing thioglycolate waves. The A layer and exocuticle contain high concentrations of cystine residues [110] and are therefore highly reactive to reducing agents. The endocuticle, on the other hand, contains relatively little cystine [110]. Therefore the primary reaction of the reducing agent in the cuticle will be with the stiff, resistant A layer and exocuticle regions. These stiff cuticle layers will be softened, allowing for structural rearrangement. On reoxidation, the macromolecules of the A layer and exocuticle layers will be rehardened to a new configuration. Thus, a combination of cortical and cuticle changes occurs in permanent-waving to provide a new "permanent" shape to the keratin fiber.

One might also anticipate greater cuticle contribution to a permanent wave in fine hair as compared to coarse hair, because Wolfram and Lindemann [111] have provided evidence for a greater ratio of cuticle to cortex in fine hair. Greater cuticle contribution to waving of fine hair occurs because thin and coarse hair contain the same number of cuticle layers, with essentially the same thickness, and fine hair therefore, contains a greater proportion of cuticle to cortex than coarse hair.

Assuming a 4-μm-thick cuticle, for coarse, 100-μm-diameter hair fibers, the cuticle would constitute only 15% of the fiber cross section. However, for thin, 40-μm hair fibers, the cuticle would be 36% of the total fiber cross section. Fine hair is known to be resistant to waving, and this may result from its high cuticle content. The cuticle may play a lesser role than the cortex in permanent waving. Nevertheless, it is difficult to conceive that the cuticle does not play an important role in waving reactions that involve reduction, shaping, and rehardening of the high-sulfur A layers and exocuticle layers of the cuticle cells.

Hair-swelling agents, such as concentrated solutions of alkali metal halides [112] or aqueous solutions of reducing agents [113], are capable of promoting supercontraction as well as setting. The types of reagents that promote supercontraction suggest that hydrogen bond breakage occurs during supercontraction. Burley [114] has shown that disulfide–mercaptan interchange is also important for supercontraction. In addition, keratin fibers, while undergoing supercontraction, suffer a loss in birefringence, and the alpha X-ray diagram disappears [115,116], which suggests structural change in the filamentous regions of the cortex. The alpha keratin is thus rearranged to a less organized structure. Therefore, supercontraction, with the exception of the driving force, and the final molecular orientation are very much related mechanistically to the process of setting.

In the first edition of this book, I proposed that one might consider the curling (waving) of a human hair fiber as a combination of setting and supercontraction. A bent hair fiber that is treated with an aqueous solution of a reducing agent is undergoing concomitant extension (setting) and compression (which should be more analogous to supercontraction than to setting). If one perceives the waving of human hair in this manner, one may then apply the testing procedures and mechanisms for these two phenomena to arrive at a picture of the waving process.

Wortmann and Kure [1] have gone beyond that proposal and have been able to provide a model that explains the set behavior obtained in the permanent waving of human hair in terms of the bending stiffness of single hairs during the reduction and the oxidation reactions. Wortmann and Kure propose a distribution of Young's moduli from the hair surface to the center of the fiber and a diffusion-controlled breakdown during reduction. This model is simple, yet elegant, and is highly satisfactory in spite of the fact that it does not consider the two-phase composite nature of the cortex of human hair. Feughelman [3] has extended this model by proposing a model for setting a bent fiber and taking into account the two-phase composite nature of the cortex of keratin fibers.

In conclusion, it should be pointed out that even though the criterion for permanent set in the wool industry is markedly different from that in the hair-waving industry (boiling water versus neutral-pH shampoos near room temperature), much can be gained with regard to understanding hair waving by using the testing procedures employed for wool and by drawing analogies to the mechanisms of setting and supercontraction.

Swelling: During and After Waving

The microscopic method (change in volume) [117] and the centrifugation method (change in weight) have been used for studying the swelling of human hair by aqueous solutions of mercaptans. Both the rate and the extent of the swelling of human hair by mercaptan solutions are highly pH dependent and increase dramatically with increasing pH above neutrality [27,118]. In fact, at high pH, using a high solution-to-hair ratio, swelling in excess of 300% is possible with thioglycolic acid using relatively short reaction times [118]. The swelling of human hair in aqueous mercaptan solutions is a direct reflection of the chemical reactions occurring inside the fibers and can therefore be described in terms of the reactivity considerations outlined in the section on the kinetics of this reaction.

Shansky [119] has studied the swelling action of hair fibers during reduction, rinsing, and chemical neutralization, that is, a simulated cold wave process. During the reaction with mercaptan, the swelling action is extensive. On rinsing, swelling continues but at a reduced rate. These actions have been attributed to osmotic forces arising from the rapid decrease in salt concentration outside the hair compared to inside the hair on rinsing. During neutralization, swelling is reversed.

Hair fibers that have been reduced and reoxidized approach the original fiber diameter. Indeed, Eckstrom [117] has suggested that the milder the conditions of reduction, the closer the fiber will return to its original dry-state diameter on neutralization. Fiber diameter determined in the wet or swollen state is sensitive for detecting changes produced by damaging treatments such as permanent waves or bleaches. The swelling action of permanent-waved hair [27] and bleached fibers [29] is greater than in unaltered hair and has been used to estimate the relative extent of alteration produced by reduction and reoxidation [28]. For additional discussion on hair swelling, see Chapters 1 and 8.

Permanent Waving of Human Hair

Nessler is reputed as a key figure in the invention of the permanent wave during the early 1900s [120]. The first permanent waves were concentrated solutions of alkali (5%–15%) or alkaline sulfites [120,121], which were reacted with hair at elevated temperatures. In these treatments, high temperatures were achieved by using either curling irons, chemical heating pads, or electric heaters [122].

Current permanent waves are greatly superior to the early hot waves and do not require elevated temperature, thus the designation "cold waves." Cold waves became successful during World War II and have not changed substantially for nearly 40 years. These products are based on mercaptans or sulfites, the most common of these being thioglycolic acid, which is generally employed at a concentration of approximately 0.6 N and a pH of 9 to 9.5.

The sulfite wave has made a comeback, with more than one manufacturer now offering a sulfite-based product. Sulfite waves employ a pH near 6 and a hydrogen peroxide neutralizer. This type of wave generally claims to provide a wave that does not "frizzle" the hair, that is, is gentle to the hair and can be used on any type of hair, damaged or undamaged [123]. This image is consistent with the fact that sulfite is a weaker reducing agent than thioglycolate (see the discussion on the reduction of hair by sulfite earlier in this chapter).

Thioglycolate and sulfite waves are the primary reducing agents used in home permanent waves today, although as indicated earlier glycerylmonothioglycolate (GMT) is being used in the professional field in commercial acid waves. This type of wave requires a covering cap and the heat of a dryer to accomplish sufficient reduction to provide a satisfactory permanent wave [24].

Cold Wave Formulations and Making Cold Wave Products

A typical thiol permanent wave will consist of two compositions. The first, a reducing solution (often called a waving lotion), will have a composition similar to that shown in Table 3–2.

The emulsifier/wetting agents (steareth-20) are first melted and then added to oxygen-free water, under an inert atmosphere, at about 50° while stirring. After cooling to room temperature, about 3% concentrated ammonium hydroxide is added followed by, thioglycolic acid with stirring. The other ingredients are added and the pH adjusted with ammonium hydroxide or the preferred form of alkalinity.

TABLE 3–2. Thiol Permanent Wave Waving Lotion.

Ingredient	Percent
Thioglycolic acid	6.0
Steareth-20	2.5
Fragrance	0.5
Ethylene diamine tetraacetic acid	0.2
Colors	0.02
Water	q.s.[a]
Ammonium hydroxide	(to pH 9.3)

[a] q.s., add water to 100%.

TABLE 3–3. Neutralizer.[a]

Ingredient	Percent
Hydrogen peroxide (30%)	7.0
Polysorbate-40	2.5
Phenacetin	0.5
Water	q.s.[b]
Phosphoric acid (85%)	(to pH 4)

[a] Heat water to 75°C, melt polysorbate-40, and slowly add to water with stirring. Cool to room temperature, then add peroxide, phosphoric acid, and preservative.
[b] q.s., add water to 100%.

TABLE 3–4. Waving Lotion for Softwave Formulation.[a]

Ingredient	Percent
Ammonium bisulfite	4
Ammonium sulfite	3
Laureth-23	2.5
Fragrance	~0.5
Water	q.s.[b]
Ammonium hydroxide	(to pH 8)

[a]The neutralizer described for the ammonium thioglycolate wave can also be used with a bisulfite system.
[b]q.s., add water to 100%.

Note the following precautions for making thiol perms. One should use a vessel lined with materials such as glass, plastic (polyethylene or Teflon or other inert plastic), or stainless steel. One should avoid contact with most metals, because thiols react with many metals to form colored salts. Note that salts of thioglycolic acid may be handled in stainless steel, but thioglycolic acid may not. Also, heat should be minimized whenever thioglycolic acid is present. Exposure to air and oxygen should be avoided; for example, use oxygen-free water and package with a minimum of headspace, because thiols are sensitive to air oxidation. The neutralizer components are given in Table 3–3.

A milder waving lotion, sometimes called a softwave, can be made in the following manner. Add the laureth-23 to water at 70°C with stirring. Cool to room temperature, dissolve the bisulfite and sulfite, then add the fragrance and adjust the pH to 8 with ammonium hydroxide. As with the thiol wave, oxygen-free water should be used and the mixing done in an inert atmosphere (Table 3–4).

Acid Wave

Acid waves are generally based on GMT although some bisulfite systems are sold as acid waves. For a GMT wave, the waving lotion itself consists of two parts because GMT is not stable for long periods of time in water (Table 3–5).

To make the acid wave, for part I of the waving lotion add glycerol thioglycolate to glycerine (oxygen-free) in an inert atmosphere taking the same precautions as described previously for the thiol wave. For part II of the waving lotion, dissolve the sulfonate and the neodol in water and then add the remaining ingredients in the order listed in the foregoing for-

TABLE 3–5. Waving lotion for GMT wave.

Ingredient	Percent
Part I	
Glycerol thioglycolate (GMT) (75%)	77.3
Glycerine (oxygen-free dry glycerine)	22.7
Part II	
Urea	4.1
Neodol 91-8	1.0
Dodecyl benzene sulfonate	0.5
Triethanolamine	0.8
Potassium sorbate	0.35
Ammonium carbonate	0.2
Disodium EDTA	0.2
Water (oxygen free)	q.s.[a]

[a] q.s., add water to 100%.

TABLE 3–6. Depilatory cream or lotion.

Ingredient	Percent
Part I	
Mineral oil (heavy)	5.0
Emulgade 1000NI (cetearyl alcohol and ceteareth-20)	5.0
Part II	
Water (oxygen-free)	q.s.[a]
Part III	
Sodium thioglycolate	3.5
Calcium thioglycolate	3.0
Calcium hydroxide	~1.5 (to pH 11.5)
Fragrance	<1.0

[a] q.s., add water to 100%.

mula. Immediately before application to the hair, mix part I and part II of the waving lotions.

The following table (Table 3–6) lists depilatory ingredients.

To make this depilatory, melt the Emulgade into the mineral oil (part I). Heat the water (part II) to 75°C, then add part I to part II and continue heating and stirring for about 10 min. Cool to 40°C, and while stirring, add the individual ingredients of part III, adding calcium hydroxide last. Homogenize. Precautions as described for the thiol wave, such as the exclusion of oxygen and metals from the system, must also be exercised for making a thiol depilatory.

Other cold-wave formulations, straighteners, and depilatories are described by Gershon et al. [123], and methods for the qualitative and quantitative analysis of ingredients in cold-wave lotions have been described by Walker [124,125]. Product ingredient labels provide the most up-to-date qualitative information on these types of products.

Properties of Cold-Waved Hair

The chemical changes produced in hair by permanent waving, as indicated by amino acid analysis, are quantitatively small and do not reflect the vast structural changes that have taken place in the fibers during a permanent wave. Small decreases in cystine [77,126] and corresponding increases in cysteic acid [77,126] and in cysteine [77] have been reported. Small quantities of mixed disulfide [127], sorbed thioglycolic [77], and dithidiglycolic acids [127] have also been detected in hair that has undergone cold-waving treatments. Zahn et al. [128] have demonstrated small quantities of intermediate oxidation products of cystine in permanent-waved hair. For additional details of the chemical changes occurring in hair that has undergone permanent waving, see Chapter 2.

The wet tensile properties (through 30% extension) of hair are decreased by permanent waving; however, the dry tensile properties remain virtually unchanged. The torsional behavior of hair that has been permanent-waved is also changed. Bogaty [129] has shown that waved hair is more rigid in the dry state yet less rigid in the wet state than unwaved hair, and Schwartz and Knowles [130] have shown that the frictional resistance of human hair is increased by permanent waving, evidence of changes (damage) in the cuticle of hair. Increased fiber friction results in more difficult combing of hair that has undergone permanent-wave treatment. For additional details on the changes in these properties, see Chapter 8.

The swelling capacity of permanent-waved hair is increased in proportion to the damage rendered by the waving process [117]. Increased swelling is evidence of cortical damage to hair. Greater swelling produces a substantial increase in the chemical reactivity of hair toward those reactions in which diffusion is rate limiting. Because most of the whole-fiber chemical reactions that human hair undergoes are diffusion controlled, permanent waving can markedly alter the chemical character of human scalp hair.

The Nature of the Cold-Wave Process

The Reduction Step

A very important factor in cold-waving hair on heads is the solution-to-hair ratio, which is limited by the capillary spaces between the fibers, and the amount of solution absorbed into the fibers before solution runoff occurs. Assuming a solution to hair ratio of 2:1 for a twofold addition of reducing solution to hair, a 0.6 M mercaptan solution, and a favorable equilibrium constant (for thioglycolate at alkaline pH), there is insufficient mercaptan for total reduction of the disulfide bonds in hair. Randebrook and Eckert [131] and Reed et al. [132] have suggested that only about 20% of the cystine in hair is reduced during an average thioglycolate permanent-wave treatment. Presumably, less reduction occurs for an average sulfite wave.

During the reduction step, a highly reduced zone proceeds into the cuticle and eventually into the outer regions of the cortex, leaving an inner zone of unreduced hair. The relative quantities of reduced versus unreduced fiber depend on the reducing agent (thioglycolate versus sulfite), its concentration, the solution-to-hair ratio, pH of the reaction medium, time of reaction, fiber diameter, and the condition of the hair; these variables for the most part have already been considered in this chapter. For a more detailed discussion of the waving process, see the article by Gershon et al. [123].

The relatively high cleavage of cystine residues and the resultant high concentration of cysteinyl residues produced from the reaction of thioglycolic acid or sulfite with hair permit molecular reorientation to occur through a disulfide–mercaptan interchange pathway. The reduction occurs primarily in the high-sulfur regions of the fibers, that is, the A layer and exocuticle of the cuticle and the matrix of the cortex, permitting molecular reorientation and structural changes to occur in both cuticle and cortex (as described in the section on setting and supercontraction).

Rinsing

Cessation of the reduction reaction and removal of most of the reducing agent is the prime function of this step. The continued increase in swelling during rinsing from osmotic forces has already been described.

Creep Period

After rinsing, the hair is often wrapped in a towel and maintained in the desired configuration for a given period of time (up to 30 min). This step has been called the "creep period" and was introduced into the waving process in the early 1950s [133]. Continued molecular reorientation through disulfide–mercaptan interchange and secondary bond formation (other than covalent bond formation) occurs during this step, and because secondary bonds can contribute to wave stability [134], this step is important to the total permanent-wave process.

Neutralization

Neutralization or reoxidation is accomplished primarily through chemical means, mild oxidation for thioglycolate waves or mild oxidation or even mild alkali for sulfite waves. Neutralization rapidly decreases the mercaptan content in the fibers, decreasing the probability of disulfide–mercaptan interchange, and thereby stabilizes the permanent wave.

Safety Considerations for Permanent Waves

As for other reactive hair products, the primary safety concerns for permanent waves generally arise from misuse or failure to comply with product usage instructions. Skin irritation, hair breakage, oral toxicity, sensitization, and scarring alopecia either have been reported in the literature or are referred to in the warning instructions for home permanent-wave products.

An old but relevant review of the toxicity of thiol home permanent waves and neutralizer solutions was published by Norris in 1964 [135]. Thioglycolates are moderately toxic yet comparable to bisulfite. Sodium thioglycolate has an LD_{50} of 148 mg/kg (ip in rats) [136] versus 115 mg/kg (iv in rats) for sodium bisulfite [137].

Thioglycolate waving lotions can irritate skin [138]; however, irritation in home use is rare and may in part be related to the alkalinity of the system [139]. Among the different thioglycolate salts, monoethanolamine thioglycolate is reported to be less irritating to skin than ammonium thioglycolate [140]. Although ammonium thioglycolate has been reported as having a low sensitization potential [141], a few incidents of sensitization reaction have been reported by hairdressers with frequent contact [141].

Hair breakage and some permanent hair loss from misuse of these products have been reported by Bergfeld [142] and attributed to scarring alopecia. Bergfeld did not specify the extent of hair loss observed; however, concludes that side effects from home permanent-waving products are minimal if consumers are aware of their hair damage and any inherent skin diseases and if they comply with the product usage instructions [142].

References

1. Wortmann, F.J.; Kure, N. J. Soc. Cosmet. Chem. 41:123 (1990).
2. Salce, L.; Savaides, A.; Schultz, T. In 8th International Hair-Science Symposium, Kiel, Germany (German Wool Res. Inst. Publ. Abstracts.) (1992).
3. Feughelman, M. J. Soc. Cosmet. Chem. 42:129 (1991).
4. Cleland, W. Biochemistry 3:480 (1964).
5. Fruton, J.S.; Clark, H.T. J. Biol. Chem. 106:667 (1934).
6. Kolthoff, I.M.; et al. J. Am. Chem. Soc. 77:4733 (1955).
7. Patterson, W.I.; et al. J. Res. Natl. Bur. Stand. 27:89 (1941).
8. Weigmann, H.D.; Rebenfeld, L. Text. Res. J. 36:202 (1966).
9. Wickett, R.R. J. Soc. Cosmet. Chem. 34:301 (1983).
10. Wickett, R.R.; Barman, B.G. In 4th International Hair Science Symposium, Syburg, W. Germany, German Wool Res. Inst. Publ. Abstracts. (1984).

11. Wickett, R.R.; Barman, B.G. J. Soc. Cosmet. Chem. 36:75 (1985).
12. Haefele, J.W.; Broge, R.N. Proc. Sci. Sect. Toilet Goods Association, No. 36, p. 31 (1961).
13. O'Donnell, I.J. Text. Res. J. 24:1058 (1954).
14. Leach, S.J.; O'Donnell, I.J. Biochem. J. 79:287 (1961).
15. Thompson, E.O.P.; O'Donnell, I.J. Biochim. Biophys. Acta 53:447 (1961).
16. Middlebrook, W.R.; Phillips, H. Biochem. J. 36:294 (1942).
17. Carter, E.G.H.; et al. J. Soc. Dyers Col. 62:203 (1946).
18. Schoeberl, A.; Tausent, H. Proc. Int. Wool Text. Res. Conf. C:150 (1955).
19. Foss, O. In Organic Sulfur Compounds, Vol. 1, Kharasch, N., ed. Pergamon Press, New York (1959).
20. Eldjorn, L.; Pehl, A. J. Biol. Chem. 225:499 (1957).
21. Weigmann, H.D. J. Polymer Sci. A-1(6):2237 (1968).
22. Kubu, E.T.; Montgomery, D.J. Text. Res. J. 22:778 (1952).
23. Herrmann, K.W. Trans. Faraday Soc. 59:1663 (1963).
24. Edman, W.W.; Klemm, E.J. Cosmet. Toiletries 94:35 (1979).
25. Valko, E.I.; Barnett, G. J. Soc. Cosmet. Chem. 3:108 (1952).
26. Zviak, C.; Ronet, A. U.S. patent 3,271,258 (1966).
27. Heilingotter, R. Am. Perfumer 66:17 (1955).
28. Klemm, E.J.; et al. Proc. Sci. Sect. Toilet Goods Association, No. 43 (May 1965).
29. Edman, W.W.; Marti, M.E. J. Soc. Cosmet. Chem. 12:133 (1961).
30. Menkart, J.; Speakman, J.B. J. Soc. Dyers Col. 63:322 (1947).
31. Gould, E.S. Mechanism and Structure in Organic Chemistry, p. 259. Holt, Rinehart & Winston, New York (1959).
32. Haefele, J.W.; Broge, R.W. Proc. Sci. Sect. Toilet Goods Association, No. 32, p 59 (1959).
33. Wolfram, L.J. In Hair Research, Orfanos, C.E.; Montagna, W.; Stuttgen, G., eds., pp. 486–491. Springer-Verlag, Berlin (1981).
34. Heilingotter, R. Am. Perfumer 69:41 (1957).
35. Heilingotter, R.; Komarony, R. Am. Perfumer 71:31 (1958).
36. Rieger, M. Am. Perfumer 75:33 (1960).
37. Deadman, L.L. Br. patent 798,674 (1958).
38. U.S. patent 3,064,045 (1962).
39. Danehy, J.P. In The Chemistry of Organic Sulfur Compounds, Vol. 2, Kharasch, N.; Meyers, F. eds., p. 337. Pergamon Press, New York (1966).
40. Zahn, H.; et al. J. Text. Inst. 51:T740 (1960).
41. Clark, H.T. J. Biol. Chem. 97:235 (1932).
42. Reese, C.E.; Eyring, H. Text. Res. J. 20:743 (1950).
43. Elsworth, F.F.; Phillips, H. Biochem. J. 32:837 (1938).
44. Elsworth, F.F.; Phillips, H. Biochem. J. 35:135 (1941).
45. Volk, G. Proc. 3rd Int. Wool Text. Res. Conf. II:375 (1965).
46. Wolfram, L.J.; Underwood, D.L. Text. Res. J. 36:947 (1966).
47. Wolfram, L.J. In Hair Research, Orfanos, C.E., Montagna, W.; and Stuttgen, G., eds., pp. 491–494, Springer-Verlag, Berlin (1981).
48. Sneath, R. L. 8th International Hair Science Symposium of the German Wool Research Institute, Kiel, Germany, German Wool Res. Inst. Publ. Abstracts (1992).

49. Albrecht, L.; Wolfram, L.J. J. Soc. Cosmet. Chem. 33:363 (1982).
50. Barry, R.H. In Cosmetic Science and Technology, Sagarin, E., ed., p. 463. Interscience, New York (1962).
51. Asquith, R.S. Cosmet. Toiletries 95:40 (1980).
52. Speakman, J.B. J. Soc. Dyers Col. 52:335 (1936).
53. Asquith, R.S.; Speakman, J.B. Proc. Int. Wool Text. Res. Conf. C:302 (1955).
54. Parker, A.J.; Kharasch, N. Chem. Rev. 59:583 (1959).
55. Kharasch, N. In Organic Sulfur Compounds, Vol. 1, Kharasch, N., ed., p. 392. Pergamon Press, New York (1961).
56. Ziegler, K. Proc. 3rd Int. Wool Text. Res. Conf. II:403 (1965).
57. Feughelman, M. J. Soc. Cosmet. Chem. 42:129 (1991).
58. Speakman, J.B.; Whewell, C.S. J. Soc. Dyers Col. 52:380 (1936).
59. Horn, M.; et al. J. Biol. Chem. 138:141 (1941).
60. Danehy, J.P.; Hunter, W.E. J. Org. Chem. 32:2047 (1967).
61. Swan, J.M. Proc. Int. Wool Text. Res. Conf. Australia C:25 (1955).
62. Earland, C.; Raven, D.J. Nature 191:384 (1961).
63. Danehy, J.P.; Kreuz, J.A. J. Am. Chem. Soc. 83:1109 (1961).
64. Asquith, R.S.; et al. J. Soc. Dyers Col. 90:357 (1974).
65. Tolgyesi, E.; Fang, F. In Hair Research, Orfanos, C.E.; Montagna, W.; Stuttgen, G., eds., pp. 116–122. Springer-Verlag, Berlin (1981).
66. Asquith, R.S.; Garcia-Dominguez, J. J. Soc. Dyers Col. 84:155 (1968).
67. Asquith, R.S.; Carthew, P. J. Text. Inst. 64:10 (1973).
68. Asquith, R.S. Cosmet. Toiletries 95:40 (1980).
69. Goddard, D.R.; Michaelis, L. J. Biol. Chem. 106:605 (1934).
70. Cuthbertson, W.R.; Phillips, H. Biochem. J. 39:7 (1945).
71. Bajpai, L.S.; et al. J. Soc. Dyers Col. 77:193 (1961).
72. Jenkins, A.D.; Wolfram, L.J. J. Soc. Dyers Col. 79:55 (1963).
73. Wolfram, L.J. Proc. 3rd Int. Wool Text. Res. Conf. II:393 (1965).
74. Carr, E.M.; Jensen, W. Ann. N.Y. Acad. Sci. 116(II):735 (1961).
75. Bogaty, H.; Brown, A.E. U.S. patent 2,766,760 (1956).
76. Elod, E.; Zahn, H. Textil Praxis, p. 27 (1949).
77. Zahn, H.; Gerthsen, T.; Kehren, M.L. J. Soc. Cosmet. Chem. 14:529 (1963).
78. Whitman, R. Proc. Sci. Sect. Toilet Goods Association, 18:27 (1952).
79. Head, R.C. U.S. patent 2,633,447 (1953).
80. Reed, R.E.; et al. U.S. patent 2,564,722 (1951).
81. Sagal, J. Jr. Text. Res. J. 35:672 (1965).
82. Bell, T.E. U.S. patent 2,774,355 (1956).
83. Sanford, D.; Humoller, F.L. Anal. Chem. 19:404 (1947).
84. Harris, M.; Brown, A. Symp. Fibrous Proteins, Soc. Dyers Colquest Publ., p. 203 (1946).
85. Hall, K.E.; Wolfram, L.J. J. Soc. Cosmet. Chem. 28:231 (1977).
86. Walker, G.T. Seifen-Oele-Fette-Wachse 89:147 (1963).
87. Leach, S.J. Aust. J. Chem. 13:547 (1960).
88. Schoberl, A. J. Text. Inst. 51(T):613 (1960).
89. Geiger, W.B.; et al. J. Res. Natl. Bur. Stand. 29:381 (1942).
90. Burley, R.W.; Horden, F.W.A. Text. Res. J. 27:615 (1957).
91. Madaras, G.W.; Speakman, J.B. J. Soc. Dyers Col. 70:112 (1954).
92. Negishi, M.; Arai, K.; Nagakura, I. J. Appl. Polymer Sci. 11:115 (1967).

93. Wall, R.A. Can. patent 795,758; U.S. patent 3,472,243 and 3,472,604 (1969).
94. Robbins, C.R.; Crawford, R.; Auzuino, G. J. Soc. Cosmet. Chem. 25:407 (1974).
95. Wolfram, L.J. J. Soc. Cosmet. Chem. 20:539 (1969).
96. Ingram, P.; et al. J. Polymer Sci. 6:1895 (1968).
97. Campbell, D.; et al. Polymer Lett. 6:409 (1968).
98. Wolfram, L.J.; Diaz, P. 4th Int. Hair Sci. Symp., Syburg, W. Germany, German Wool Res. Inst. Publ. Abstracts, (1984).
99. Robbins, C.R.; Reich, C. Unpublished work.
100. Stam, P.B.; Katy, R.F.; White, Jr., H.J. Text. Res. J. 22:448 (1952).
101. Scott, G.V.; Robbins, C.R. J. Soc. Cosmet. Chem. 29:469 (1978).
102. King, G. J. Text. Inst. 41:T135 (1950).
103. Brown, A.E.; et al. Text. Res. J. 20:51 (1950).
104. Rebenfeld, L.; et al. Text. Res. J. 33:79 (1963).
105. Jenkins, A.D.; Wolfram, L.J. J. Soc. Dyers Col. 80:65 (1964).
106. Farnworth, A.J. Text. Res. J. 27:632 (1957).
107. Weigmann, H.D.; et al. Proc. 3rd Int. Wool Text. Res. Conf. II:244 (1965).
108. Milligan, B.; et al. Proc. 3rd Int. Wool Text. Res. Conf. II:239 (1965).
109. Menkart, J.; et al. Proc. 3rd Int. Wool Text. Res. Conf. II:253 (1965).
110. Swift, J.; Bews, B. J. Soc. Cosmet. Chem. 27:289 (1976).
111. Wolfram, L.J.; Lindemann, M. J. Soc. Cosmet. Chem. 22:839 (1971).
112. Alexander, P. Ann. N.Y. Acad. Sci. 53:653 (1951).
113. Speakman, J.B. Nature 132:930 (1933).
114. Burley, R.W. Proc. Int. Wool Text. Res. Conf. D:88 (1955).
115. Haly, A.R.; Smith, J.W. Biochim. Biophys. Acta 44:180 (1960).
116. Skertchly, A. J. Text. Inst. 51:528 (1960).
117. Eckstrom, M.G., Jr. J. Soc. Cosmet. Chem. 2:244 (1951).
118. Powers, D.H.; Barnett, G. J. Soc. Cosmet. Chem. 4:92 (1953).
119. Shansky, A. J. Soc. Cosmet. Chem. 14:427 (1963).
120. DeNavarre, M.G. Am. Perfumer 46:49 (1944).
121. Kalisch, J. Drug Cosmet. Ind. 49:156 (1941).
122. Suter, M.J. J. Soc. Cosmet. Chem. 1:103 (1949).
123. Gershon, S.D.; Goldberg, M.A.; Rieger, M. In Cosmetics Science and Technology, Sagarin. E., eds., pp. 583–627. Interscience, New York (1963).
124. Walker, G.T. Soap Perfumery Cosmet., p. 816 (Aug. 1954).
125. Walker, G.T. Soap Perfumery Cosmet., p. 929 (Aug. 1954).
126. Robbins, C.R.; Kelly, C.H. J. Soc. Cosmet. Chem. 20:555 (1969).
127. Gerthsen, T.; Gohlke, C. Parfeum. Kosmet. 45:277 (1964).
128. Zahn, H.; et al. 4th Int. Hair Sci. Symp., Syburg, W. Germany, German Wool Res. Inst. Publ. Abstracts (1984).
129. Bogaty, H. J. Soc. Cosmet. Chem. 18:575 (1967).
130. Schwartz, A.; Knowles, D. J. Soc. Cosmet. Chem. 14:455 (1963).
131. Randebrook, R.; Eckert, L. Fette Seifen Anstrichmittel 67:775 (1965).
132. Reed, R.; et al. J. Soc. Cosmet. Chem. 1:109 (1948).
133. Brown, A.E. U.S. patent 2,688,972 (1954).
134. Bogaty, H. J. Soc. Cosmet. Chem. 11:333 (1960).
135. Norris, J.A. Bibra Info. Bull. 3:471 (1964).
136. Freeman, H.; Rosenthal, C. Fed. Proc. 11:347 (1952).

137. Hoppe, J.D.; Goble, F.C. J. Pharmacol. Exp. Ther. 101:101 (1951).
138. Behrman, H.T. J. Soc. Cosmet. Chem. 2:228 (1951).
139. Ishahara, M. In Hair Research, Orfanos, C.E.; Montagna, W.; Stuttgen, G., eds., p. 536. Springer-Verlag, Berlin (1981).
140. Whitman, R.; Brookins, M. Proc. Sci. Sect. Toilet Goods Association, 25:42 (1956).
141. Bourgeois-Spinasse, J. In Hair Research, Orfanos, C.E.; Montagna, W.; Stuttgen, G., eds., p. 544. Springer-Verlag, Berlin (1981).
142. Bergfeld, W.F. In Hair Research, Orfanos, C.E.; Montagna, W.; Stuttgen, G., eds., p. 507. Springer-Verlag, Berlin (1981).

4

Bleaching Human Hair

The composition of amino acid residues in bleached hair and in hydrolysates of oxidized keratin fibers is described in Chapter 2 and also in publications by Zahn [1,2], Robbins et al. [3–6], Maclaren et al. [7,8], and Alter and Bit-Alkis [9]. Although, several questions remain unanswered, especially with regard to the structures and reactions of hair pigments [10–14], general features about the chemical structure of hair and its reactions with bleach products have been relatively well described using the language of physical-organic chemistry.

The objectives of this chapter are to describe bleach product compositions and their formulation and to review the chemistry of both chemical and photochemical bleaching, that is, the oxidative degradation of hair pigments and the accompanying oxidative degradation of the proteins of human hair, and to describe the chemical nature of bleached hair.

Hair-Bleaching Compositions

Bleaching compositions were described several years ago by Cook [15], and complete formulations were listed by Wall years ago in the book edited by Sagarin [16]. However, for the most reliable up-to-date qualitative information on bleaching compositions one should examine product ingredient labels. Hydrogen peroxide is the principal oxidizing agent used in bleaching compositions, and salts of persulfate are often added as "accelerators" [15]. The pH of these systems generally ranges from 9

TABLE 4–1. Hair lightener base.

Ingredient	Percentage
Oleic acid	20
Dodecyl benzene sulfonate (50%)	2
Neodol 91-2.5	2
Concentrated ammonium hydroxide	6
Sodium sulfite	0.5
Deionized water	69.5

TABLE 4–2. Lotion developer.

Ingredient	Percentage
Hydrogen peroxide (30%)	17
Dodecyl benzene sulfonate (50%)	16
Nonoxynol-9	6
Cetyl alcohol	3
Stearyl alcohol	2
Phosphoric acid	1
Water	55

to 11, and stabilizers (e.g., sequestrants) are often used to reduce the rate of decomposition of the peroxide.

A maximum hair-lightening product for either stripping or frosting hair will generally consist of three different units: the hair lightener base (alkalinity), the lotion developer (containing the peroxide), and the booster powder or accelerator containing salts of persulfate. It is prepared just before use by adding approximately 100 g of the lotion developer to about 50 g of the hair lightener base and then adding one to two packets of the booster powder (~12–16 g per packet) (Table 4–1).

To formulate this hair lightener base, add the sulfonate and neodol to the water while stirring at room temperature. Add the sodium sulfite and then the alkalinity followed by the oleic acid, stirring continuously as the system thickens. The developer ingredients are listed in Table 4–2.

The lotion developer is made by dissolving the sulfonate and nonoxynol in water. Heat to 65°C and add the melted cetyl and stearyl alcohols while stirring; cool and add the phosphoric acid and the hydrogen peroxide. The booster powder is listed in Table 4–3.

Premix the silica, the silicate, the sulfate and the EDTA; add the persulfates one at a time and mix until uniform.

For permanent hair dyes in which small shade changes to a lighter color are involved, bleaching is also employed. For these systems of permanent dyes, extra peroxide is formulated into the creme developer for the necessary bleaching action. For formulas of this type, see the section of Chapter

TABLE 4–3. Booster powder (accelerator).

Ingredient	Percentage
Potassium persulfate	27
Sodium silicate	26
Ammonium persulfate	25
Silica	20
Sodium lauryl sulfate	1.8
Disodium EDTA	0.2

TABLE 4–4. Spray–in hair lightener.

Ingredient	Percentage
Water	q.s.[a]
Hydrogen peroxide (30%)	10
Hydroxyethyl cetyldimonium phosphate	1
Polysorbate-20	1
Quaternium-80	0.5
Benzoic acid	0.3
Disodium EDTA	0.2
Fragrance	0.2 to 0.5

[a] q.s., add water to 100%.

6 that describes the formulation of permanent hair dyes. A spray-in hair lightener is listed in Table 4–4.

While stirring, dissolve the polysorbate into water followed by the hydroxyethyl cetyldimonium phosphate. Add the quaternium-80, benzoic acid, and disodium EDTA followed by the fragrance.

Reactions of the Proteins of Human Hair with Bleaches

Chemical Oxidation of the Disulfide Bond

The primary purpose in bleaching human hair is to lighten the hair, and this is most readily accomplished by oxidation. However, because of the severe reaction conditions required for destruction of the chromophore of hair pigments, side reactions with the hair proteins occur simultaneously. Wolfram [12] has provided evidence that hydrogen peroxide, the principal component of hair bleach systems, reacts faster with melanin than with hair proteins. However, because hair is primarily proteinaceous and contains a large percentage of oxidizable groupings (e.g., disulfide bonds), oxidation of the hair matrix and of the cuticle also occurs during the bleaching of human hair.

Zahn [1] first demonstrated that the primary reaction of oxidizing agents with the proteins of human hair occurs at cystine. Small amounts of degradation also occur to the amino acid residues of tyrosine, threonine, and methionine during severe bleaching [5]. The main site of attack, however, is at the disulfide bonds of the cystyl residues in the fibers. Robbins and Kelly [5] have shown that 15% to 25% of the disulfide bonds in human hair are degraded during "normal" bleaching. However, as much as 45% of the cystine bonds may be broken during severe "in practice" bleaching. This amount of damage may occur while frosting hair, or while bleaching hair from black or brown-black to light blond.

The kinetics of the oxidation of cystyl residues in hair by hydrogen peroxide has not been reported, although there is evidence to suggest that this reaction is a diffusion-controlled process. Harris and Brown [17] reduced and methylated keratin fibers and demonstrated that the wet tensile properties decrease almost linearly with the disulfide content. Alexander et al. [18] arrived at this same conclusion after oxidizing wool fiber with peracetic acid. A similar phenomenon has been observed by Robbins for hair that has been oxidized with alkaline hydrogen peroxide. (Some of these data are described in Chapter 8). Therefore, one may conclude that the percentage loss of the wet tensile properties that occurs during bleaching (e.g., the decrease in the 20% index [20]) is an estimate of the percentage of cystine linkages that are broken. For a more complete discussion of the effects of bleaching on the tensile properties of hair, see Chapter 8.

Edman and Marti [19] described the change in the 20% index of hair fibers as a function of treatment time in 6% hydrogen peroxide, at 32°C using a 25:1 solution-to-hair ratio at a pH of 9.5. When their data are plotted versus the square root of time, the resultant straight line indicates a diffusion-controlled process (Figure 4–1). These data have been applied to an equation developed by Crank [20] describing diffusion from a stirred solution of limited volume into a cylinder of infinite length.

$$C_T/C_\infty = 2\left[\frac{2}{\sqrt{\pi}}(Dt/a^2)^{1/2} + \cdots\right]$$

The term C_T is the 20% index at time t and represents the amount of cleaved disulfide at time t; C_∞ is the 20% index at time 0 and represents the total amount of disulfide before oxidation; and a represents the fiber radius, assumed to be 40 μm. Considering these assumptions, one obtains an approximate diffusion coefficient of 1.8×10^{-9} cm^2/min. This diffusion coefficient is of the anticipated magnitude, suggesting that the oxidation of the disulfide bond in hair by alkaline hydrogen peroxide is a diffusion-controlled reaction.

Two types of mechanisms have been proposed for the oxidative degradation of disulfide bonds [21,22]: a sulfur–sulfur (S–S) fission process, and a

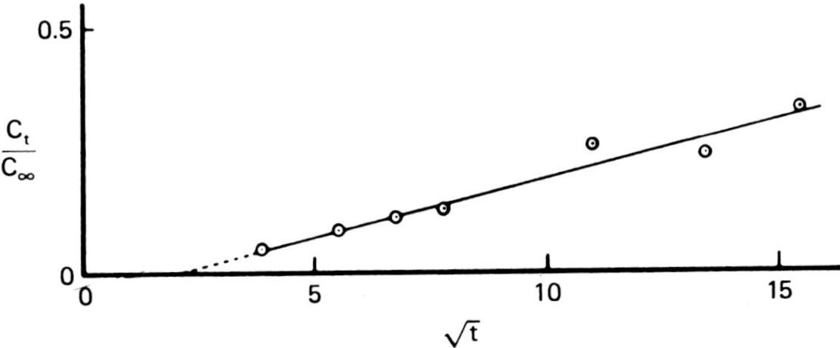

FIGURE 4–1. Rate of cleavage of cystine cross-links estimated from tensile properties [21]. (Reprinted with permission of the *Journal of the Society of Cosmetic Chemists.*)

S–S FISSION

$$R\text{-}S\text{-}S\text{-}R \rightarrow R\text{-}SO\text{-}S\text{-}R \rightarrow R\text{-}SO_2\text{-}S\text{-}R \rightarrow [R\text{-}SO_2\text{-}SO\text{-}R] \rightarrow R\text{-}SO_2\text{-}SO_2\text{-}R \rightarrow 2\,R\text{-}SO_3H$$

C–S FISSION

$$R\text{-}S\text{-}S\text{-}R \rightarrow R\text{-}S\text{-}S\text{-}OH + R\text{-}OH \rightarrow R\text{-}S\text{-}SO_2H \rightarrow R\text{-}S\text{-}SO_3H \rightarrow R\text{-}SO_3H + H_2SO_4$$

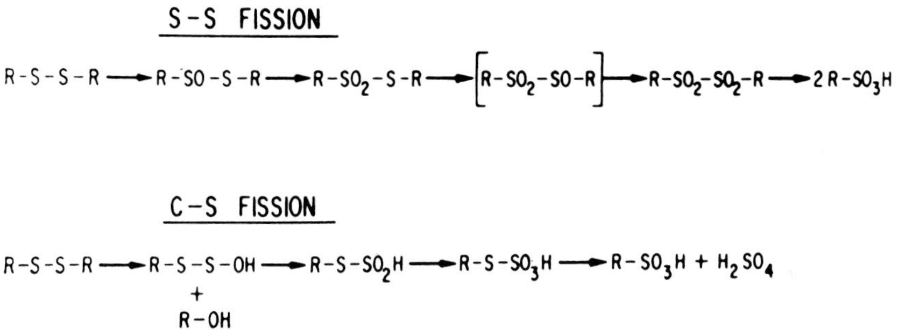

FIGURE 4–2. Schemes for disulfide fission [21]. (Reprinted with permission of the *Journal of the Society of Cosmetic Chemists.*)

carbon–sulfur (C–S) fission process. These two mechanistic schemes are summarized by Figure 4–2. Table 4–5 defines the different functional groups involved in these two oxidative schemes. Figure 4–2 shows that if the oxidation of cystine in hair proceeds totally through S–S fission, then 2 moles of sulfonic acid should be produced per mole of reacted disulfide. However, if the reaction goes totally through C–S fission, then only 1 mole of sulfonic acid can be produced from each mole of disulfide that reacts. Nachtigal and Robbins [4] have shown that this ratio is greater than 1.6 for frosted hair, suggesting that this reaction occurs largely through the S–S fission route. Second, if this reaction occurs through the C–S fission route, the alcohol produced would be a seryl residue and on hydrolysis would produce significantly larger quantities of serine in bleached hair hydrolysates than in hydrolysates of unbleached hair. However, this is not the case, because Robbins and Kelly [5] have shown that in samples of hair bleached on heads with commercial bleaching products the amount of serine is equal to or less than that of unbleached hair. Thus, the oxidative cleavage of the disulfide bond that occurs during the chemi-

TABLE 4–5. Functional groups involved in the oxidation of disulfides and mercaptans.

Disulfide Oxides	
–S–S–	Disulfide
–SO–S–	Monoxide
$-SO_2-S-$	Dioxide
$-SO_2-SO-$	Trioxide
$-SO_2-SO_2-$	Tetroxide
Sulfur Acids	
–SH	Mercaptan
–S–OH	Sulfenic Acid
$-SO_2H$	Sulfinic Acid
$-SO_3H$	Sulfonic Acid

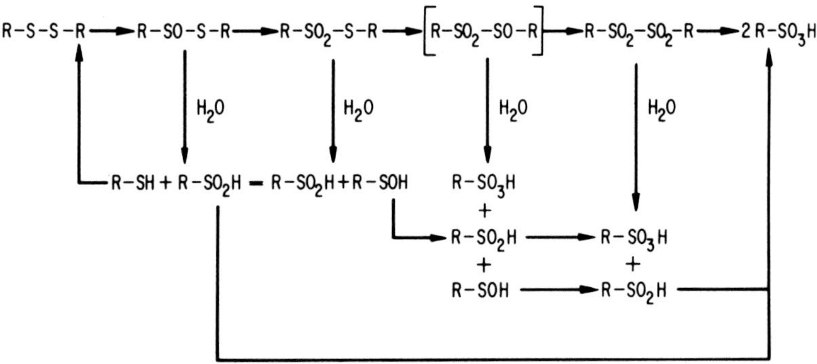

FIGURE 4–3. S–S fission of disulfides in an aqueous alkaline oxidizing medium [21]. (Reprinted with permission of the *Journal of the Society of Cosmetic Chemists.*)

cal bleaching of human hair by current bleach products is predominately an S–S fission process.

Because the bleaching of human hair is carried out in an aqueous alkaline oxidizing medium, hydrolysis of the cystine oxide intermediates (Figure 4–3) should be competitive with their oxidation. In fact, disproportionation of the cystine oxides [22] may also occur, adding to the complexity of the total reaction scheme; however, the net highest oxidation state of a disulfide under S–S fission conditions is sulfonic acid. This is illustrated by the oxidation and hydrolysis reactions of the cystine oxides summarized in Figure 4–3. Note that disulfide trioxides have never been isolated but are thought of as possible intermediates in this scheme, because both disulfide dioxides and disulfide tetroxides of pure compounds [22] have been isolated by oxidation in an acidic medium. Cystine monoxide and dioxide are sensitive to alkaline hydrolysis [23,24] but have been isolated from aqueous acidic oxidations [24]. The tetroxide should be even more sensitive to alkaline hydrolysis [22]. Although the importance of hydrolysis relative to oxidation for each of the cystine oxides is not known, it is certain that hydrolysis should be increasingly important with increasing pH. At the pH of current bleach products (pH 9–11), the rate of hydrolysis of these species should be competitive with oxidation, thereby decreasing the probability of existence of these species in major quantities in bleached hair.

Intact hair from bleaching experiments using alkaline hydrogen peroxide and peroxide-persulfate has been examined by both infrared spectroscopy [3,9] and electron spectroscopy for chemical analysis [6]. Evidence for intermediate oxidation products of cystine (the monoxide through tetroxide) could not be found. However, one could not conclude that small quantities of these species do not exist in bleached hair. The primary conclusions from these spectroscopic studies are (1) the principal end product formed from the oxidation of cystine during chemical bleaching of hair with either alkaline peroxide or alkaline peroxide-persulfate is cysteic acid, and (2) the cleavage of cystine appears to proceed primarily through the S–S fission route.

Zahn and coworkers [2], using two-dimensional gel electrophoresis, have separated up to 62 isolated protein spots from human hair. From the fluorogram of bleached hair, these scientists identified cystine oxides (monoxide and dioxide). Although, the exact quantities of these intermediate oxidation products versus cysteic acid were not reported, the quantities were indicated to be small relative to the cysteic acid content [2].

To summarize the end products of the oxidation of cystyl residues in hair from chemical bleaching, sulfonic acid is the principal established end product of the oxidative cleavage of the disulfide bond that occurs during the chemical bleaching of human hair with current hair bleach products [3,9]. The mercaptan content of bleached hair is less than that of unbleached hair [4]. The intermediate oxidation products of cystine, the disulfide monoxide, dioxide, trioxide, and tetroxide, do not exist as significant end products of hair bleaching [3,6,9]. Nevertheless, evidence has been presented demonstrating low levels of cystine oxides in bleached hair [2].

Considering all the species from the oxidation of disulfides described in Figure 4–3, sulfinic acid is the only species of even moderate stability [25] remaining to be examined. Sulfenic acids are notoriously unstable [26], and disulfide trioxides and disulfide tetroxides are even more sensitive to alkali than are the dioxides and the monoxides.

Oxidation of Other Amino Acid Residues

Robbins and Kelly [5] have examined bleached and unaltered hair by hydrolysis and amino acid analysis. Their results for severely bleached

TABLE 4–6. Effect of bleaching on the amino acid residues in hair [5].

Amino acid	Micromoles/gram dry hair		
	Nonfrosted	Frosted	Difference (%)[a]
Half-cystine	1,509	731	−50
Methionine	50	38	−24
Tyrosine	183	146	−20
Lysine	198	180	−10
Histidine	65	55	−15

[a] Only those amino acids found to be 10% or lower in bleached hair are included in this table.

hair, summarized in Table 4–6, suggest that methionine, tyrosine, lysine, and histidine, in addition to cystine, are degraded to the greatest extent (tryptophan could not be evaluated in this study).

These results are consistent with the relative sensitivities of these species to oxidation. Cystine and its reactions with oxidizing agents have already been described. Methionine is also sensitive to oxidation and is probably oxidized to its sulfoxide and possibly to methionine sulfone. Tyrosine, with its electron-rich phenolic ring, is also sensitive to oxidation. The amine salts of lysine and histidine should be resistant to oxidation, although the free amines of these species may be slowly oxidized in the bleach medium.

Hydrolysis or the Action of Alkali

Because bleaching compositions are usually formulated between pH 9 and 11, the hydrolysis of peptide and amide bonds and the formation of lanthionyl residues in hair are possible side reactions in bleaching. The hydrolysis of amide groups of the residues of aspartic and glutamic acids, in addition to the formation of cysteic acid residues, will increase the ratio of acidic to basic groups in the fibers; that is, amide hydrolysis will decrease the isoelectric and isoionic points of the fibers.

$$-\overset{\overset{O}{\|}}{C}-\underset{\underset{\underset{\underset{NH_2}{|}}{C=O}}{|}}{\underset{CH_2}{|}}{CH}-NH- \; + \; {}^{\ominus}OH \longrightarrow -\overset{\overset{O}{\|}}{C}-\underset{\underset{\underset{\underset{O^{\ominus}}{|}}{C=O}}{|}}{\underset{CH_2}{|}}{CH}-NH- \; + \; NH_3$$

Amide of aspartic acid residue

Peptide bonds are the major repeating structural unit of polypeptides and proteins. Hydrolysis of peptide bonds can occur at high pH and is most likely to occur during frosting or bleaching from black or brown-black to light blond, during which processes long reaction times and higher concentrations of alkali and oxidizing agent are employed.

$$-C(=O)-CH(R)-NH-C(=O)-CH(R)-NH-C(=O)-CH(R)-NH- \xrightarrow{^{\ominus}OH} {}^{\ominus}O-C(=O)-CH(R)-NH-C(=O)-CH(R)-NH- + -C(=O)-CH(R)-NH_2$$

Alkaline hydrolysis of a peptide bond

Harris and Brown [17] have shown that the wet tensile properties of keratin fibers are related to the disulfide bonds, whereas the dry tensile properties are influenced more by peptide bonds [18]. In an examination of frosted hair, Robbins found a 4% to 8% decrease in dry tensile properties (see Chapter 8 for details), which suggests that some peptide bond hydrolysis may occur during severe bleaching conditions. However, because the frequency of peptide bonds is nearly an order of magnitude greater than that of the disulfide bonds, the percentage of peptide bonds broken during bleaching must be relatively small.

The formation of lanthionyl residues in alkaline media is described in Chapter 3. Note that if lanthionine is formed during hair bleaching, its sulfoxide and sulfone are also possible oxidation products.

Oxidation of Hair Proteins by Sun and Ultraviolet Light

Light radiation attacks both hair proteins and hair pigment. The emphasis in this section is on the photochemical degradation of the proteins of hair. Later in this chapter, in the discussion on hair pigments, the effects of light on melanins are considered.

Sunlight and ultraviolet light have been shown by Beyak et al. [27] to decrease the wet tensile properties of human hair. Beyak relates these effects to the total radiation to which the hair is exposed, rather than to any specific wavelength. However, more recently, hair protein degradation by light radiation has been shown to occur primarily in the wavelength region of 254 to 400 nm. Hair proteins have been shown to absorb light primarily between 254 and 350 nm [28].

Several amino acids of hair absorb light in this region, and these amino acids are the most subject to degradation by light. The following amino acids have been shown to be degraded by weathering actions (primarily light radiation) on wool fiber [29,30]: cystine and methionine (the sulfur-containing amino acids); the aromatic and ring containing amino acids phenylalanine, tryptophan, histidine, and proline; and the aliphatic amino acid leucine. Pande and Jachowicz [31] used fluorescence spectroscopy to monitor the decomposition of tryptophan in hair. These scientists demonstrated that photodegradation to tryptophan does occur in hair, and they speculated that tryptophan photodamage may lead to, or make other amino acids more sensitive to, photodegradation.

Oxidation at the amide carbon has also been shown to occur both in wool [32] and in hair [6], producing carbonyl groups (this is called photo-yellowing of wool). However, the primary reaction in the weathering of human hair involves degradation of the disulfide bond of cystine residues.

Robbins and Kelly [33] analyzed amino acids of both proximal and distal ends of human hair and showed significantly more cysteic acid in tip ends. They attribute this change to weathering actions, specifically to ultraviolet radiation. This same study also found significant changes in tyrosine and histidine similar to the weathering effects in wool fiber as well as changes in the lysine content.

Robbins and Bahl [6] examined the effects of sunlight and ultraviolet radiation on disulfide sulfur in hair via electron spectroscopy for chemical analysis (ESCA) [6]; both UV-A (320–400 nm) and UV-B (290–320 nm) radiation were shown to oxidize sulfur in hair. The primary oxidation occurs closer to the hair fiber surface, producing a steep gradient of oxidized to less oxidized hair from the outer circumference of the hair to the fiber core.

The ESCA binding energy spectra (S 2p sulfur) for weathered hair and hair exposed to an ultraviolet lamp in the laboratory are similar to each other but are different from spectra of chemically bleached hair (alkaline hydrogen peroxide). Similar binding energies suggest similar end products and similar mechanisms of oxidation. As described earlier in this chapter, the mechanism for peroxide oxidation of pure disulfides and for disulfide residues in hair is believed to proceed through the S–S fission route. On the other hand, for irradiation of pure cystine, existing evidence suggests the C–S fission route as the scheme for photochemical degradation of cystine and for other pure disulfides [22] (see Figure 4–2).

The evidence from ESCA suggests that both the chemical and the photochemical degradation of cystine in hair are similar to that of pure disulfides [6]; that is, for chemical degradation, S–S fission occurs, while for photochemical degradation the C–S fission route is followed. For the S–S fission route, the main end product is sulfonic acid, and the C–S fission route produces mainly S-sulfonic and sulfonic acids [22]. However, S-sulfonic acid is ultimately degraded by light to sulfonic acid [22]. The ESCA spectra suggest that cystine S-sulfonate and cysteic acid are both formed in weathered (tip) ends of hair and in hair exposed to ultraviolet light, while cysteic acid is the primary end product formed from the oxidation of cystine in hair during chemical bleaching [6]. These results suggest that the mechanism for the radiation-induced degradation of cystine follows the C–S fission pathway and is different from the chemical oxidation of cystine that proceeds through the S–S fission route.

Chlorine Oxidation of Hair

Treatment of hair with chlorine or bromine water was first observed by Allworden [34], who noted that bubbles or sacs form at the surface of the fibers during treatment with chlorine water. This oxidizing system diffuses across the epicuticle and degrades protein beneath the epicuticle, producing smaller, water-soluble species too large to migrate out of the hair. As a result, swelling occurs beneath the epicuticle from osmotic forces, producing the characteristic Allworden sacs (see Figure 1–22).

The qualitative observations of the Allworden reaction are produced by relatively large concentrations of chlorine or bromine water. The effect of chlorine water on hair, at the parts per million level, has been investigated by Fair and Gupta [35] in an attempt to assess the effects of chlorine in swimming pools on hair. In this study, hair effects were measured by following changes in hair fiber friction. In general, the effect of chlorine was to increase the coefficient of fiber friction and to decrease the differential friction effect in hair. Changes in hair friction were observed even at parts per million levels of chlorine. Effects increased with the number of treatments and as pH decreased from 8 to 2.

The actual oxidizing species present depends on pH and is either Cl_2 or HOCl. Although the chemistry of these interactions was not examined, one would expect disulfide bond cleavage and peptide bond fission similar to the effects shown for the reaction of chlorine and wool fiber [36].

Hair Pigment Structure and Chemical Oxidation

The principal pigments of human hair are the brown-black melanins (eumelanins) and the less prevalent red pigments, the pheomelanins. The latter pigments at one time were called trichosiderins. For this discussion, the brown-black pigments of hair are referred to as melanins and the yellow and red pigments as pheomelanins.

In scalp hair, the pigments reside within the cortex and medulla [37] in ovoid or spherical granules that generally range in size from 0.2 to 0.8 μm along their major axis [38]. Hair pigments are not normally found in the cuticle of scalp hair (see Figure 1–2), and pigment granules generally constitute less than 3% of the total fiber mass, as estimated by the residue weight after acid hydrolysis [39].

Methods used for pigment granule isolation usually involve dissolving the hair from the granules [13,38,40–43]. Funatsu and Funatsu [43] have found that the general composition of melanin granules consists of pigment, protein, and minerals. Flesch [13] reports a similar general composition for the pheomelanin-containing granules. Schmidli et al. [44,45], after acid or alkaline hydrolysis of hair, were able to isolate melanin combined with protein, suggesting that melanin exists in combination with protein in the granules; this is sometimes referred to as melanoprotein.

Because the pigment granules of human scalp hair are located in the cortical cells and the medulla, it is reasonable to assume that pigment degradation by chemical means is a diffusion-controlled process. However, evi dence supporting this contention is not available at this time. Determining the rate-controlling step in this process is in fact a large-order task. It is difficult to follow quantitatively the loss of pigment in hair, and further, two important side reactions consume oxidizing agent: the previously described oxidation of amino acid residues [5] and, in addition, the association of dibasic amino acid residues of hair with many oxidizing agents including hydrogen peroxide and persulfate [46,47].

Melanins (Eumelanins)

The current knowledge of the structure of melanic pigments, the preponderant pigments of human hair, has been described by Nicolaus [10] and Mason [11]. Two theories are under consideration for the structure of

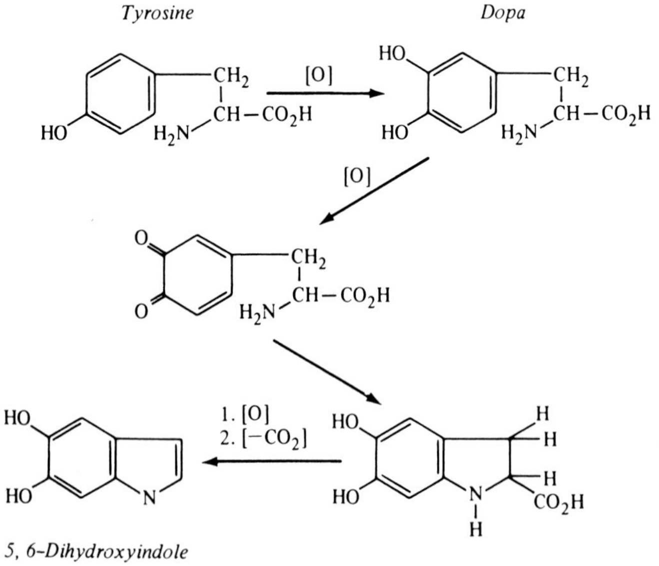

Tyrosine *Dopa*

5, 6-Dihydroxyindole

FIGURE 4–4. Raper's scheme for formation of 5,6-dihydroxyindole from tyrosine.

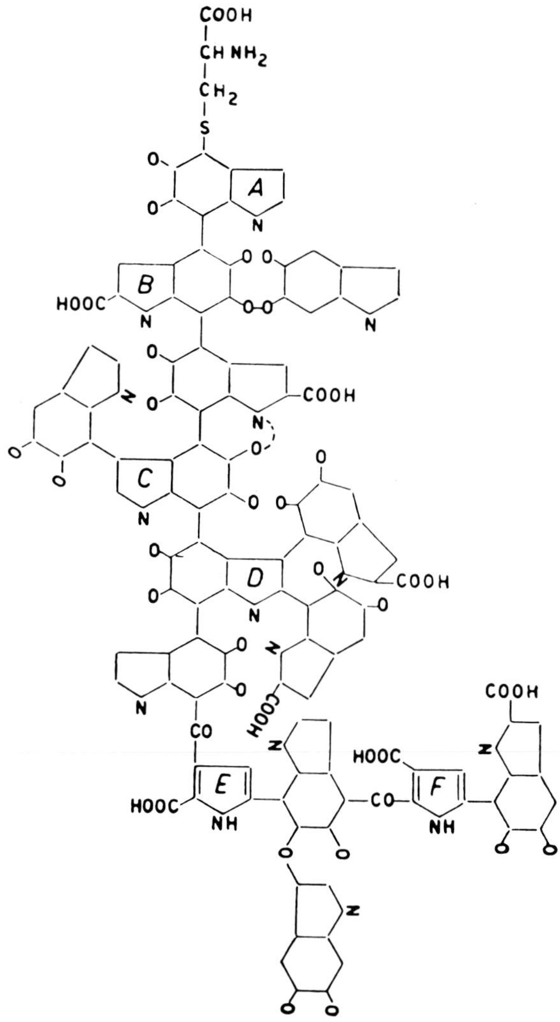

FIGURE 4–5. Part of the complex random polymer proposed for melanin by Nicolaus.

melanin, and both consider structures involved in Raper's scheme (Figure 4–4) [48,49] as intermediates in the formation of all melanic pigments.

Nicolaus [50] has proposed that melanin is a complex random polymer (Figure 4–5), formed from several species of the Raper scheme. Nicolaus' random polymer proposal is of special interest, because other polymers in nature, such as proteins, carbohydrates, and nucleic acids, contain regular repeating structural units and the structure he proposes deviates from these other naturally occurring polymers.

In contrast to the random polymer proposal of Nicolaus, Mason has suggested that melanin is a homopolymer of 5,6-dihydroxyindole. This proposal is consistent with, and in a sense biased by, the regularity of other natural polymers.

The Nicolaus group has isolated several pyrole carboxylic acids and indole derivatives from degradation studies of melanins [11,50–52]. Among these degradation products are the following pyrole carboxylic acids:

dicarboxylic acids of pyrole: 2,2; 3,5; 2,5;
tricarboxylic acids of pyrole: 2,3,4; 2,3,5; and
tetracarboxylic acid of pyrole: 2,3,4,5.

Binns and Swan [53] have also isolated similar pyrole carboxylic acids after the oxidative degradation of melanins.

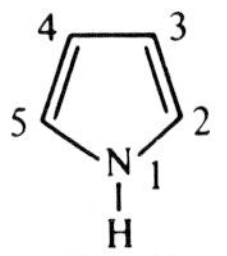

Pyrole

The Nicolaus group has also isolated 5,6-dihydroxyindole, 2-carboxy-5,6-dihydroxyindole, and 5,6-dihydroxy-4,7-dicarboxy-indole in their degradative studies of melanins.

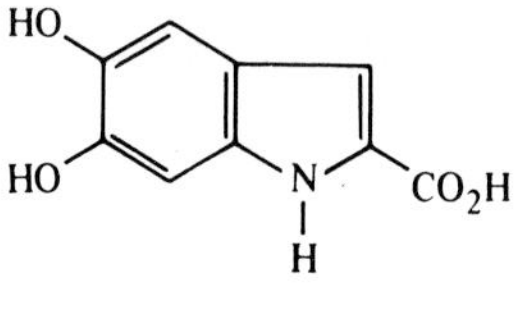

2-Carboxy-5,6-dihydroxyindole

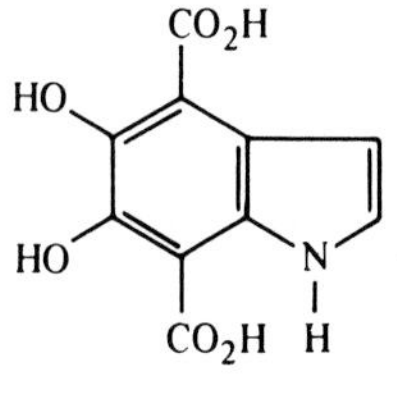

5,6-Dihydroxy-4,7-dicarboxyindole

Several of these isolated structures could not have resulted from the degradation of a melanin homopolymer. Therefore, these degradation products provide major support for the Nicolaus random polymer hypothesis. The low yields of isolated pyrolic acids from these degradation studies (generally $<1\%$) may be explained on the basis that these fragments are themselves sensitive to oxidation. However, a second explanation suggests that a multiplicity of sites in the pigment macromolecules are susceptible to attack by oxidizing agents, producing many unrecovered fragments. Mason [11] has offered a rebuttal to the evidence provided by the degradation studies of Nicolaus. Regardless of the differences between these two hypotheses, both suggest that the indole quinone grouping or its reduced form is a major repeating unit of melanins.

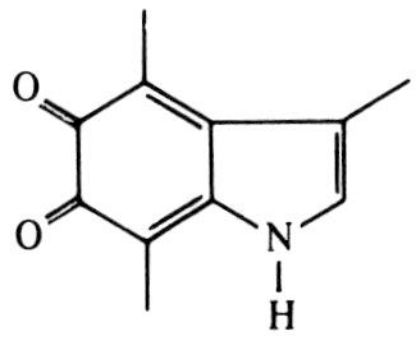

An indole quinone unit 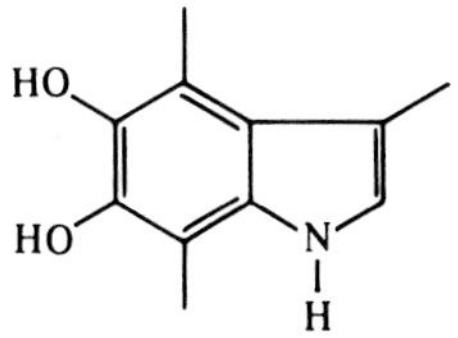 *A reduced indole quinone unit*

Wolfram and coworkers [12,54] have studied the oxidation of human hair with and without pigment and the oxidation of melanin granules isolated from human hair. These scientists found that hair with pigment degrades hydrogen peroxide at a measurably faster rate than hair without pigment. As melanin represents less than 2% of the hair, this result suggests a faster reaction of peroxide with hair pigment than with hair protein.

For the reaction of peroxide with hair containing no pigment versus hair containing pigment, the initial reaction rates are similar (through 10 min). However, longer reaction times (30–90 min) produce marked differences in the reaction rates. The initial rates are due to reaction with the surface and cuticle proteins and are expected to be similar, because pigment is not present in the cuticle. However, as the reaction continues and the pigment becomes involved, peroxide is degraded faster by the pigment-containing hair.

Treatment of isolated melanin granules (from hair) with a large number of reagents, including thioglycolic acid, persulfate, permanganate, or perchlorate (at different pH values), failed to provide a detectable physical change in the granules. However, treatment of the granules with alkaline hydrogen peroxide produces disintegration and dissolution of the granules. Wolfram also found that the pH of dissolution is at a maximum near the pK of hydrogen peroxide (pH 11.75). Furthermore, the dissolved pigment produces an intensely colored solution that fades on further reaction.

Wolfram [54] examined the effects of different oxidizing agents on their ability to decolorize soluble melanin and found the following order of efficiency: permanganate > hypochlorite = peracid > peroxide. This finding suggests that the melanin pigments within the granules are not accessible to oxidizing agents, and the granules must be degraded, perhaps even solubilized, before extensive decolorization of the pigment chromophore can occur. Further, the first step (dissolution of the pigment granules) is a relatively specific reaction requiring oxidation at specific sites. Hydrogen peroxide is not as strong an oxidizing agent as permanganate or peracetic acid, but it is actually more effective for dissolving the granules than either of these other two oxidizing species. Once the granules are dissolved, reactions to degrade the chromophoric units of melanin can proceed more readily. Because the melanin chromophoric units contain many different sites that are susceptible to oxidation, the rate of the second step (degradation of the pigment chromophore) proceeds faster with the stronger oxidizing agents, for example, permanganate > hypochlorite = peracid > peroxide, which is the order for decolorization of the solubilized melanin found by Wolfram.

Although persulfate is not a stronger oxidant than hydrogen peroxide, mixtures of persulfate and peroxide provide a more effective bleaching system than peroxide alone. Persulfate oxidation of fungal melanins has been studied by Martin et al. [55], who showed persulfate to be a selective oxidizing agent, releasing only those portions of melanins containing primarily fatty acids and phenolic compounds. Presumably, persulfate and peroxide are both somewhat selective in their attack on melanins; they attack different portions or sites on the melanin macromolecules, and thereby facilitate solubilization of melanin so that the more potent peroxide can degrade it in solution. Thus, one might conclude that persulfate and peroxide complement each other in terms of their ability to bleach melanin pigment and therefore to bleach human hair.

Wolfram and Hall [54] have also isolated several products from the reaction of alkaline hydrogen peroxide with melanin pigments, including proteinaceous species up to 15,000 daltons. These scientists have further developed a procedure for the isolation of melanoprotein and determined the amino acid composition of melanoprotein from Oriental hair. They found fewer cystine linkages in melanoprotein than in whole fiber and a larger percentage of ionizable groups; that is, approximately 35% more dibasic amino acid residues and 15% more diacidic groups.

Pheomelanins

Pheomelanins, the yellow-red pigments, are lighter in color than the brown-black pigments of human hair. However, both pheomelanins and eumelanins occur as granules in melanocytes in hair. Prota and Nicolaus [14,56] have proposed that red hair pigments are formed by a modification of the eumelanin pathway described earlier. The pheomelanin pathway, however, involves the interaction of cysteine with dopaquinone.

Figure 4–6 summarizes Prota's [56] description of a common metabolic pathway for formation of all melanins and shows how pheomelanin formation relates to eumelanin biosynthesis.

In that scheme, tyrosine is first oxidized to dopaquinone, the key intermediate for the formation of all melanins. If virtually no cysteine is present, then eumelanins are the primary products that arise through cyclization and oxidative polymerization, as described in the section on eumelanins.

Cyclization and oxidative polymerization of 5-*S*-cysteinyldopa and its isomers (present in smaller quantities) forms the high molecular weight, yellow-red pheomelanins. A partial structure proposed by Prota [56] for pheomelanins is depicted in Figure 4–7.

If cysteine is present, however, it adds to the reactive *o*-quinone unit of dopaquinone, forming 5-*S*-cysteinyldopa (isomers of this species are also

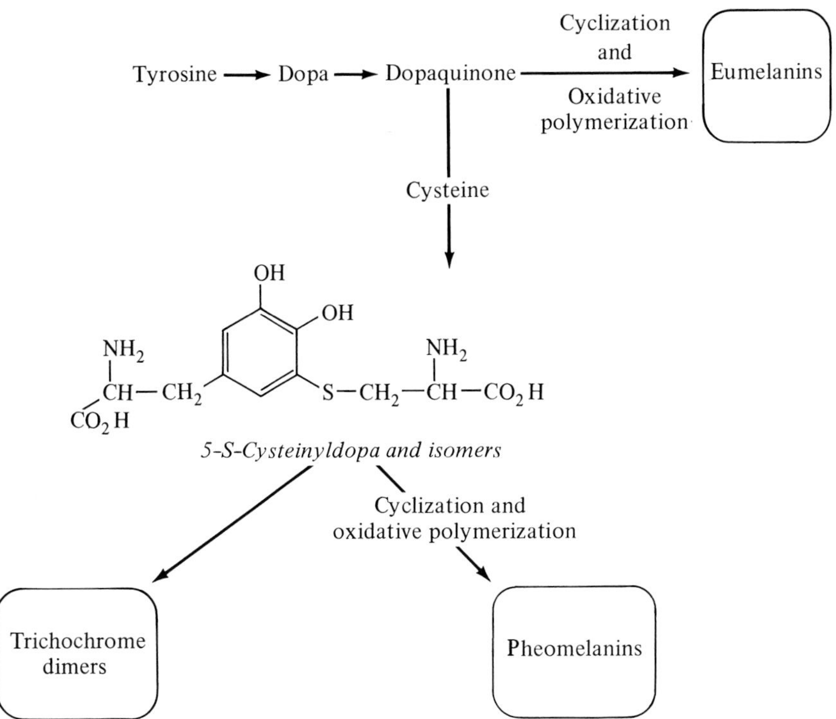

FIGURE 4–6. Prota's mechanism for formation of eumelanins, pheomelanins, and trichochromes.

FIGURE 4–7. Partial structure proposed by Prota for the primary unit of pheomelanins.

formed). Cyclization and dimerization produce the yellow-to-violet trichochrome pigments (trichochrome F and C), formerly reported by Flesch and Rothman as iron-containing pigments, which they called trichosiderins [57,58].

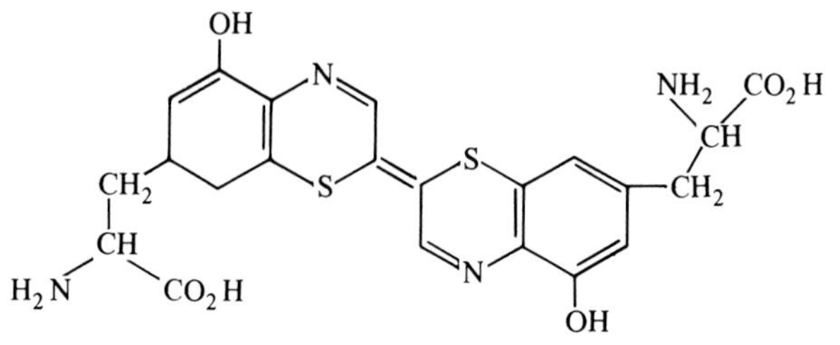

Trichochrome F

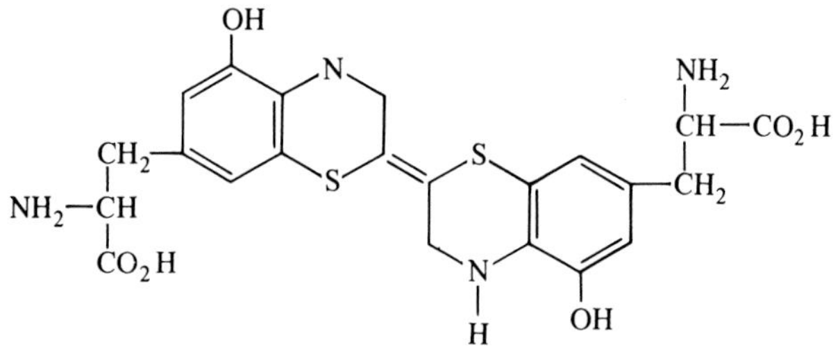

Trichochrome C

The two melanins are formed from a common metabolic pathway, which involves highly reactive intermediates. It is highly probable, therefore, that both types of pigments (eumelanin and pheomelanin), and even mixtures of these two pigments and the trichochromes, may be formed in the same hair, depending primarily on the amount of cysteine present in the melanocyte. Wolfram and Albrecht [59] have also shown that hue differences in hair not only result from chemically different pigments, but also from differences in the degree of aggregation and dispersion of the eumelanin pigment.

Both eumelanins and pheomelanins are polymeric and are believed to be formed from a common metabolic pathway. Both these polymeric pigments contain polypeptide chains with similar amino acids [60]. The red hair melanin contains more sulfur (as 1,4-benzothiazine units) than the brown-black melanins.

The chemical degradation of pheomelanins by oxidizing agents has not been reported. However, there are undoubtedly several similarities with regard to the chemical bleaching of eumelanins and pheomelanins. However, Wolfram and Albrecht [59] have shown that pheomelanin in hair is more resistant to photodegradation than are the brown-black eumelanins. The aromatic rings of both melanic structures are of high electron density, and consequently they are both sensitive to attack by oxidizing agents, as already demonstrated for the brown-black melanins.

Photochemical Degradation of Melanins

Photochemical degradation of hair results in attack on both hair proteins and hair pigment. Photochemical degradation of hair proteins occurs primarily near 254 to 350 nm, the primary absorbance region of unpigmented hair [28]. Although several amino acids are degraded by light, the primary degradation occurs at cystine. Launer [29] and Inglis and Lennox [30] have provided evidence for photochemical degradation to other amino acids including methionine, histidine, tryptophan, phenylalanine, and leucine.

The mechanism for photochemical degradation of cystine is believed to be the C–S fission mechanism (see the section on photochemical degradation of hair proteins described earlier in this chapter). Hair pigments function to provide some photochemical protection to hair proteins, especially at lower wavelengths where both the pigments and the proteins absorb light (254–350 nm). Hair pigments accomplish this protection by absorbing and filtering the impinging radiation and subsequently dissipating this energy as heat. However, in the process of protecting the hair proteins from light, the pigments are degraded or bleached.

Eumelanin ring opening may result from ionic reaction (chemical degradation) or from free radical reaction (photochemical degradation), and Slawinska and Slawinski [61] have suggested that these two mechanistic schemes may have some common intermediates. The ionic pathway probably begins by nucleophilic attack of the peroxide anion on the *o*-quinone

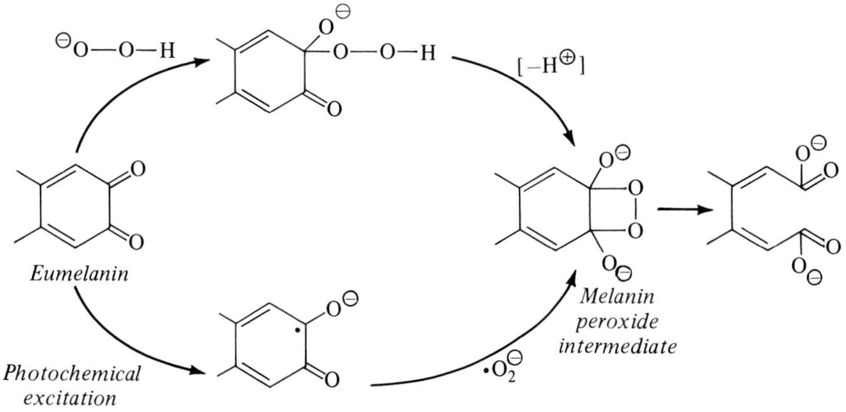

FIGURE 4–8. Proposed mechanisms for degradation of melanins [61].

grouping. Slawinska and Slawinski suggested that photochemical degradation of melanin occurs through a similar peroxide intermediate. The first step in the photochemical degradation of the eumelanin chromophore probably involves excitation to a radical anion and then attack by the oxygen radical anion on the *o*-quinone grouping (Figure 4–8). Ring opening of the six-membered ring indolequinone species then follows.

A related scheme may be involved for the photochemical degradation of pheomelanins, as Sarna et al. [62] have shown that pheomelanins are very similar to eumelanins with regard to their susceptibility to photooxidation although Arakindakshan Menon et al. [60] reported that pheomelanins are more easily induced to an excited state than the brown-black eumelanins. On the other hand, data by Wolfram and Albrecht [59] suggest that pheomelanin in hair is more resistant to photodegradation than melanin.

Physical Properties of Bleached Hair

The gross or bulk chemical changes produced in hair by bleaching, described in the previous sections of this chapter, consist of primary attack on cystine with degradation to some of the other amino acids. The primary changes are produced in those morphological regions of the hair containing higher concentrations of cystine (the A layer and exocuticle of the cuticle and the matrix of the cortex) and the other degraded amino acids. Generally, greater changes occur closer to the surface than at the core of the fibers, and often greater changes can be detected in the tip ends of hair than in the root ends, because of longer exposure.

The swelling capacity of bleached hair increases with increasing oxidation, and because of this relationship, Klemm et al. [63] have developed a swelling test to assess hair damage by bleach treatments. Both wet bending stiffness (by the hanging fiber method) and wet stretching stiffness (estimated from Hookean slopes) decrease with bleaching [64]. The wet tensile properties through 30% extension are decreased by both chemical bleaching [65] and by photochemical oxidation [29]; however, the dry tensile properties remain virtually unchanged (see Chapter 8 for additional details.).

Safety Considerations for Hair Bleaches

The primary safety concerns with hair bleaches, as with most hair care products, arise from misuse or failure to comply with the usage instructions. Skin irritation, hair breakage, oral toxicity, sensitization, and scarring alopecia either have been reported [66] from use (misuse) of hair bleaches or are mentioned on the warning labels of these products.

Bergfeld [66] has reviewed adverse effects of hair cosmetics recorded at the Cleveland Clinic Dermatology Department over a 10-year period.

Effects attributed to hair bleaches were simple skin irritation and hair breakage. However, neither sensitization reactions nor complex toxic symptoms were reported by Bergfeld for hair bleaches.

Bourgeois-Spinasse [67] indicates a few incidents of allergic manifestations caused by ammonium persulfate powder; however, most bleaches today use potassium persulfate as the primary bleach accelerator. Permanent hair loss was also reported by Bergfeld [66] following misuse of hair bleaches and attributed to scarring alopecia, although Bergfeld did not specify the extent of hair loss observed. However, Bergfeld concludes that side effects from hair bleaches are minimal if the consumer is aware of damaged hair, any inherent skin disease and complies with the product usage instructions.

References

1. Zahn, H. J. Soc. Cosmet. Chem. 17:687 (1966).
2. Zahn, H.; 4th Int. Hair Sci. Symp., Syburg, West Germany, German Wool Res. Indust. Publ. Abstracts November (1984).
3. Robbins, C. Text. Res. J. 37:811 (1967).
4. Nachtigal, J.; Robbins, C. Text. Res. J. 40:454 (1970).
5. Robbins, C.; Kelly, C. J. Soc. Cosmet. Chem. 20:555 (1969).
6. Robbins, C.; Bahl, M. J. Soc. Cosmet. Chem. 35:379 (1984).
7. Maclaren, J.A.; Leach, S.J.; Swan, J.M. J. Text. Inst. 51T:665 (1960).
8. Maclaren, J.A.; Savige, W.E.; Sweetman, B.J. Aust. J. Chem. 18:1655 (1965).
9. Alter, H.; Bit-Alkis, M. Text. Res. J. 39:479 (1969).
10. Nicolaus, R.A. In Advances in Biology of Skin, The Pigmentary System, Vol. 8, Montagna, W.; Hu, F.; eds., p. 313. Pergamon Press, New York (1966).
11. Mason, H.S. In Advances in Biology of Skin, The Pigmentary System, Vol. 8, Montagna, W.; Hu, F.; eds., pp. 293–312 (and references therein). Pergamon Press, New York (1966).
12. Wolfram, L.J. J. Soc. Cosmet. Chem. 21:875 (1970).
13. Flesch, P. J. Soc. Cosmet. Chem. 19:675 (1968).
14. Prota, G.; Piattelli, M.; Nicolaus, R.A. Rend. Accad. Sci. Fis. Mat. (Naples) 35:1 (1968).
15. Cook, M. Drug Cosmet. Ind. 99:47, 154 (1966).
16. Wall, F.E. In Cosmetics: Science and Technology, Sagarin, E.; ed., pp. 479–530. Interscience, New York (1957).
17. Harris, M.; Brown, A.E. J. Soc. Dyers Col.; Symp. Fibrous Proteins, Soc. Dyers Colquists, Publ. 203–206 (1946).
18. Alexander, P.; Hudson, R.F.; Earland, C. In Wool, Its Chemistry and Physics, 2d Ed., pp. 63:289. Franklin Publ. Co., New Jersey (1963).
19. Edman, W.; Marti, M. J. Soc. Cosmet. Chem. 12:133 (1961).
20. Crank, J. The Mathematics of Diffusion, p. 71. Clarendon Press, Oxford, UK. (1967).
21. Robbins, C.R. J. Soc. Cosmet. Chem. 22:339 (1971).
22. Savige, W.E.; Maclaren, J.A. In The Chemistry of Organic Sulfur Compounds, Vol. 2, Kharasch, N.; Meyers, F., eds., pp. 367–402. Pergamon Press, New York (1966).
23. Lavine, T.F. J. Biol. Chem. 113:583 (1936).

24. Savige, W.E.; Eager, J.; Roxburgh, C.M. Tetrahedron Lett. 44:3289 (1964).
25. Truce, W.E.; Murphy, A.M. Chem. Rev. 48:69 (1951).
26. Kharasch, N. In Organic Sulfur Compounds, Vol. 1, Kharasch, N., ed., p. 375. Pergamon Press, New York (1961).
27. Beyak, R.; Kass, G.S.; Meyer, C.F. J. Soc. Cosmet. Chem. 22:667 (1971).
28. Arnaud, J. Int. J. Cosmet. Sci. 6:71 (1984).
29. Launer, H.F. Text. Res. J. 35:395 (1965).
30. Inglis, A.S.; Lennox, F.G. Text. Res. J. 33:431 (1963).
31. Pande, C.M.; Jachowicz, J. J. Soc. Cosmet. Chem. 44:109 (1993).
32. Holt, L.A.; Milligan, B. Text. Res. J. 47:620 (1977).
33. Robbins, C.R.; Kelly, C. 40:891 (1970).
34. Allworden, A. Angew. Chem. 29:77 (1916).
35. Fair, N.; Gupta, B.S. J. Soc. Cosmet. Chem. 33:229 (1982).
36. Makinson, K.R. Text. Res. J. 44:856 (1974).
37. Birbeck, M.; Mercer, E.H. Proc. Stockholm Conf. Electron Microsc. p. 158 (1956).
38. Gjesdal, F. Acta Pathol. Microbiol. Scand. 133:12 (1959).
39. Menkart, J.; Wolfram, L.J.; Mao, I. J. Soc. Cosmet. Chem. 17:769 (1966).
40. Laxer, G.; Whewell, C.S. Chem. Ind. (Lond.) 5:127 (1954).
41. Serra, J.A. Nature 169:771 (1946).
42. Laxer, G.; et al. Biochim. Biophys. Acta 15:174 (1954).
43. Funatsu, G.; Funatsu, M. Agric. Biol. Chem. 26:367 (1962).
44. Schmidli, B. Helv. Chem. Acta 38:1078 (1955).
45. Schmidli, B.; Robert, P. Dermatologica 108:343 (1954).
46. Heald, R.C. Schimmel Briefs, No. 382 (January 1967).
47. Breuer, M.M.; Jenkins, A.D. Proc. 3rd Int. Wool Text. Res. Conf. II:346 (1965).
48. Raper, H. S. Biochem. J. 20:735 (1926).
49. Raper, H.S. Biochem. J. 21:89 (1927).
50. Nicolaus, R.A. Rass. Med. Sper. 9:1, (1962).
51. Piattelli, M.; Fatturusso, E.; Magno, S. Tetrahedron 18:941 (1962).
52. Piattelli, M.; Fatturusso, E.; Nicolaus, R.A. Tetrahedron 19:2061 (1963).
53. Binns, F.; Swan, G.A. Chem. Ind. (Lond.) p. 396 (1957).
54. Wolfram, L.J.; Hall, K. J. Soc. Cosmet. Chem. 26:247 (1975).
55. Martin, F.; et al. Sail Sci. Am. J. 47:1145 (1983).
56. Prota, G. J. Invest. Dermatol. 75:122 (1980).
57. Flesch, P. J. Soc. Cosmet. Chem. 21:77 (1970).
58. Flesch, P.; Rothman, S. J. Invest. Dermatol. 6:257 (1945).
59. Wolfram, L.J.; Albrecht, L. J. Soc. Cosmet. Chem. 38:179 (1987).
60. Arakindakshan Menon, I.; et al. J. Invest. Dermatol. 80:202 (1983).
61. Slawinska, D.; Slawinski, J. Physiol. Chem. Phys. 14:363 (1982).
62. Sarna, T.; et al. Photochem. Photobiol. 39(6):805 (1984)
63. Klemm, E.; et al. Proc. Sci. Sect. Toilet Goods Association. 43:7 (1965).
64. Robbins, C.R.; Scott, G.V. Unpublished data.
65. Scott, G. V.; Robbins, C.R. Unpublished work.
66. Bergfeld, W.F. In Hair Research, Orfanos, C.E.; Montagna, W.; Stuttgen, G., eds., pp. 507–511. Springer-Verlag, Berlin (1981).
67. Bourgeois-Spinasse, J. In Hair Research, Orfanos, C.E.; Montagna, W.; Stuttgen, G., eds., pp. 543–547. Springer-Verlag, Berlin (1981).

5

Interactions of Shampoo and Conditioning Ingredients with Human Hair

According to legend, the word "shampoo", is derived from a Hindustani word meaning "to squeeze." Shampoos have a long and varied history. However, hair conditioners were not widely used until the mid-twentieth century following the introduction of "cold" permanent wave-type products that exacerbated combing problems and damaged the hair.

The primary function of shampoos is to clean both the hair and the scalp of soils and dirt, while the primary function of hair conditioners is to make the hair easier to comb. Secondary benefits such as preventing flyaway hair, making the hair "shine", and protecting the hair from further damage are also important functions of hair conditioners. Shampoos also have important secondary functions such as dandruff control, minimizing irritation (baby shampoos), and conditioning (including both the primary and the secondary functions of conditioners), and these functions have become even more important with the advent of 2-in-1 shampoos. Even fragrance impact, character, and preference have created new market segments or have become primary reasons for the purchase of some shampoos and conditioners.

Shampoos and hair conditioners have generally been perceived as products that do not damage hair. However, there is increasing evidence that these products, particularly shampoos, can contribute to hair damage through abrasive/erosive actions both during and after the shampoo process and through the slow but gradual dissolution of the important nonkeratinous components of the cell membrane complex and the endocuticle. For a detailed discussion of this subject see the section on hair damage near the end of this chapter.

The principal function of hair conditioning products involves combability, which depends on lubrication of the fiber surface; this is accomplished by the sorption or binding of lubricating or conditioning ingredients to the hair fiber surface. Thus, the most important interactions for both shampoos and conditioners are those that occur at or near the fiber surface or near the first few cuticle layers (see Figure 1–7). Of course if the hair surface is badly damaged and the cortex is exposed, then shampoos and conditioners interact with exposed cortex too.

The first section of this chapter is concerned with shampoo and conditioner formulations and with procedures for making these same products. The control of product viscosity and important parameters concerned with product stability for shampoos, hair conditioners, and other types of hair care products are also discussed. The second section describes the different types of soil found on hair, their origins, and the ease or difficulty in their removal. Methods to evaluate hair cleaning, the perception of hair cleaning, and shampoo lather as it relates to cleaning are then described. The third section is concerned with the attachment and the affinity of surfactant/conditioning-type molecules to hair, including the theory of sorption, both surface adsorption (which is more concerned with shampoos and hair conditioners) and whole-fiber studies including fiber diffusion.

Diffusion or penetration of chemicals into hair is more concerned with permanent waves, hair straighteners, and hair dyes; however, recent evidence that shampoos over time can damage the nonkeratinous pathways for entry into hair means that diffusion is also important to shampoos. Damaging effects to hair caused by shampooing and rubbing actions as occur in hair grooming such as in shampooing, drying, combing, brushing, and styling hair, and safety concerns for shampoos, are also considered. A brief introduction to the subject of dandruff concludes this chapter.

General Formulations for Shampoos and Conditioners

Shampoos consist of several types of ingredients, which usually include many of the following types of components:

Primary surfactant for cleaning and foaming
Secondary surfactant for foam and/or viscosity enhancement
Viscosity builders: gums, salt, amide
Solvents to clarify the product or to lower the cloud point

Conditioning agents
Opacifier for visual effects
Acid or alkali for pH adjustment
Colors (D&C or FD&C colors) for visual effects
Fragrance
Preservative
UV absorber (usually for products in a clear package) to protect the product against UV light.
Specialty active ingredients, such as antidandruff agents, conditioning agents, etc.

Hair conditioners on the other hand are very different compositionally from shampoos. Conditioners are usually composed of many of the following types of ingredients:

Oily or waxy substances including mineral oil, long-chain alcohols or triglycerides or other esters, including true oils and waxes, and silicones or fatty acids
Cationic substances consisting of monofunctional quaternary ammonium compounds or amines or polymeric quaternary ammonium compounds or amines
Viscosity builders
Acid or alkalies for pH adjustment
Colors
Preservative.

Temperature and Aging Stability

There are no standard aging or stability tests in the cosmetic industry. Each company has developed its own set of standards to assess product stability at higher-temperatures and to use high-temperature aging as a means to project longer term aging effects, thus accelerating product introductions. The best approach is to test the product at multiple temperatures because in some cases, for example, for some emulsions, a product can be more stable at a higher than at a lower temperature.

Freeze-thaw or temperature cycling is also important, especially in temperate or colder climates, because we have to know if a phase change occurs when the product is frozen or taken to a lower temperature and then returned to room temperature, will the appearance and product performance be restored? If precipitation or a permanent phase change occurs at lower temperatures, such problems can sometimes be addressed by improving the solvency of the system by adding solvents, or even by adding fluoride salts, hydrotropes, urea, or other such solubilizing additives.

The aging conditions listed in Table 5–1 are useful to evaluate a hair care product for its quality before sale.

TABLE 5–1. Aging conditions for hair product evaluation.

Temperature aging	Time
50° (122°F)	3 months
40° (104°F)	3–6 months
25° (77°F)	1 year
25° (77°F)	In sunlight (if clear package)
5° (40°F)	3 months
−20° (−4°F)	Freeze/thaw (lower colder climates)

Obviously, in many cases one cannot affort to wait 1 year for completion of aging studies before going to market. In such cases, 3 to 6 months of aging under these conditions is helpful in making judgement about product stability, especially if one has additional longer term aging data with related formulations.

I also recommend aging the product in glass and the actual package in which it is to be sold. Thus, if a problem arises during aging, one can determine if the problem is in the formulation itself or a reaction of the formulation with the package.

Color Stability

Color instability can be caused by many factors, such as the degradation of color additives or of another formula component through chemical interaction with formula components or trace contaminants of components, or by ultraviolet radiation. This section is concerned with the latter problem, stabilization of the system to light radiation.

For hair products that are sold in clear packaging, light stability is usually a major concern. For example, exposure to light may cause the dyes used in the product to fade, fragrance components to degrade, or other additives to fade or decompose. Structurally, a common source of this problem is unsaturated groups in the structure of the sensitive component.

The easy solution, to use an opaque container, may not be compatible with the marketing plan. An alternative solution is to add ultraviolet absorbers to the product to absorb degrading radiation and thus to inhibit, retard, or prevent product degradation. Benzophenone-2 or benzophenone-11 are usually the preferred agents, because of their broad-spectrum protection (Table 5–2).

Benzophenone-2 is usually preferred over benzophenone-11 because it is a single component, whereas benzophenone-11 is a mixture of benzophenone-6, benzophenone-2, and other tetrasubstituted benzophenones. Most of these ultraviolet absorbers can be used in concentrations of about 0.05% to 0.2% for effective protection against degradation by ultraviolet light.

TABLE 5–2. Most effective wavelengths of ultraviolet absorbers (nm).

Benzophenone-2	290–350
Benzophenone-4	285
Benzophenone-8	355
Benzophenone-9	333
Benzophenone-11	290–355

Preservation Against Microbial Contamination

Preservation of consumer products against microbial contamination is important because such contamination can lead to product degradation and, in the worst case scenario, to the spread of disease. Thus, it is necessary to preserve consumer products against microbial contamination at the time of manufacture and to ensure that our products are preserved for a reasonable period of time thereafter.

Some formulations are inherently more difficult to preserve than others. In general, the more water in a product the more difficult it is to preserve. In addition, some ingredients, for example, plant extracts, vitamins, and some nonionic detergents, are more difficult to preserve than others.

Formaldehyde, specifically formalin, is perhaps the single most effective preservative for shampoos and conditioners; however, because of its reputation for sensitization, which actually occurs well above levels used in consumer products, it is not used in most countries. One convenient way to classify preservatives is as those that release formaldehyde and those that do not. In the former group, we have Germaben II, which is one of the more effective preservatives, Germall 115, Germall II, and Glydant. Germaben II, often used in shampoos and conditioners at a level of approximately 0.5% of the product, is a mixture of diazolidinyl urea (which releases formaldehyde) and parabens in propylene glycol. Germall 115, another effective preservative, is actually imidazolidinyl urea, and can be made more effective by the addition of parabens; approximately 0.05% methyl paraben and 0.1% propyl paraben is highly effective in the presence of this preservative. Germall II, which is diazolidinyl urea, is another effective preservative, but not as effective as Germaben II because of the parabens in Germaben II. Glydant, which is actually DMDM hydantoin (1,3-Dimethylol-5,5-Dimethyl Hydantoin), is often used in the vicinity of 0.5% of the product and is another effective preservative. It too is made more effective by the addition of parabens.

Among the more commonly used preservatives that do not release formaldehyde are parabens, Dowicil 200, and Kathon CG. A commonly used mixture of parabens consists of 0.1% methyl paraben and 0.7% propyl paraben. This mixture of parabens is moderately effective alone, but is more effective in combination with other preservatives. The European Economic Community (EEC) prohibits the use of parabens above 0.8%. Dowicil 22 has the CTFA designation Quaternium-15, and is sometimes used at concentrations between 0.05% and 0.2%; it can be used in combination with parabens to enhance their preservative capacity. Kathon CG, a mixture of methyl chloroisothiazolinone and methyl isothiazolinone, is another useful preservative for cosmetic hair products.

Benzyl alcohol, sodium benzoate, sorbic acid, and even sequestrants are used as adjuncts for the preservation of hair care products. For example, ethylene diaminetetracetic acid (EDTA) is effective against *Pseudomonas* and should be considered in systems where *Pseudomonas* is a problem, but it should not be considered as the only preservative for a shampoo or a hair conditioner.

Viscosity control in Shampoos and Conditioners

To control the viscosity of many shampoos, salt is added to the surfactant system; the interaction between salt and the long-chain surfactants tends to form a lamellar or liquid crystalline-"type" structure that helps to stabilize or control the consistency of the product. If one plots the salt concentration versus the viscosity in such a system, one typically finds an optimum for the maximum viscosity (Figure 5–1). Above this optimum salt concentration, additional salt decreases the viscosity. In developing such a system in which viscosity is controlled by salt addition, it is recommended that one select the appropriate salt concentration on the ascending part of the

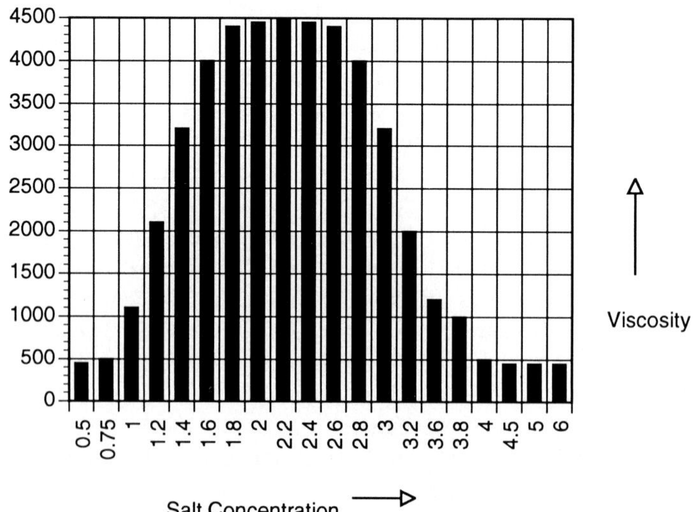

FIGURE 5–1. Relationship of the salt content to the viscosity in surfactant systems (shampoos).

viscosity–salt concentration curve. The selection of surfactant, amide, and other components are critical to viscosity–salt concentration control in such a system. Further, impurities such as salt contaminants in surfactants must be carefully controlled to obtain the appropriate viscosity when salt control is used.

Polymeric gums such as methyl cellulose or hydroxy ethyl cellulose have also been used in shampoos to help control viscosity. In such systems, the salt concentration is also helpful to viscosity control. Solvents such as propylene glycol, glycerine, carbitols, or other alcohols are sometimes used in shampoos to help solubilize or to clarify product or to lower cloud-clear points. Such ingredients often tend to lower product viscosity and are sometimes used for this purpose also.

Structures, Formulation Examples, and Procedures for Making Shampoo and Conditioning Ingredients

The principal primary surfactant used in the United States for shampoos is ammonium lauryl sulfate, although sodium or ammonium laureth sulfate (with an average of 2 or 3 moles of ethylene oxide) is the current leader in many other countries. These two surfactants are used alone or blended together for shampoos because of their fine ability to clean sebaceous soil, and perhaps even more importantly, because of their excellent lathering properties.

$CH_3(CH_2)_{11}OSO_3^{\ominus}NH_4^{\oplus}$ — *Ammonium lauryl sulfate*

$CH_3(CH_2)_{11}O(CH_2CH_2O)_xSO_3^{\ominus}Na^{\oplus}$ — *Sodium laureth sulfate*

Alpha olefin sulfonate has been used to a limited extent in lower priced shampoos. This surfactant is represented by the following structure:

$$R-CH_2-CH=CH-CH_2-SO_3^{\ominus}Na^{\oplus}$$

$$R-CH=CH-CH_2-CH_2-SO_3^{\ominus}Na^{\oplus}$$

$$R-CH_2-\underset{\displaystyle OH}{\underset{|}{C}H}-CH_2-CH_2-SO_3^{\ominus}Na^{\oplus}$$

$$R-\underset{\displaystyle OH}{\underset{|}{C}H}-CH_2-CH_2-CH_2-SO_3^{\ominus}Na^{\oplus}$$

Alpha olefin sulfonate consists of a mixture of these four surfactants in about equal quantities. The commercial shampoo material is 14 to 16 carbon atoms in chain length; therefore, $R = 10$ to 12 carbon atoms.

A carbon chain length of 12 to 14 carbon atoms or a coco-type distribution (derived from coconut oil) that is approximately 50% C-12 is generally used for the primary surfactant in shampoos, because of the excellent foam character and surface-active properties provided by such chain lengths.

TABLE 5–3. Example of a cleaning shampoo.

Ingredient	Percentage
Ammonium lauryl sulfate	15
Cocamide DEA	2
Cocamidopropyl betaine	2
Glycerin	1
Fragrance	0.7
Preservative (Germaben II)	0.5
Citric acid	0.3
Ammonium chloride (to adjust viscosity)	up to 2%
Colors	0.02
Water	q.s.[a]

[a] q.s., add water to 100%.

Longer or shorter chain length surfactants are used only in specialty systems.

Secondary surfactants are used for foam modifiers, for added cleaning, or even for viscosity enhancement. The principal secondary surfactants used in shampoos are amides such as cocodiethanolamide, lauric diethanolamide, or cocomonoethanolamide. Betaines are excellent foam modifiers, and cocamidopropylbetaine is the most widely used betaine in shampoos today (see Table 5–3).

The pH is usually adjusted with a common acid such as citric acid or even mineral acid, and buffers such as phosphate or other ordinary inexpensive materials are also used for pH control. Preservation against microbial contamination is necessary, as was discussed.

Baby shampoos and some light conditioning shampoos employ nonionic surfactants such as PEG-80 sorbitan laurate and amphoteric surfactants such as cocoamphocarboxyglycinate or cocoamphopropylsulfonate to improve the mildness of anionic surfactants and at the same time to improve cleaning and lather performance (Table 5–4).

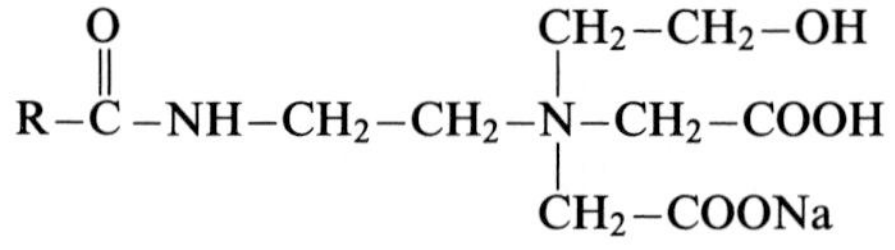

$$R{-}\overset{\overset{\displaystyle O}{\|}}{C}{-}NH{-}CH_2{-}CH_2{-}\underset{\underset{\displaystyle CH_2{-}COONa}{|}}{\overset{\overset{\displaystyle CH_2{-}CH_2{-}OH}{|}}{N}}{-}CH_2{-}COOH$$

Cocoamphocarboxyglycinate

$$R{-}\overset{\overset{\displaystyle O}{\|}}{C}{-}NH{-}CH_2{-}CH_2{-}CH_2{-}\underset{\underset{\displaystyle CH_3}{|}}{\overset{\overset{\displaystyle CH_3{}^{\oplus}}{|}}{N}}{-}CH_2{-}\underset{\underset{\displaystyle OH}{|}}{CH}{-}CH_2{-}SO_3^{\ominus}$$

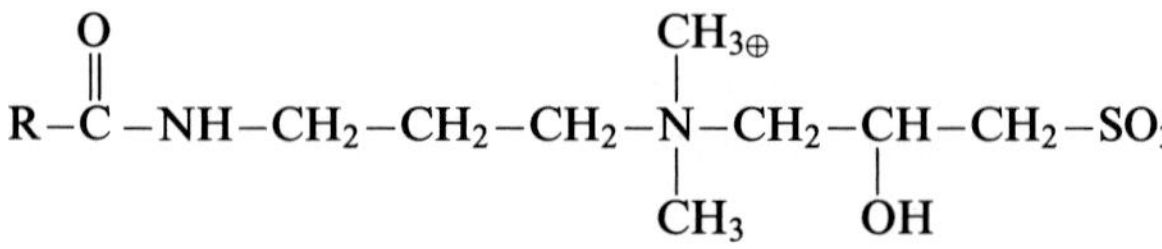

Cocoamidopropylhydroxysultaine

Conditioning agents for shampoos are varied and may generally be classified as lipid type, soap type, or salts of carboxylic acids, cationic type

TABLE 5–4. Example of baby shampoo.

Ingredient	Percentage
PEG-80 sorbitan laurate	12
Sodium trideceth sulfate	5
Lauroamphoglycinate	5
Laureth-13 carboxylate	3
PEG-150 distearate	1
Cocamidopropyl hydroxysultaine	1
Fragrance	1
Preservative (Germaben II)	0.5
Colors	0.02
Water	q.s.[a]

[a] q.s., add water to 100%.

TABLE 5–5. Example of a light conditioning shampoo.

Ingredient	Percentage
Ammonium lauryl sulfate	8
Sodium laureth-2 sulfate	6
Cocamide DEA	3
Polyquaternium-10	1
Sodium phosphate buffer	0.4
Fragrance	1
Ethylene glycol distearate	0.6
Preservative (Germaben II)	0.5
Sodium chloride (to adjust viscosity)	up to 2%
Colors	0.02
Water	q.s.[a]

[a] q.s., add water to 100%.

including cationic polymers, or silicone type including dimethicone or amodimethicones (See structures in later discussion of hair conditioners). An example of a light conditioning shampoo is given in Table 5–5.

Opacifiers such as ethylene glycol distearate or soap-type opacifiers are often used in conditioning shampoos for visual effects, that is, to provide the perception that something is deposited onto the hair for conditioning it.

"Two-in-one" (2-in-1) shampoos are usually higher in conditioning than ordinary conditioning shampoos and normally contain a water-insoluble dispersed silicone as one of the conditioning agents. Conditioning shampoos containing water-insoluble dispersed silicones are generally better for conditioning unbleached hair than other conditioning shampoos, but are not as effective for bleached hair as for unbleached hair. The making procedure is also more complex, and the particle size of the active ingredient is critical to its effectiveness. This type of system is also difficult to

TABLE 5–6. Example of a 2-in-1 conditioning shampoo.

Ingredient	Percentage
Ammonium lauryl sulfate	10
Sodium laureth-2 sulfate	6
Dimethicone	2.5
Ammonium xylene sulfonate	2
Glycol distearate	2
Cocamide MEA	2
Fragrance	1
Thickening gum (hydroxy ethyl cellulose)	0.3
Stearyl alcohol (Germaben II)	0.3
Preservative	0.5
Colors	0.03
Water	q.s.[a]

[a] q.s., add water to 100%.

stabilize. The formula in Table 5–6 is stabilized by a combination of the long chain acylated agent, for example, glycol distearate and the thickening gum. Although, Grote et al [1] described thickeners as optional components, in our experience with this type of acylated suspending agent thickeners are essential to long-term product stability.

Introduction to Making Procedures for Clear Shampoos and Emulsion Products

The simplest procedure is for a clear solution product with no gums or water-insoluble solids in the formulation. In this case, heat is usually not required to make the product. This procedure may be considered as consisting of four steps.

1. Dissolve the surfactants in water with stirring. Note the order of addition may be important. In general, add the foam modifier last.
2. Add the fragrance, color solutions, and preservative and stir until a uniform solution is obtained.
3. Adjust the pH with either acid or alkalinity.
4. Add salt for the final viscosity adjustment.

Note that whenever possible the final step in product manufacture should be viscosity adjustment to allow for optimum mixing and for maximum energy conservation. It may be useful or necessary to dissolve the fragrance or an oily component in a small amount of concentrated surfactant before adding it to the aqueous phase.

If solid amides are used as foam modifiers, then heating above the melt point may be necessary to either dissolve or emulsify such an ingredient. If gums are used, it may be necessary to dissolve the gum in a small amount of water before adding it to the detergent phase. In any case, when polymeric gums are used one should consult and follow the manufacturer's directions for dispersing/solubilizing the gum into the formulation.

Most conditioners and conditioning shampoos, such as 2-in-1s, are oil-in-water emulsions and are more complex to make than the simple clear shampoo just described. The following procedure can be used to make most oil-in-water emulsion products:

1. Dissolve the water-soluble ingredients in deionized water while stirring and heat if necessary (this is part A).
2. If necessary, heat the oil-soluble components to melt the solids. These ingredients may be added together or separately. The order of addition is often critical (this is part B). When adding part B or its components to part A, heat part A to approximately 10°C above the melting point of the solids. Add part B or its components to part A while stirring.
3. Continue stirring for at least 10 to 15 min and then add the remaining water.
4. Cool; add preservative, fragrance, and colors.
5. Adjust pH and then the viscosity.

The speed of agitation, type of mixer, rate of cooling, and order of addition are all important to produce consistent emulsion products that are stable and provide high performance. In the case of 2-in-1 conditioning shampoos containing water-insoluble silicones, the silicone will generally be added after the fatty components once the emulsion has been formed. Three examples of hair conditioner formulations and making procedures are described in the next section. These should provide a better feel for how to make and formulate emulsion hair products than the general outline just described.

This discussion is obviously a cursory introduction into shampoo and conditioner making procedures. For more details on emulsions, their structure, stability, and formation, see the review by Eccleston [2] and the references therein. For additional details on the making of shampoos and conditioners, consult formularies and recent literature from cosmetics courses such as offered by The Society of Cosmetic Chemists, and The Center for Professional Advancement. For additional details on product compositions, consult references 1 to 3, product ingredient labels, and the books by Hunting [4,5].

Hair Conditioners

Creme rinses and most hair conditioners are basically compositions containing cationic surfactant in combination with long-chain fatty alcohol or other lipid components. Distearyl dimethylammonium chloride and cetyl trimethylammonium chloride are typical cationic surfactants used in many of today's hair-conditioning products. Amines such as dimethyl stearamine or stearamidopropyl dimethylamine are other functional cationics used in these products.

$$CH_3-\overset{\displaystyle CH_3}{\underset{\displaystyle [CH_2-(CH_2)_{16}-CH_3]_2}{\overset{|}{\underset{|}{N^{\oplus}}}}} \quad Cl^{\ominus}$$

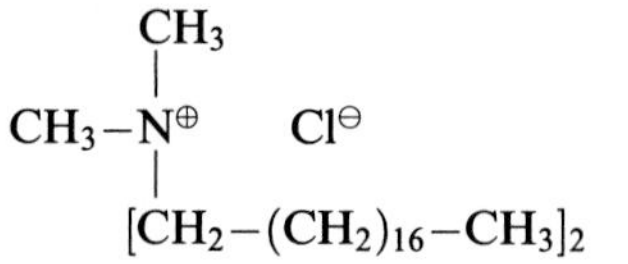

Distearyldimethylammonium chloride

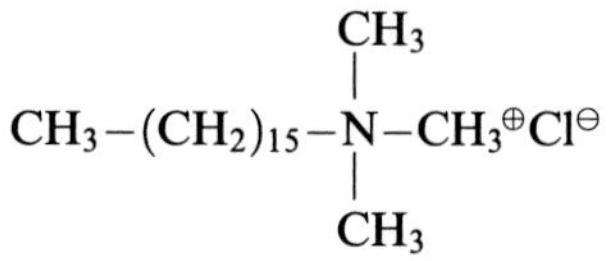

$$CH_3-(CH_2)_{15}-\overset{\displaystyle CH_3}{\underset{\displaystyle CH_3}{\overset{|}{\underset{|}{N}}}}-CH_3{}^{\oplus}Cl^{\ominus}$$

Cetyltrimethylammonium chloride

Typical lipids used in these products are cetyl alcohol or stearyl alcohol, glycol distearate, or even silicones such as dimethicone or amodimethicones.

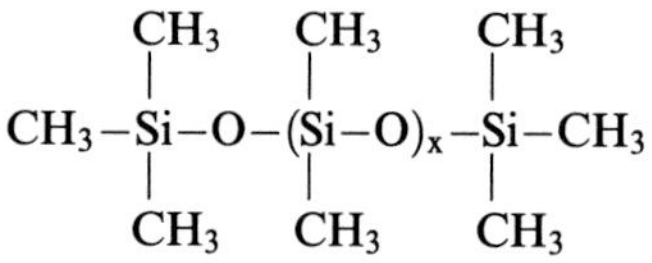

$$CH_3-\overset{\displaystyle CH_3}{\underset{\displaystyle CH_3}{\overset{|}{\underset{|}{Si}}}}-O-(\overset{\displaystyle CH_3}{\underset{\displaystyle CH_3}{\overset{|}{\underset{|}{Si}}}}-O)_x-\overset{\displaystyle CH_3}{\underset{\displaystyle CH_3}{\overset{|}{\underset{|}{Si}}}}-CH_3$$

dimethicone

Some Hair Conditioner Formulations and Making Procedures

An example of a good, simple, yet effective formulation for a creme rinse/conditioner is listed in Table 5–7.

The making procedure for this type of hair conditioner is the one described for oil-in-water-emulsion conditioning shampoos. The CTAC is first dissolved in one-half the water and heated to 60°C (part I). The cetyl alcohol is heated to melt and added to part I while stirring. Cool to 40°C and add fragrance, preservative, and citric acid. The gum is hydrated in the remaining water, and after it is in solution is slowly added to the product with stirring.

If one examines conditioners in the marketplace one also finds more complex conditioners, many of which differ for the sake of using ingredient names rather than for real product performance. An example of such

TABLE 5–7. Example of a simple hair conditioner.

Ingredient	Percentage
Cetyl trimethyl ammonium chloride (CTAC)	1
Cetyl alcohol	2.5
Thickening gum (hydroxy ethyl cellulose)	0.5
Fragrance	0.2
Preservative (Germaben II)	0.5
Citric acid	0.2
Water	95.1

TABLE 5–8. Example of a more complex hair conditioner.

Ingredient	Percentage
Cetyl alcohol	1
Stearyl alcohol	1
Hydrolyzed animal protein	<1
Stearamidopropyl dimethyl amine	<1
Cetearyl alcohol	<1
Propylene glycol	<1
Keratin polypeptides	<1
Aloe	<1
Tocopherol	<1
Panthenol	<1
Preservative	<1
Fragrance	<1
Water	q.s.[a]

[a] q.s., add water to 100%.

TABLE 5–9. Example of a "deep" hair conditioner.

Ingredient	Percentage
Part I	
Cetyl alcohol	6.0
Stearamidopropyl dimethyl amine	1.5
Mineral oil, heavy	0.5
Propylene glycol	1.0
Part II	
Citric acid	0.2
Dicetyl dimethyl ammonium chloride	1.0
Germaben II	0.5
Fragrance	0.4
Water	q.s.[a]

[a] q.s., add water to 100%.

a product is detailed in Table 5–8. The "kitchen-sink" hair conditioner in Table 5–8 would be made according to the same procedure described for making oil-in-water-emulsion conditioning shampoos. Hype compounds such as proteins, placenta extract, vitamins (tocopherol), provitamins (panthenol), etc. that are almost always nonfunctional, such as many of those described in the formulation in Table 5–8, are commonly used in conditioning products.

To make the deep hair conditioner described in Table 5–9, first melt the oil phase, cetyl alcohol, and stearamidopropyl dimethyl amine in the presence of mineral oil and propylene glycol and heat to 80°C. Add citric acid to water and heat to 80°C as the dicetyl dimethyl ammoniumchloride is added to the aqueous phase. Add the oil phase (Part I) to the aqueous phase, stir for about 20 min, then cool; add in the preservative, colors, and fragrance.

Cleaning Hair

Shampoos are formulated under several constraints because a hair-cleaning system must contact the scalp. These constraints include the following. Cleansing ingredients must be safe, thus requiring low toxicity, low sensitization potential, and low skin and eye irritation potential. Low temperatures (20–44°C) are used during shampooing. Short cleaning or reaction times (minutes) are also employed. Low substantivity of detergent for hair is preferred, except for conditioning, where adsorption is necessary (see the section on the sorption or binding of ingredients to hair). Essentially no degradation of the hair substrate by the cleansing system is allowable. The cleansing system should be capable of removing a variety of different soils without complicating interactions between shampoo ingredients and the soils.

The most common test criterion used to assess cleaning efficiency of shampoo products relates to the amount of soil left on the hair surface after shampooing. However, the rheological and other physical properties of the soil have recently been shown to be important also. Specific properties of hair fibers versus assemblies, attributes of the product (fragrance, lather, and viscosity), and the rate of resoiling are also relevant to hair-cleaning efficacy. The next section of this chapter is concerned primarily with the different types of soil found on hair, their origins, and their removal by existing surfactant systems.

Hair Soils and Detergency Mechanisms

Hair soils may be classified as one of four different types:

1. Lipid soils are the primary hair soil and are principally sebaceous matter. For a more complete description of the chemical composition of sebaceous soil, see Chapter 2.
2. Soils from hair products or hair preparations represent another important group consisting of a variety of different cationic ingredients, polymers, and lipophilic ingredients. Lipid soils are not only of sebaceous origin, but also from hair product formulations. They may exist on hair surfaces either as neutral lipid (oils, waxes, silicones) or as calcium-bridged fatty acids or neutralized or free fatty acids. Fatty acids are not used as frequently as in the past in hair conditioning products, but are rapidly being replaced by silicone or lipid-type conditioning agents. As a

result, these same ingredients, lipids and silicones, are becoming more important as hair soils.

3. Protein soils are from the skin, but probably in most cases do not constitute a serious soil removal problem.
4. Environmental soils vary, and consist of particulate matter from air (hydrocarbons and soot) and minerals from the water supply.

Soils from Hair Products

A variety of different soils from hair products may be found on hair surfaces, and it is essential for a good shampoo to remove these soils without complicating interactions between the surfactant and the soil. Hair products provide lipid-type soils, cationic soils, polymeric soils, and metallic ions or fatty acids that can bridge metallic ions to hair.

Neutral lipids are found in many different types of hair products, including some conditioners, pomades, men's hair dressings, etc. Monofunctional cationic ingredients such as stearalkonium chloride and cetyl trimethyl ammonium chloride are the primary active ingredients of creme rinses and other hair-conditioning products, and the increased usage of such products during the past decade makes this soil type even more common. Nevertheless, use of dialkyl quaternary ammonium ingredients such as dicetyl dimethyl ammonium chloride are becoming more common in hair conditioner usage.

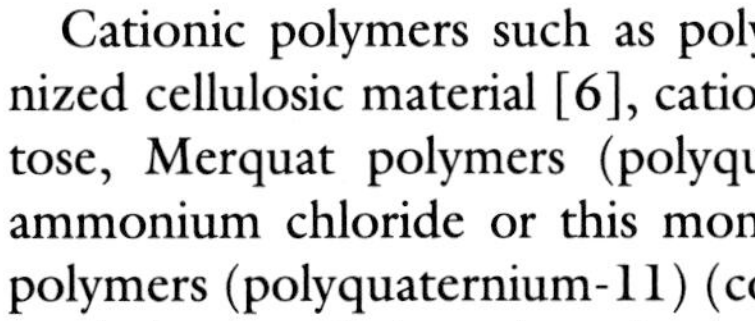

$$C_6H_5-CH_2-\overset{\displaystyle CH_3}{\underset{\displaystyle CH_2-(CH_2)_{16}-CH_3}{\overset{|}{\underset{|}{N}}}}-CH_3^{\oplus}\ Cl^{\ominus}$$

Stearalkonium chloride

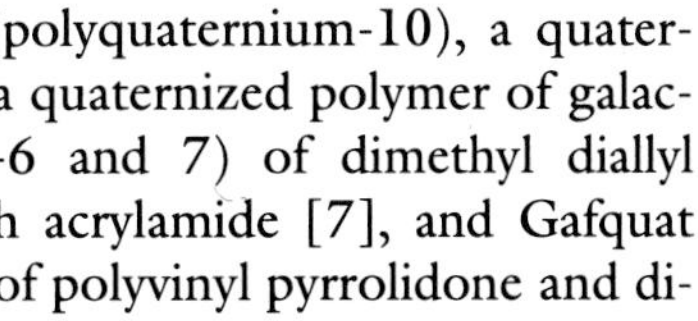

$$CH_3-(CH_2)_{15}-\overset{\displaystyle CH_3}{\underset{\displaystyle CH_3}{\overset{|}{\underset{|}{N}}}}-CH_3^{\oplus}\ Cl^{\ominus}$$

Cetyl trimethyl ammonium chloride

Cationic polymers such as polymer JR (polyquaternium-10), a quaternized cellulosic material [6], cationic guar, a quaternized polymer of galactose, Merquat polymers (polyquaternium-6 and 7) of dimethyl diallyl ammonium chloride or this monomer with acrylamide [7], and Gafquat polymers (polyquaternium-11) (copolymer of polyvinyl pyrrolidone and dimethylaminoethyl methacrylate) [8] have all been used and are currently being used in conditioning shampoos, setting lotions, or mousses. (See Chapter 7 for additional details regarding cationic polymers in hair care products.)

Neutral and acidic polymers such as polyvinyl pyrrolidone, copolymers of polyvinyl pyrrolidone and vinyl acetate, copolymers of methyl vinyl ether and half-esters of maleic anhydride, etc. are all used in hair-styling and hair-setting products (see Chapter 7 for additional details).

Fatty acids such as lauric, myristic, or palmitic have been commonly used as conditioning ingredients in conditioning shampoos, although as indicated these are now being used less frequently than in previous years. Fatty acids interact with calcium and magnesium ions of the water supply and deposit on the hair. It is believed that at least part of this type of conditioning agent binds to the hair through metal ion bridges [9]. The "harder" the water the greater the amount of deposition of fatty acid conditioner on the hair [10]. Thus the primary sources of calcium-bridged fatty acids on hair are conditioning shampoos and bar and liquid soap products that react with metal ions in the water supply. In moderate to high hard-water areas, fatty acids from sebum may also be a source for metal ion-bridged fatty acid on the hair fiber surface.

Environmental Soils

Hair is an excellent ion-exchange system. Therefore, other metallic ions such as copper (+2) [11] can adsorb to hair, especially after frequent exposure to swimming pool water. It has been suggested that metallic ions such as chromium, nickel, and cobalt may also bind to hair from swimming pool water [11]. Sorption of metallic ions like calcium or magnesium occurs even from low concentrations in the water supply rather than from hair products. However, fatty acids present in hair products enhance the adsorption of most of these metallic ions to the hair surface, as described. Heavy metals such as lead and cadmium have been shown to accumulate in hair from air pollution [12], and other metals such as zinc are available from antidandruff products (from the zinc pyrithione active ingredient).

Other soils that shampoos must remove are proteinaceous matter arising from the stratum corneum, sweat, and other environmental sources. We have already described metallic ion contamination from the water supply, swimming pools, and sweat. In addition, particulate soils from the environment include hydrocarbons, soot, and metal oxide particles, which should also be removed by shampoos [13].

Detergency Mechanisms

Although mechanical action is involved in cleaning hair, this may be assumed to be constant for any given person. Therefore, detergency mechanisms are the only viable approach to improve hair cleaning. Detergency mechanisms [14] generally consider soils as either oily (liquid soils) or particulate (solid soils). The removal of oily soils involves diffusion of water to the soil–fiber interface and rollup of the soil, which generally determines the rate of soil removal, although solubilization, emulsification, and soil penetration are also involved. Rollup of oil on a fiber surface is caused by interfacial tensions of oil on fiber, $\pounds_{fo}$; water on fiber, $\pounds_{fw}$; and between oil and water, $\pounds_{ow}$; the oily soil rolls up when the combination of these interfacial tensions (R) is positive in the expression:

$$R = \pounds_{fo} - \pounds_{fw} + \pounds_{ow} \times \cos\phi$$

In other words, to produce oily soil rollup, the detergent must make the fiber surface more hydrophilic [14]. Thus, the removal of lipid soil from hair is dominated by the hydrophilicity of the fiber surface, and anything that can be done to make the fiber surface more hydrophilic, such as bleaching or washing with anionic surfactants in water, should facilitate oily soil removal. This is one of the reasons why damaged hair, which is generally more hydrophilic at the surface, is so sensitive to oil removal.

The removal of particulate soil is not controlled by the hydrophilicity of the fiber surface. Particulate soil removal depends on the bonding of the particle to the surface and the location of the particle [14]. When the soil particle consists of nonpolar components, for example, waxes or polymeric resins such as hair spray resins, its adhesion depends mainly on van der Waals forces. Unless very high molecular weights are involved, the removal of such soils is oftentimes easier than for cationic polymers for which adhesive binding includes a combination of ionic and van der Waals forces.

Many hair soils fit into these two distinct classes, and the mechanism for their removal is easy to understand. However, other soils (e.g., conditioners containing cationic surfactant plus oily substances or some plasticized resins) are intermediate in classification, and their removal probably involves either a combination of mechanisms or a more complicated cleaning mechanism.

Methods to Evaluate Clean Hair

Several methods have been described to evaluate the ability of different shampoos or detergents to clean soil from the hair [11,13,15–23]. Most of these methods have been developed to evaluate the removal of lipid soil from the hair [13,15,17,18]. Some of these methods are soil specific [17] or are more sensitive with specific soil types [13,20], while others work for most soils [19].

Hair-cleaning methods may be classified according to the following categories: chemical and physical properties, microscopic methods, or subjective or sensory evaluation procedures. Chemical or physical methods may involve either direct analysis of the hair itself [13,19] or analysis of hair extracts [16,17]. For direct analysis of hair, chemical methods such as electron spectroscopy for chemical analysis (ESCA) [21] or infrared spectroscopy may be used. Physical methods such as fiber friction [20], light scattering [19], or examination of interfiber spacings [13], on the other hand, are less soil specific than chemical methods; these have the ability to investigate a variety of soil types, but sometimes distinguish less well in so doing.

Microscopic methods have also been used to evaluate hair cleanliness [16]. However, sensory evaluation of hair greasiness on hair swatches

[22] and subjective assessments from half-head and consumer tests are also useful. The latter evaluations are, in a sense, the final word in the estimation of cleanliness by shampoos. Most procedures involve evaluation of either single soils (primarily hair lipid) or short-term effects of different products. One area of concern that has received relatively little attention is long-term effects that might result from gradual buildup, or from gradual interactions between different hair products such as silicone-containing products, or between hair lipid and different hair products. This is an important area for future research for shampoos.

Cleaning Efficiency of Shampoos

To evaluate shampoo efficiency, one must consider the different soil types separately and then together, and also attempt to distinguish between cleaning soil from hair and the deposition of ingredients from the shampoo formulation itself.

Cleaning Lipid Soil from Hair

For efficiency in removing lipid soil from hair, the literature does not provide a consensus. For example, Shaw [16] concludes that a one-step application of anionic shampoo removes essentially all the hair surface lipid and that differences in cleaning efficiencies cited (for different surfactants [16]) reflect differences in the amount of internal hair lipid removed. To support this conclusion, Shaw cites results from scanning electron micrographs (SEM) of hair washed with anionic surfactant (monoethanolamine lauryl sulfate) compared with SEM photographs of hair washed with ether (Figure 5–2). In addition, Shaw cites in vitro studies showing that various shampoos remove 99% of an artificial lipid mixture deposited on the hair.

Robbins [24] independently arrived at a similar conclusion suggesting that shampoo surfactants in a normal two-step shampoo operation are very effective in removing "surface" lipid; because of their limited penetration into hair, however, they are less effective but capable over time of removing internal lipid.

Table 5-10 offers some evidence for the effectiveness of current shampoos for removing a sebaceous-type soil from wool fabric in moderately soft water (80 ppm hardness). These data show that a cocomonoglyceride sulfate (CMGS) shampoo at only 5% concentration (5% shampoo and 1% active ingredient (A.I.)) under these laboratory conditions approaches the efficiency of boiling chloroform in a 4-h Soxhlet extraction for removing lipid soil (Spangler synthetic sebum; see Chapter 2) from wool swatches. On the other hand, the soap-containing shampoo of Table 5–10 is not very effective in removing this lipid from hair. The normal usage concentration of shampoos is 20% to 25%. The solution-to-hair ratio in normal usage is lower than in this experiment, but this probably does not make a

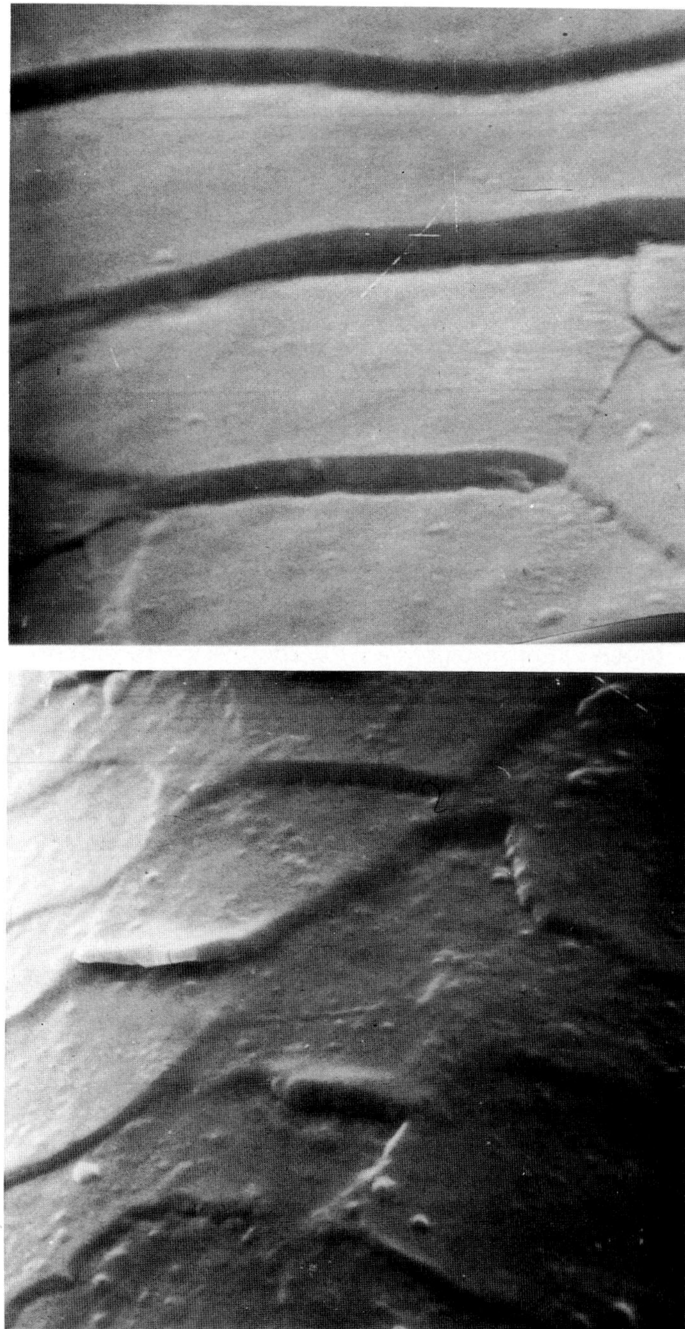

FIGURE 5–2. Scanning electron micrographs (SEMs) of hair fiber surfaces illustrating sebaceous soil versus a fiber cleaned with sodium lauryl sulfate solution.

TABLE 5–10. Shampoo versus chloroform extraction of animal hair.

	Percent of sebum removed[a]	
Shampoo (%)	CMGS formula	Soap-containing formula
0.5	63	18
1.0	76	32
2.0	86	53
5.0	91	77

[a] These percentage values were obtained by extracting these same wool swatches in boiling chloroform for 4 h in a Soxhlet apparatus after shampooing and comparing the residue weight versus total soil deposited. Thus, for practical purposes boiling in chloroform for 4 h is 100% sebum removal.

Test procedure: Wool swatches were soiled with synthetic sebum and weighed to determine the amount of soil deposited. The swatches were then washed in a tergitometer with CMGS and soap-containing (dry hair) shampoos at varying concentrations. The swatches contained approximately 100% sebum. The temperature was 105°–110°F; time was 30 sec; and a 200:1 solution-to-wool ratio and water rinsings of 2 to 30 s were used.

substantive difference. Wool swatches were used in this experiment instead of hair, because most shampoos are too effective to provide distinctions in removing sebum from hair under these conditions.

Another experiment compared the total amount of extractable lipid from hair after washing with two different shampoos in a half-head test. These shampoos were selected because they displayed differences in their ability to remove sebum from hair in the laboratory. This experiment was performed twice, using five subjects per test. Both on-head tests show no significant difference in the amount of extractable hair lipid after shampooing the hair with these two shampoos (Table 5–11). This result suggests that the laboratory test is more sensitive for detecting differences in sebum removal than the half-head test. It also suggests that both shampoos are removing most of the surface hair lipid in the on-head procedure.

If large differences in cleaning efficiency really exist between most shampoos in consumer usage, then other variables such as lather or fragrance would not likely have a large impact on the consumer's assessment of cleaning efficiency. However, it is well known that variables such as fragrance or lather do have a large impact on the consumer's perception of cleaning efficiency by different shampoos (see the section on the perception of cleaning). The foregoing results are consistent with the conclusion that current cleaning shampoos are very efficient for removing "surface" lipids from the hair.

Statements contrary to the conclusions of Shaw and Robbins exist in the scientific literature. Schuster and Thody [25] state that "shampooing" with sodium lauryl sulfate is an ineffective means of removing hair lipid. Thompson et al. [17] report from in vitro testing that anionic surfactants

TABLE 5–11. Sebum found on hair clippings after half-head shampooing with shampoos of different laboratory sebum-removing potential.

Test	Formula	Sebum removal in lab (%)[a]	Amount lipid extracted using alcohol (4h)[b]
#1	TEALS liquid	54	3.0
	Tas-6-1065	82	3.0
#2	TEALS liquid	54	4.2
	Tas-6-1179-A	88	4.2

[a] Values obtained using synthetic sebum and 0.5% surfactant and the procedure described in Table 5–10.

[b] Test procedure: Hair clippings were taken from both sides of heads after half-head shampooing (two applications of shampoo) on five subjects per test. Clippings were combined into two sets, keeping treatments separate. Each set was randomized, divided into three equal portions (~5.5 g each), and extracted in a Soxhlet apparatus for 4 h with ethanol. The lipid extract is expressed as a percentage of the dry weight of the hair for an average of triplicate determinations.

remove polar components of sebum more readily than nonpolar components (paraffin waxes), the implication being that nonpolar components of sebum are not efficiently removed from hair by normal anionic shampoos. Clarke et al. [26] have shown that laureth-2 sulfate is one of the most effective surfactants for removing virtually all sebaceous components from hair. However, lauryl sulfate is not as effective for removing fatty acids in the presence of water hardness. It is nevertheless, highly effective for removing other sebaceous components from hair. Shorter chain length surfactants are, as expected, less effective for removing lipid components from hair.

The effect of temperature on the selective removal of sebum components from hair was also compared by Clarke et al. [27] for sodium laureth-2 sulfate and ammonium lauryl sulfate. Laureth sulfate was found to be the more effective detergent at both 21°C (70°F) and 43°C (110°F). Surfactant efficacy decreased with temperature, providing a slightly greater selectivity in component removal at the higher temperature than at the lower temperature.

The ability of anionic surfactants to remove hair lipid is dependent on surfactant structure, concentration, agitation, temperature, time, and other variables including other soils on the hair. In addition, detergents such as sodium lauryl sulfate do not penetrate rapidly into hair and should not be expected to remove the same amount of lipid from hair at the same rate as a penetrating lipid solvent like ethanol.

Under optimum conditions such as in vivo shampooing, anionic surfactants are nearly as effective as chloroform or ether for removing surface-deposited lipid. In most of the tests described in the literature, care was taken to exclude conditioning products containing cationics and cationic polymers or silicones, setting resins, and hard water to provide more control over the experiments. Obviously these variables must be included before we can arrive at a full understanding and a consensus about the efficiency of anionic shampoos for cleaning hair lipid from the surface of hair.

Other soils have not been studied so extensively; however, Robbins et al. [28] have shown that C-12 alkyl sulfates or alkyl ether sulfates, the traditional shampoo surfactants, do not remove cationic surfactants from hair effectively under certain conditions. However, shorter chain length anionics such as deceth-2 sulfate are more effective for removing cationics. In addition, alkyl ether sulfates are more effective for removing fatty acid soils in the presence of water hardness than alkyl sulfates [28].

Surface Versus Internal Lipid

Human hair contains both surface lipid (dirt) and internal lipid, as indicated in Chapter 2. Much of the internal lipid is structural material, and is difficult to remove by ordinary shampooing [16,21] because of the slow penetration rates of surfactants. However, there is increasing evidence that structural lipid, probably the inert β-layers of the cell membrane complex (the major pathway for entry into the fibers; see Figure 1–21) can be removed over time by shampooing [29,30], (see Chapter 1).

The total amount of lipid extractable from hair can be as high as 9% of the weight of the hair [31], when a penetrating hair-swelling solvent like ethanol is used on hair that has not been shampooed for 1 week. A sizable fraction of the "total" hair lipid is not removed by shampooing or by extraction with a nonpenetrating, low-boiling lipid solvent like ether. We have obtained as much as 4.2% ethanol-extractable matter from hair cut from heads immediately after shampooing two times with a triethanol ammonium lauryl sulfate (TEALS) shampoo [31] (see Table 5–11).

Curry and Golding [32] have shown that the rate of extraction of lipid from hair by solvents is very slow, and even after 100 Soxhlet cycles with ether (four successive extractions), a significant amount of lipid can be obtained by additional extraction. As indicated earlier, Shaw [16], using SEM techniques, reported that washing hair with either ether or shampoos in a one-step application removes virtually all the surface lipid from hair and that differences in cleaning efficiencies of surfactants relate to the amounts of internal hair lipid removed. Shaw found that one-step shampooing removes approximately 50% of the ether-extractable matter.

Koch et al. [9] report that repeated shampooing removes 70% to 90% of the ether-extactable lipid, and that enzymatic hydrolysis of hair after ether extraction followed by extraction of the residual membranes yields "internal lipid." Koch found the composition of this internal lipid to be somewhat similar to that of surface hair lipid (see Chapters 1 and 2). Koch

TABLE 5–12. Amount of hair lipid in oily versus dry hair after shampooing.[a]

	Average amount of lipid recovered:	
	Weight (g)	Percent weight of hair
Dry hair	0.164	3.6
Oily hair	0.161	3.6

[a] Test Procedure: Six panelists were selected for this test, three having dry hair and three oily hair, as judged by both beauticians and the panelists themselves. Hair clippings were taken from heads after shampooing with a TEALS shampoo, using the usual two-step application procedure. The clippings were combined from all three dry-hair and all three oily-hair panelists. They were randomized into three replicates (sets) per sample, and Soxhlet extracted in triplicate for 4 h with ethanol.

therefore concludes that internal lipid of hair must in part originate from the sebaceous glands (see the section entitled Cell Membrane Complex in Chapter 1).

It is difficult to isolate only surface lipid or only internal lipid; however, it is certain that both surface and internal lipids exist in hair. Koch suggests that the external lipid may be extracted by boiling ether saturated with water followed by ethereal hydrochloric acid. The former solvent removes neutral surface lipid; the latter solvent removes calcium-bridged fatty acids attached to the hair surface. He suggests that surface lipid so defined is removed under conditions that simulate the "strongest shampooing conditions imaginable."

This definition of surface lipid by Koch probably provides a high estimate for surface hair lipid. Another definition is the amount of lipid removed by a double application of an anionic surfactant. This latter definition probably provides a more realistic estimate for surface hair lipid. However, if one accepts this latter definition, then the amount of lipid left in hair after shampooing represents internal lipid and may be estimated by solvent extraction (ethanol) after shampooing.

Table 5–12 summarizes data from an experiment conducted to determine if the quantity of internal hair lipid differs in dry (chemically unaltered hair) versus oily (chemically unaltered) hair. Immediately after shampooing twice with a TEALS shampoo, hair clippings were taken from three oily-haired panelists and three dry-haired panelists and extracted with boiling ethanol. The results suggest similar quantities of internal hair lipid in these six hair samples.

This test result suggests that the amounts of internal lipid in dry and oily hair are similar. Therefore, the primary differences between dry and oily hair lipid are in the amount and the composition of the surface lipids. In summary, the current literature suggests that human hair contains lipid at or near its surface and that it also contains internal lipid. The surface lipid provides many of the negative physical characteristics attributed to oily (greasy) hair, while some of the interior lipid will slowly diffuse to the surface on successive washings (shampooings) or extractions. Furthermore, this internal hair lipid is similar (but not exactly the same) in composition to the external hair lipid. Hair also contains bound or structural internal lipid that is presumably resistant to shampooing. Further details on the composition of hair lipid are described in Chapters 1 and 2.

The Transport of Hair Lipid

After shampooing, the surface of hair is relatively free of lipid, or at least its concentration is considerably reduced. Sebum (produced by the sebaceous glands) and epidermal lipid (produced by the cells of the horny layer of the scalp) are transferred to the hair because of its greater surface area and absorptive capacity. Creeping of sebum along the hair has also been suggested [33], although Eberhardt [34] has shown that creeping does not occur along single hair fibers. Eberhardt suggests that transport occurs primarily by mechanical means, that is, by contact of hair with scalp (pressure of pillows and hats), rubbing (combing and brushing), and hair-on-hair contact.

Distribution of sebum along the fibers by combing and brushing is very important, and wicking as occurs in textile assemblies might also be involved [35,36]. The net result is that the rate of accumulation of lipid is fastest for oily hair, and after the lipid accumulates beyond a given level it interferes with the appearance and overall aesthetics of the hair, causing fibers to clump or to adhere together, thus producing limp hair.

The composition of the lipid itself may influence its transport, because ingredients that either lower the surface tension of the sebum or increase its fluid nature (make it less viscous) can facilitate transport and even increase the perception of oiliness. In addition, other ingredients left behind on the hair surface, such as conditioning agents, may exacerbate oiliness in an analogous manner.

Hair characteristics such as fineness, degree of curvature, and length are also relevant to the transport of lipid and to the influence of lipid on hair assembly properties. For example, fine, straight hair will provide optimum characteristics for transport of sebum. This type of hair will also provide the maximum amount of hair clumping by a given amount of lipid, thus it will appear more oily than curly hair. In contrast, curly-coarse hair will tend to inhibit transport and also to minimize the influence of tress compacting. Among all hair properties, increasing fiber curvature provides the greatest influence against the cohesive forces of hair lipid and the resultant compacting of tresses (limpness) [37].

Cationic Soils

Dye staining tests [38] on wool fabric or hair swatches (containing cationic) and ESCA studies on hair containing monofunctional cationic surfactant [21] show that a single washing of hair with an anionic detergent does not remove all the quaternary ammonium compound from hair. Radiotracer studies of cotton fabric containing presorbed sodium lauryl sulfate (SLS) indicate that SLS sorbs to the fabric in an equimolar quantity to the deposited quaternary ammonium compound [39]. Robbins et al. [40] found that by presorbing anionic surfactant (SLS) to hair and then treating with cationic (dodecyl trimethyl ammonium bromide), the presorbed anionic enhanced the adsorption of the cationic to the hair. These results suggest that monofunctional cationics are resistant to removal by anionic surfactant because they form adsorption complexes on hair.

Robbins et al. [28] have shown that washing monofunctional cationic surfactants like cetyl trimethyl ammonium chloride from hair with normal alkyl sulfates or alcohol ether sulfates does not remove all the cationic from the hair, and in addition the anionic detergent can build up with the cationic. Adsorption complexes formed in this manner adsorb to hair with the potential for building-up. Shorter chain length surfactants like deceth-2 or deceth-3 ether sulfate do not build up in the same manner. However, this type of buildup generally levels off after five to six treatments. In addition, hair matting has been reported in vivo and attributed to the adsorption of cetyl trimethyl ammonium bromide on hair [41].

Certain cationic polymers have been reported to build up on hair [8], and even low charge density cationic polymers like polymer JR (a quaternized cellulasic polymer) have been reported to be resistant to removal from hair surfaces by anionic surfactant [42,43]. For example, in one study 3% sodium lauryl sulfate, after 1 min, removed 50% of the polymer JR from the hair and nearly 70% in 30 min. However, some strongly bound cationic polymer was still attached to the hair and resistant to removal by anionic surfactant after 30 min.

Hannah et al. [43] have shown that polymer JR deposits on hair in the presence of excess sodium lauryl sulfate. This deposited complex is highly substantive to hair and resists removal from the hair surface by either water or 3% sodium lauryl sulfate solution. Therefore, adsorption complexes of polymeric cations also resist shampooing from hair.

Polyethyleneimine, a high charge density cationic polymer, appears to be even more strongly bound to hair than polymer JR, and has been shown to be resistant to removal by anionic surfactant [44]. For example, PEI-600 was sorbed onto hair and tested for desorption toward a 10% shampoo system. After 30 min, less than 20% of the PEI was removed and only about 30% PEI was removed after 6 h. For additional details on the adsorption and removal of cationic polymers from hair, see Chapter 7 and the references therein.

The foregoing results clearly show that cationic soils are resistant to removal by anionic surfactant, and it appears that anionic surfactants are not capable of completely removing high charge density cationic polymers from hair.

Other Soils

The original hair spray lacquers of the 1950s were more difficult to remove from hair than the anionic and neutral polymers of today's hair-setting products. However, no systematic study of the ease or difficulty in removing these ingredients from hair could be found. Gloor et al. [45] have examined the influence of hair spray on reoiling; however, no systematic study of the effects of hair spray on the ease of removal of hair lipid has been reported.

Calcium-bridged fatty acid may be deposited onto hair even in shampoos containing anionic surfactant such as ammonium lauryl sulfate. It is also well known that acid rinses may be used to remove calcium-bridged fatty acid from hair, and anionic sulfate surfactants appear to remove fatty acid deposits from hair. However, copper (cupric ion) adsorbs strongly to hair and is reported to be resistant to removal by anionic surfactant [11].

Published literature regarding the efficacy of anionic surfactant systems for removing particulate soils such as soot, hydrocarbons, etc. could not be found. One may speculate that two important variables with this kind of soil are particle size and the type of chemical bonding between the particle and the hair. As particle size decreases to less than about 1 μm, the resistance to removal should increase. However, when van der Waals attractive forces are the primary adhesive forces, removal probably occurs. When hydrogen bonding or ionic bonding are involved, the particle will be more resistant to removal.

Rate of Reoiling of Hair

Breuer [13] has described the kinetics for reoiling of hair in terms of sebum production and sebum removal. He derived the following expression to describe the rate of reoiling:

$$m = A/K(1 - e^{-Kt})$$

where m = amount of sebum on the hair (at any time after cleaning), t = time after cleaning (min), A = production rate of sebum (12.5×10^5 ng/min), and K = rate constant for sebum removal.

Using experimental data, this expression was solved numerically by Breuer suggesting that in a 4-day period approximately 65% of the sebum that is produced is lost from the head by rubbing against objects such as pillows, combs, or brushes. Breuer concludes that shampoo and postshampoo treatments influence the reoiling rate of hair. As indicated earlier, anionic surfactants alone do not stimulate the rate of refatting [16], although selenium sulfide has been reported to increase the rate of sebum production [15,46], and zinc pyrithione [15] and climbasole, two other antidandruff agents, have been shown to behave similarly by increasing hair greasiness [15].

Perceptions in Cleaning Hair and the Subjective Testing of Shampoos

With the advent of 2-in-1 shampoos, a new era has begun in cosmetic science. Differences in the performance between different conditioning shampoos are relatively large and can be detected in laboratory tests, in half-head tests, and even in consumer tests on sample sizes smaller than $n = 100$. To the cosmetic scientist, this is a pleasant situation. We can now turn our attention to real product performance. We can truly work to create products that are really better, not only in the laboratory, but products that consumers will discern as better. This situation was created by a combination of new technology with consumers who are becoming willing to accept different standards of performance for shampoos, and I believe this same situation exists for other opportunities in hair care in the future.

The situation is not as clear for cleaning shampoos. However, with the new soils that we are leaving behind on hair for superior conditioning, body, and style control, perhaps new performance opportunities in hair cleaning will also be created in the future.

Nevertheless, the following discussion is useful for all product types when the differences between real product performance become relatively small, a situation that will probably occur for 2-in-1 shampoos or high-performance conditioning shampoos within this decade.

Questions regarding the removal of sebaceous soil and other soils from hair are fundamental to the action of shampoos. However, another fundamental question is: Which is more important to the sale of shampoos—the actual abilities of different shampoos to remove soil from hair, or factors relevant to the perception of cleaning such as lather, viscosity, fragrance, etc? Laboratory or in vitro tests are critical to provide an understanding of shampoo behavior. However, subjective tests are ultimately involved to evaluate the consumer's response to the total product. The next section describes some of the more common subjective tests used in shampoo development and raises some important questions.

Shampoo Performance

The evaluation of overall shampoo performance is determined by the hair effects that the product provides and by the properties of the shampoo itself (properties that do not directly influence hair effects). A helpful distinction defines hair effects as all performance attributes of the shampoo evaluated after rinsing, and shampoo properties as all performance attributes noted or evaluated before and during the rinse step.

Hair Effects and Discernibility versus Perception

For this discussion, discernibility is considered as the objective (not necessarily numerical) ability of the users of a product to isolate and to discriminate between effects on hair without being influenced by related stimuli such as fragrance, lather, viscosity, etc. Perception on the other hand is the subjective response to a hair property, and this response is influenced by all related stimuli including the hair property itself, advertising and label copy, and all relevant shampoo properties.

The question of whether or not a hair effect is discernible to a given percentage of consumers is relevant to the understanding of the perception of a product and to understanding why a product does or does not sell well. However, its answer will often be in doubt. This is because it is difficult for consumers or panelists to be objective and to isolate and measure performance properties without being influenced by other product properties; and, for pragmatic (financial) reasons, insufficient blind tests are generally conducted to determine discernibility, because the bottom line is sales not objective understanding. It is for this reason that many executives question the relevance of testing performance in isolation. Thus, judgment is involved to answer questions of discernibility and subjectivism often interferes in its evaluation and interpretation.

Different Test Procedures

Some of this author's conclusions relevant to different types of shampoo (product) tests are described here. In general, objective discernibility of a hair effect often becomes progressively less important as one proceeds from laboratory to sales tests. This is because subjective perceptions involving psychologically related stimuli become more important as one moves from the laboratory (where experimental control isolates discernibility from perception) to sales testing.

Laboratory Tests

Certain laboratory tests (tress combing, fiber friction, light scattering, sebum-removing ability, etc.) can be more sensitive than consumers' evaluations (see the section on Methods to Evaluate Hair Cleaning, and Chapter 8). However, the most severe constraint with laboratory testing is that laboratory measurements are often only a portion of the related consumer assessment. For example, fiber friction is only a small part of how the hair feels to a consumer or how easily her hair combs, and hair combing is only one part of hair conditioning.

Half-Head Tests with Evaluations by Trained Cosmetologists

These tests are side-by-side comparisons and can be more precise than most assessments by consumers (who rely on memory comparisons) for discerning most important hair effects. It can be argued that half-head tests generally involve short-term effects, and they may be misleading with regard to long-term effects.

Blind Product Tests

The standard 2-week crossover blind product test with a large panel size ($n \simeq 300$) is a relatively sensitive means for discerning whether or not product differences exist between different shampoos. On the other hand, long-term effects from buildup or product interactions may be either not detected or further complicated by the 2-week crossover design. One further difficulty even in short-term evaluations is in understanding the meaning of the differences detected in this type of test procedure. It is very easy to take the conservative stance and to rule out a product if it loses in a blind product test. Yet, I often wonder how many excellent products never reached the marketplace bacause of an inconsequential loss in a blind product test.

The overall data of a blind product test are usually more consistent and more sensitive than the majority of the individual panelists (Table 5–13). This table summarizes data from a blind test in which a baby shampoo was compared with a high-foaming TEALS-based shampoo containing the baby shampoo fragrance and color. This test was actually two tests run back to back, comparing these two products for four 2-week intervals. Among the 73 panelists, for all attributes other than lather and fragrance, fewer than 14 panelists were consistent in their ratings. The consistency obtained in overall preference and in all hair effect attributes was not beyond that expected by random chance. These results show that differences do exist between these two shampoos, but only about one-third of this panel could discriminate sufficiently to repeat their lather and fragrance choices between these two products. Fewer than 15% of these panelists could distinguish between any hair effect differences between these two products, that is, could duplicate their choice for any hair effect.

There is a significant lather preference but not a significant fragrance preference. Only a small percentage of the panelists were capable of duplicating their choice for hair effects; for example, for flyaway and luster, 9

TABLE 5–13. Adult shampoo versus baby shampoo (blind, back-to-back tests).[a]

Specific Attribute	Probability, Test 1 ($n = 75$)	Probability, Test 2 ($n = 73$)	Total, consistent ratings of 73
Overall preference	0.85	0.66	16
Lather	0.99	0.88	23
Ease of rinsing	0.62	0.99	11
Cleaning efficiency	0.88	0.95	13
Feel of wet hair	0.50	0.88	12
Ease of combing (wet)	0.44	0.77	11
Feel of dry hair	0.72	0.99	10
Ease of combing (dry)	0.44	0.44	10
Flyaway	0.89	0.83	9
Luster	0.77	0.84	8
Fragrance	0.55	0.62	23
Softness	0.72	0.74	10

[a] Note: This experiment suggests that most of the users individually do not clearly discriminate between these two shampoos, although the tests taken together clearly show differences. The repeat scores for overall preference are not beyond that expected by random chance, yet the overall probability that the TEALS product is preferred is at a p value of .95.

and 8 panelists, respectively, were able to duplicate their choice. However, the overall test results suggest a difference for flyaway and luster ($>p = .98$ and .96, respectively). Only a small percentage of these panelists appear to be capable of discerning hair effect differences between these two products, although the statistics for the overall test results suggest a difference for example in flyaway and luster (overall $p = .98$ and .96, respectively), but only 9 and 8 of these panelists were able to duplicate their choices. These results lead one to question whether large or meaningful differences exist between hair effects of this baby shampoo and adult shampoo. In light of the consistency ratings, one questions how discerning consumers (panelists) as individuals really are to shampoo performance attributes.

Blind tests with larger groups have been run comparing a related TEALS shampoo formulation versus this same baby shampoo in which the sebum-removing capabilities of these two products in laboratory tests were equivalent. Once again, an attempt was made to match the color and fragrance of these two products. The TEALS product was clearly superior in various laboratory foam tests and in foam property evaluations in half-head testing. The panelists as a group significantly preferred the TEALS product for cleaning efficiency, foaming properties, and for overall performance as well as for several hair effect attributes. Apparently, the superior foaming character of the TEALS system provided a "halo" effect that subconsciously reflected in the cleaning efficiency evaluation and in several of the hair effect attribute evaluations.

Identified Product Tests

Discernibility is very difficult to interpret from identified product tests, because perceptions from label copy, fragrance, lather, and other stimuli often influence and even overwhelm true performance differences. Several years ago, we tested a protein-containing shampoo, both blind and identified, 3 months after its national introduction. The blind test scores showed that in spite of color, form, and fragrance variables, panelists could not discern between the hair effects of that protein shampoo and the hair effects of another leading competitive brand. However, panelists exposed to concepts, label copy, and product names in an identified test provided highly significant wins for this same protein-containing shampoo in hair effect scores, against their favorite brands, but not for the competitive brand. These panelists ($n \simeq 300$) in a projectable identified consumer test were so taken in by the concept and label copy of the protein-containing shampoo that they simply repeated the concept and label copy in their performance ratings. This suggests that in some instances, in identified consumer tests, particularly in the event of a popular concept, the true performance attributes of shampoo products may be ignored and even overwhelmed by the impact of the concept. This situation is most likely to occur when the differences in the performance attributes between the products is small.

Sales Tests

Performance properties in sales tests are even more subject to the influence of psychologically related stimuli than any other type of test. However, because a sales test is long term, longer term performance benefits or negatives will have some bearing on the test outcome. Thus, a sales test can provide some index of longer term performance, particularly if an effort is made to control other variables. Unfortunately, this is difficult to do, and the cost of a sales test is considerably greater than that of an identified consumer test.

The conclusions from the descriptions of different tests suggest that cleaning shampoos are more successful in the marketplace for concept and for advertising execution than for real differences in hair effect benefits. This same situation does not exist for 2-in-1 shampoos. This suggests that the cleaning differences provided by performance (hair effect) differences for the shampoos that are currently in the marketplace may indeed be real. However, they are relatively small and subtle, or the sale of these-products would not be so influenced by psychological stimuli and advertising. On the other hand, the current differences between conditioning offered by conditioning shampoos are relatively large. When subtle hair effect differences are complemented by quality advertising execution and other psychological stimuli (sensory effects), an opportunity for a marketplace success exists. When larger hair effect differences are created, however, an even greater opportunity exists in the marketplace particularly when the hair effect differences are complemented by good advertising execution. When larger hair effect differences are provided, then the need for psychological reinforcement or sensory effects is not as great as in the former situation when only subtle hair effect differences between competing products are provided.

Shampoo Lather

The lathering potential of shampoos does not directly influence the physical behavior of hair fibers. However, as indicated in the previous section, shampoo lather can influence the consumer's perception of hair characteristics. Therefore, a brief introduction into this important shampoo property is provided in this section.

Perhaps the most useful work leading to the present understanding of shampoo lather has been described by Neu [47] and by Hart and colleagues [48–50]. Neu pointed out that the traditional Ross–Miles [51] shampoo foam evaluation using an active concentration of about 0.1% to 0.2% is unrealistic for simulation of shampoos. He suggested lather testing at an order of magnitude greater in active surfactant concentration. Neu used a kitchen food mixer to generate shampoo lather for laboratory evaluation. The high shear rates of a food blender produce lather from surfactants that is more similar to that obtained on hair under actual shampoo ing conditions than is provided by cylinder shake test methods such as in Ross–Miles.

Hart and DeGeorge [48] used a Waring blender similar to Neu to generate shampoo lather and measured drainage rates of the lather to provide an index of lather viscosity. Hart and DeGeorge [48] distinguished between foam and lather for shampoo evaluation and pointed out that "foam" is a broad generic term consisting of "any mass of gas bubbles in a liquid film matrix," whereas lather is a special type of foam formed during shampooing and other processes and "consists of small bubbles that are densely packed," thus resisting flow. Hart's drainage test has been shown by Domingo Campos and Druguet Toutina [52] to produce results that relate to actual in-use salon testing.

Some useful conclusions from Hart's work are as follows: A synthetic sebum load generally lowers lather quality. This is consistent with the well-known observation that the second shampoo application lathers better than the first shampoo application when more sebaceous soil is encountered. Hart also demonstrated that the traditionally known "foam booster" additives such as lauramid DEA or cocamido propyl betaine should more correctly be called lather modifiers (amides do modify lather feel and tend to make a thicker, creamier lather). However, these additives generally do not boost lather; they tend to suppress shampoo lather.

Hart and Neu have provided a useful beginning to a better understanding of shampoo lather, of lather testing, and of the effects of additives on shampoo lather. However, Hart's lather drainage rates are only one of the important components of shampoo lather relevant to the consumer's perception. Lather feel and the rate of lather generation are two other important components of shampoo lather, and as of this writing, methods for these important properties have not yet been described in the scientific literature.

Sorption or Binding of Ionic Ingredients to Hair

In this section, a summary of the latest hypothesis on the adsorption of ingredients (surfactants and hair conditioners) to hair is followed by a synopsis of the interactions of ionic ingredients of shampoos, conditioners, and ionic dyes with hair and the literature available. Because ionic dyes are sometimes used in sorption studies as model systems for surfactant molecules, some material on the sorption of ionic dyes has been included in this section.

Overview of the Binding Interactions of Shampoo and Conditioning Ingredients to Hair

Shampoo and Conditioner Ingredients

The major ingredient in most shampoos is anionic surfactant, although other surfactants—thickening agents, lather modifiers, conditioning agents, colors, and fragrance—are also usually present. Most shampoos are formulated near neutrality and are based on the anionic surfactant salts of lauryl sulfate or laureth sulfate (most commonly up to 3 moles of ethoxylation)(see the section on formulations earlier in this chapter).

Conditioners, on the other hand, are basically compositions containing cationic surfactant in combination with long-chain fatty alcohol or other lipid components. For additional details on product compositions, see the section of this chapter on formulations, and consult references 53, 54, and 55, product ingredient labels, and the books by Hunting [4,5].

The attachment of ingredients to hair fibers is fundamental to the action of conditioning agents. The amount of sorption or uptake of an ingredient by hair from an aqueous solution is governed by its attraction or binding interactions to the keratin, its hydrophilicity or binding interactions to the aqueous phase, and the diffusibility of the ingredient into the hair.

For conditioning ingredients in shampoos and hair conditioners, Robbins et al. [56] have suggested that adsorption is more critical than absorption because the conditioning ingredients are relatively large species and low temperatures are employed, in contrast to wool dyeing where

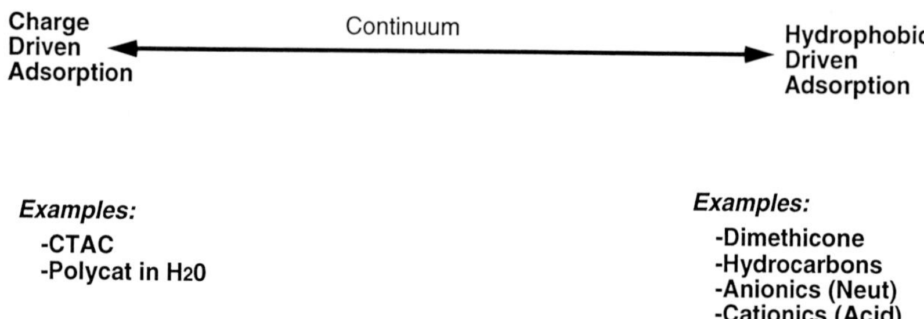

FIGURE 5–3. Schematic illustrating the hypothesis of a continuum between a charge driven and a hyrophobically driven mechanism for adsorption of conditioning agents to keratin surfaces.

diffusion is critical. Further, they have proposed a hypothesis that considers adsorption to hair in terms of a continuum between a charge driven adsorption process and a hydrophobically driven process (Figure 5–3). An example of what is essentially a purely charge-driven process is the adsorption of a water-soluble cationic surfactant such as dodecyl trimethyl ammonium chloride from an aqueous solution onto hair above the isoelectric point of hair. This adsorption process is driven by the attraction of the positively charged quaternary ammonium ion to the negatively charged hair fiber surface.

At the other end of the spectrum is the adsorption of a water-insoluble dimethicone onto hair from an anionic shampoo medium. This adsorption is driven by the fact that wet hair comes into contact with a medium in which insoluble silicone or another hydrophobic species is suspended in the system. The additional water from the wet hair and from rinsing perturbs the system, and adsorption occurs primarily because of entropy; that is, to keep the silicone suspended requires additional molecular organization (entropy), which can only be overcome by putting additional energy into the system. Thus, the silicone comes out of suspension and some of it migrates onto the water-insoluble hair. Although the primary driving force for this hydrophobic adsorption process is entropy, that is, the change in the orientation of molecules or larger species in the system, hydrophobic binding to the hair fiber surface is also involved.

This mechanism was presented as a continuum rather than as just two extremely different processes, because as we change the structures of the adsorbing species, the hair, or the solvent medium, we can see the mechanism moving toward a more charge-driven or a more hydrophobically driven process. For example, in current hair conditioners, as we change the quaternary ammonium species from a short-chain species such as dodecyl to cetyl or stearyl or even bihenyl, or as we move from monoalkyl quats to dialkyl quats, these structural changes cause the charge-driven process to take on more hydrophobic character. On the other hand, as we take water-insoluble dimethicone and add charge to it through amino groups, we decrease adsorption; however, the adsorption that takes place takes on some charge-driven character.

The binding interactions to keratin are influenced by the charge of the ingredient, its molecular size, the isoelectric point of hair, the pH of the surrounding medium, other salts or components in the formulation, and ingredients that are attached to the fiber surface. The attraction to the aqueous phase is governed primarily by its hydrophilic or hydrophobic character, which is determined by the ratio of nonpolar to polar substituent groups. If we are considering absorption, then we must consider diffusion rates, which are governed primarily by molecular size, condition of the hair, pH, and reaction temperature.

Because the isoelectric point of hair is so low, approximately 3.67 [57], its surface bears a net negative charge near neutral pH, at which most shampoos are formulated. Although anionic surfactants bind to the hair surface, the number of sites are comparatively small relative to sites for cationic ingredients. Lauryl and laureth sulfate salts and salts of olefin sulfonate are also moderately hydrophilic. They appear to rinse well (but not completely) from hair, and therefore serve as good cleaning agents. More anionic surfactant does bind to hair with decreasing pH, suggesting that low-pH shampoo formulations should leave more anionic surfactant behind after shampooing than neutral pH shampoos.

The diffusion of anionic surfactants into hair is also very slow, and it takes days for an average-size surfactant to completely penetrate cosmetically unaltered hair. Although some penetration of surfactant can and does occur, the major interactions of the surfactants of shampoos and conditioners occur at or near the fiber surface, near the first few micrometers of the periphery of the hair.

One objective of high-cleaning shampoos is to minimize sorption or deposition of its ingredients. On the other hand, the effects of conditioning shampoos and creme rinses are primarily from the adsorption of ingredients at or near the fiber surface. Soaps and surfactants, lipids, cationic ingredients, and even polymers or polymer association complexes (see Chapter 7) have been used as conditioning ingredients in shampoos and/or conditioning products. Soaps deposit their hydrophobic salts on the hair or bind by metal bridging. Cationic surfactants and polymers attach substantively to hair by ionic bonds enhanced by van der Waals attractive forces. The substantivity of most polymer association complexes probably results from their hydrophobic nature, enhanced by van der Waals forces and possibly ionic bonds.

Conditioners are analogous to conditioning shampoos in causing hair effects chiefly by the adsorption of ingredients to hair. Two of the primary active ingredients of a conditioner are stearyl benzyl dimethyl ammonium chloride (stearalkonium chloride) and cetyl trimethyl ammonium chloride (cetrimonium chloride).

More cationic than anionic surfactant binds to the hair surface above its isoelectric point, and cationics are difficult to remove by rinsing. As a result, cationic surfactants are said to be substantive to hair. Diffusion of cationic surfactants into hair is slow, similar to anionic surfactants. The more important interactions occur at or near the fiber surface (the first few micrometers of the fiber periphery), thus accounting for the low surface friction and the ability of creme rinses to make hair comb more easily (see Chapter 8). Most modern creme rinse conditioners contain a high concentration of a fatty alcohol such as cetyl alcohol or a similar fatty material in addition to a cationic surfactant. Dye binding studies suggest that these alcohols bind to hair along with the cationic ingredient (absorption maximum shifts), resulting in easier combing and more effective conditioning than by the cationic surfactant alone.

The condition of the hair also affects the uptake and the diffusion of creme rinse and shampoo ingredients. A rule of thumb is that diffusion is faster into altered or damaged hair than into unaltered hair. Bleaching (oxidation; see Chapter 4) also lowers both the isoelectric and the isoionic points of hair, thereby attracting more cationic surfactant to the hair.

Thus, although diffusion occurs more readily into cosmetically altered hair, the more important hair effects are produced by conditioner and conditioning shampoo ingredients binding at or near the fiber surface. Only in the case of severely damaged tip ends might internal binding be more important, and even here the distinction may be essentially semantic.

Transcellular and Intercellular Diffusion

Two theoretical pathways exist for diffusion into human hair [58] (Figure 5–4): (1) transcellular diffusion and (2) intercellular diffusion. The transcellular route involves diffusion across cuticle cells through both high and low cross-linked proteins. On the other hand, intercellular diffusion involves penetration between cuticle cells through the intercellular cement and other proteins that are low in cystine content (low cross-link density regions).

In the past, transcellular diffusion was the generally accepted route because of the much greater amount of surface area available for this type of penetration. Today, however, intercellular diffusion or diffusion through the nonkeratinous regions of the intercellular cement and the endocuticle (see Figure 1–19) is believed to be the preferred route for entry of most molecules (especially large ones such as surfactants or even species as small as sulfite near neutral pH).

Intercellular diffusion was first proposed as far back as 1937 by Hall

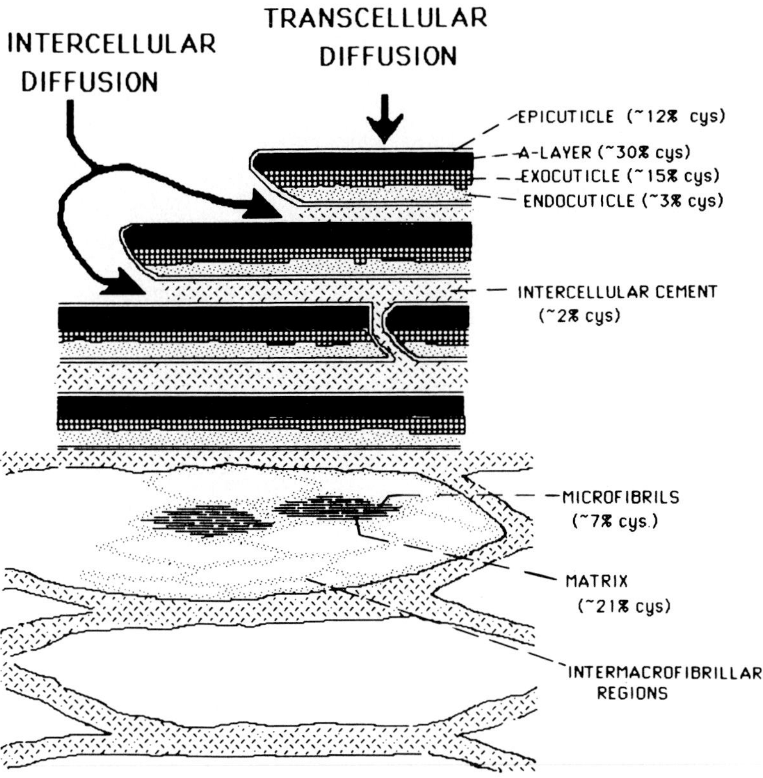

FIGURE 5–4. Schematic illustrating transcellular versus intercellular diffusion. Note the different histological regions are not drawn to scale.

[59]. It has recently been clearly demonstrated by Leeder et al. [60] for metal complex dyes; a large cationic dye (rhodamine B, 479 daltons); triphenyl pyrazine, a neutral molecule (311 daltons); and for the high molecular weight anionic oligomeric Synthappret BAP (>3,000 daltons).

Both diffusion routes probably can occur under the right circumstances, the right-sized molecule and the right solvent system. The intercellular route, however, is probably preferred in most instances, particularly for large molecules, because the low-sulfur, nonkeratinous proteins are more easily swollen than the highly cross-linked regions (see Chapter 1). However, for small molecules, transcellular diffusion under certain conditions might be the preferred route, especially if the highl; crosslinked A-layer and exocuticle are damaged.

For large metal complex dyes (>650 daltons), Leeder et al. [60] have demonstrated that intercellular diffusion of these materials occurs into wool fiber. Certain alcohols such as *t*-butanol, often considered as nonswelling solvents, have been shown by Jurdana and Leaver [61] not to

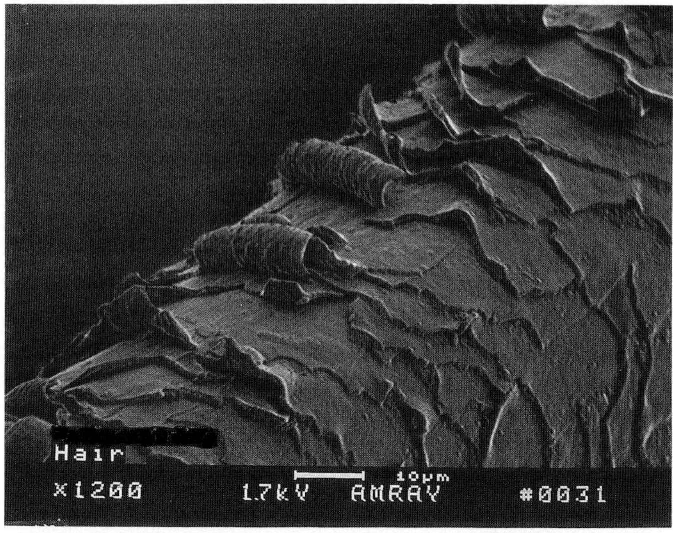

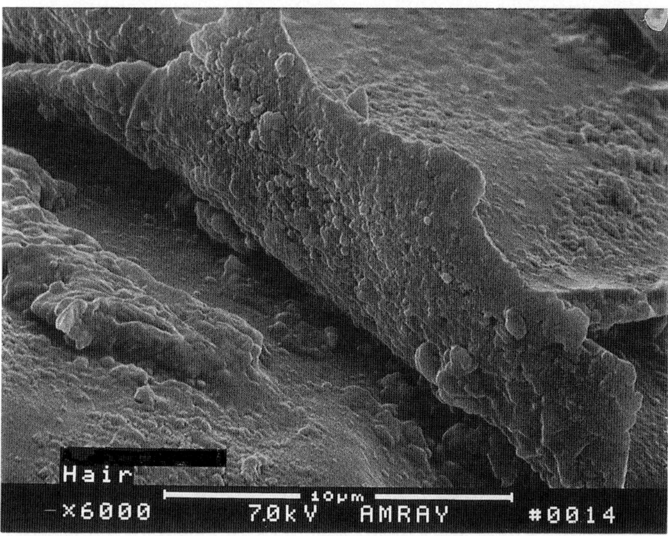

FIGURE 5–5. Lifted scales produced by penetration and deposition underneath the scales. The treatment involved alternating treatments of conditioner and shampoo ingredients.

penetrate into the cortex of wool, but to penetrate readily into the cortex of human hair via the intercellular regions.

It appears that for many dyeing processes and the penetration of other organic molecules into animal hairs, initially the surfactant, the dye or organic material, penetrates into the fibers through the cell membrane complex and into the endocuticle and intermacrofibrillar regions, and then during the later stages of reaction migrates into the more highly cross-linked exocuticle and A layer of the cuticle cells and the matrix of the cortex. Cosmetics companies have promoted the concept of penetration as positive. However, penetration into the intercellular regions of hair, underneath the scales, can degrade the nonkeratinous components [29,30] and even cause scale lifting, a form of hair damage (Figure 5–5).

Theory: Equilibria and Kinetics of Ionic Surfactant and Dye Interactions with Keratin Fibers

There are two thermodynamic quantities of pragmatic significance that characterize hair–surfactant interactions: the chemical potential (μ), and the heat of reaction (H).

The Chemical Potential (Affinity)

The chemical potential describes the tendency of solute, surfactant, or dye to move from solution to the fiber. It is analogous to a partition coefficient. Gibbs suggested the use of this parameter in place of the free energy for systems where the free energy has the disadvantage of depending on the amount of the system [62].

In actual practice, the change in the standard chemical potential ($\Delta\mu°$) is evaluated as the measure of the tendency of the solute to move from the solution to the fiber, that is, the "relative affinity" of the substance for the fiber relative to the solution phase. This parameter is generally called the affinity and is usually expressed as ion affinities instead of as molecular affinities (see Table 5–18, later in this chapter). One expression that describes the standard chemical potential, if the ion forms ideal solutions, is given here:

$$-\Delta\mu° = RT \ln(D)_f/(D)_s$$

where $R =$ gas constant (1.987 calories per degree mole), $T =$ absolute temperature ($0°C = 273.16$ K), $(D)_f =$ concentration of ion in the fiber, and $(D)_s =$ concentration of ion in solution. This equation suggests that the affinity of an ingredient for hair in an aqueous system is governed by the ratio of its binding attractions to the keratin and its hydrophilicity (the binding attractions to the aqueous phase). Perhaps the most commonly used expression for determining this parameter is the one that determines the affinities of free dye acids. When the fiber is half saturated with acid (about 0.4 mmoles acid per gram for wool and hair), the following expression applies:

$$-\Delta\mu^\circ = 4.6\ RT\ \text{pH midpoint}$$

Therefore, anion affinities may be calculated from the pH of the midpoint of the titration curve for keratin fibers and acids (see Table 5–3). For a more comprehensive treatment of this subject, including appropriate expressions for different experimental conditions, see Chapter 4 of the book by Vickerstaff [62], the paper by Lemin and Vickerstaff [63], and the paper by Han et al. [64].

The data of Table 5–18 show that anion affinities increase with increasing molecular dimensions. The same has been shown for cations. Ion affinities are generally independent of pH, and largely consist of the sum of the bond strength of the ionic attachment and van der Waals attractive forces, which can be very powerful in large molecules (see Chapter 7). Further, as van der Waals attractions increase, the hydrophobicity of the surfactant increases, further increasing the affinity of the molecule for keratin in an aqueous system.

Heat of Reaction

The heat of reaction of a surfactant or dye with a fiber is the other thermodynamic property that has practical significance. It describes the effect of temperature on equilibrium, that is, whether more or less of an ingredient combines with the fibers at equilibrium as the temperature changes. For cosmetics, the heat of reaction is not nearly as important as the chemical potential, because the change in the standard heat of reaction (ΔH°) with temperature over the narrow range of temperatures used in personal care products is comparatively small. The simplest procedure to determine ΔH° involves adsorbing a quantity of surfactant onto hair and then determining the amount of surfactant that desorbs at different temperatures. A plot of the logarithm of the concentration of desorbed surfactant (in solution) at equilibrium versus 1/temperature provides a straight line with slope of ΔH° [65]. Other methods for determining this parameter are described by Vickerstaff [66].

Oxidative Theories of Dyeing

Two primary models have been presented to account for the uptake of electrolyte by keratin fibers [66–69]. Both models consider keratin as an ion-exchange resin with positive and negative groups. The Gilbert-Rideal theory assumes that all ions are adsorbed by attachment to specific sites in the keratin, namely, the ionized carboxyl and amino groups. On the other hand, the Donnan membrane theory assumes the existence of an imaginary membrane between two phases, the solution and the fiber. The existence of a Donnan potential between the two phases then determines the partitioning of the ions between the fiber and the surrounding solution.

Both models appear to quantitatively explain the phenomenon of dyeing keratin fibers with ionic dyes, although there has been considerable controversy between supporters of each theory [62,70,71]. Oloffson [72,73] in a critical analysis of these two theories concludes that the Gilbert–Rideal theory provides the better fit to experimental data.

The objective here is to acquaint the reader with these two theories, to provide reference material if more information is desired [66–69], and to point out that most of the subsequent discussion considers interactions with specific sites in the keratin.

Kinetics of Ionic Reactions with Keratin Fibers

Summary of Reaction Steps

Reactions of hair fibers with solute in solution may be considered as a multistep process involving these phases:

1. Diffusion through solution
2. Adsorption or interaction at the fiber surface
3. Diffusion or transport into the fibers
4. Reaction at internal sites in the fibers.

Whenever diffusion through solution is rate determining, reactant concentrations are generally low, the rate is dependent on agitation, and the reaction is usually characterized by low activation energies (3–5 kcal/degree mole).

Adsorption at the "exposed" fiber surface is generally rapid for ionic ingredients, and the surface becomes filled (with respect to solute) during the first few minutes of reaction. Diffusion into the fibers is generally the rate-determining step of most hair fiber reactions and is usually characterized by higher activation energies (10–30 kcal/degree mole).

Ionic reactions are generally rapid and therefore not rate determining. However, reactions that involve breaking and formation of covalent bonds can sometimes be slower than diffusion into the fibers and therefore can be rate determining; for example, reduction of the disulfide bond by mercaptans at acidic pH.

The amount of an ingredient that penetrates into the fibers and the extent of penetration is governed by the following factors:

Reaction temperature
Molecular size
Cross-link density of the fibers
Fiber swelling
Reaction time

The rate of penetration generally increases with increasing temperature and fiber swelling, whereas it decreases with increasing cross-link density and molecular size of the penetrating species. Obviously, the extent of penetration increases with time.

Liquid water at room temperature can penetrate across the entire fiber in less than 15 min and in less than 5 min at 92°F [74], whereas more than 6 h is required for single fibers to equilibrate in a humid atmosphere, and even longer for a fiber assembly. Dyes like methylene blue (molecular weight, ~320) and orange II (molecular weight, ~350) generally require more than 1 h to penetrate through the cuticle layers to the cortex. Similar penetration times would be expected for typical anionic and cationic surfactants used in shampoos and hair conditioners.

Diffusion into Keratin Fibers

Diffusion processes may be considered as three types [75]: free or molecular diffusion, forced diffusion, and obstructed diffusion. Free or molecular diffusion applies to the transport of matter by random thermal motion. Forced diffusion involves transport by forces other than random molecular motion (e.g., pressure gradients within a fluid or electrical or magnetic fields).

Diffusion coefficients involving only free diffusion are called true or intrinsic diffusion coefficients; processes involving both free and forced diffusion are called mutual diffusion processes. Experimentally, one cannot usually evaluate free diffusion in kinetic studies on keratin fibers. Therefore, the usual practice is to apply equations derived from Fick's laws for free diffusion to data involving mutual diffusion. This provides apparent or approximate diffusion coefficients, instead of intrinsic diffusion coefficients, and compromises the fundamental significance or interpretations of these processes involving molecular motion, for example, the activation energies or entropies of activation.

In the remaining part of this book, no attempt is made to distinguish between free and mutual diffusion: the term "diffusion" is used loosely. For more comprehensive treatment of intrinsic and mutual diffusion, see the books by Crank [76] and Alexander et al. [77] and the review by Williams and Cody [75].

Fick's Laws of Diffusion

Fick's first law for unidirectional diffusion states that J, the flux (flow), is proportional to the gradient of concentration (*dc/dx*) [78].

$$J = -D(dc/dx)$$

This equation states that the flow of a substance through a surface perpendicular to its direction of movement is directly proportional to the rate its concentration changes with distance, (*dc/dx*), the concentration gradient. The proportionality constant D is the diffusion coefficient and has the dimensions of area per unit time usually expressed as cm^2/s.

Fick's second law for unidirectional diffusion may be derived from his first law [79], and it provides the fundamental differential equation of diffusion for an isotropic medium (similar properties in all directions):

$$\frac{\delta c}{\delta t} = D\frac{\delta^2 c}{\delta x^2}$$

Most kinetic studies of diffusion into keratin fibers employ equations derived from this form of Fick's law and provide approximate diffusion coefficients, assumed to be constant throughout the diffusion reaction. However, Crank [80] has provided equations for evaluating diffusion data under a wide variety of circumstances, including a variable diffusion coefficient described later in this chapter.

Experimental Approaches for Diffusion Study

The simplest experimental approach for fiber diffusion study involves periodic analysis of the decreasing concentration of solute in solution surrounding the fibers (a limited volume of solution). A second approach provides a constant concentration of solute, "infinite bath," and requires direct analysis of solute in the fibers. Diffusion equations have been developed by Crank and Hill and others for both experimental situations, and some of these are described next.

Diffusion into a Cylinder from a Solution of Limited Volume

Crank [81] described several equations for diffusion into a cylinder with changing solute concentration and a constant diffusion coefficient (D). One of these equations describes diffusion from a stirred solution of limited volume into a cylinder of infinite length:.

$$\frac{Q_t}{Q_\infty} = 2\left[\frac{2}{\sqrt{\pi}}\left(\frac{Dt}{r^2}\right)^{1/2} - \cdots\right.$$

where Q_t = amount of solute sorbed in time (t), Q_∞ = maximum sorption capacity of solute by hair, and r = fiber radius.

If a plot of Q_t/Q_∞ versus the square root of time is linear, then the latter terms of this equation (not depicted) may be neglected, and the foregoing expression applies. The approximate diffusion coefficient may then be calculated from the slope of the plot with a knowledge of the fiber radius. Weigmann [82] has shown that this equation describes the reaction of dithiothreitol with wool fiber.

A similar expression has been derived by Hill for diffusion into a semiinfinite solid [83]:

$$Q_t/Q_\infty = 2A\sqrt{Dt/\pi}$$

Hill has defined a semiinfinite solid as a tissue of irregular shape where no exact mathematical treatment is possible. Alexander and Hudson have shown that this expression applies to the diffusion of orange II dye into wool fabric [84]. A plot of Q_t/Q_∞ versus the square root of time should be linear with a slope of $2A\sqrt{D/\pi}$. The A term represents the total surface area of the fibers used in the experiment. (The variation of fiber surface area with diameter is described in Chapter 8.)

Diffusion into a Cylinder from an "Infinite Bath"

Vickerstaff [85] noted that equations describing diffusion into an infinite cylinder (e.g., hair) or into a plane slab (e.g., skin) from a constant solute concentration (infinite bath), assuming a constant diffusion coefficient, are of the following general form:

$$\frac{Q_t}{Q_\infty} = 1 - Ae^{-BK} - Ce^{-FK} - Ge^{-HK} \ldots$$

A, B, C, F, G and H are known constants; $K = Dt/r^2$ for the case of the infinite cylinder; and r equals the fiber radius [85].

In this instance, to determine the diffusion coefficient, simply carry out a sorption experiment to a fixed time (t) at a given temperature and agitation

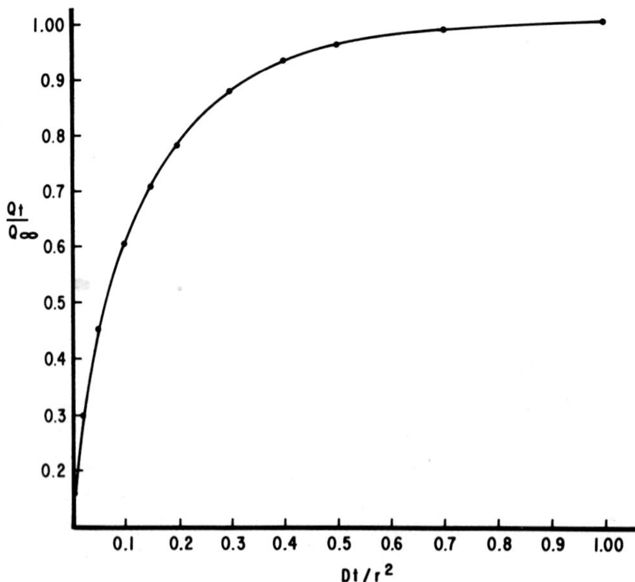

FIGURE 5–6. Diffusion into a cylinder from an infinite bath. A plot from data by Vickerstaff. For lower proportions of penetration, see the data and plots by Vickerstaff [85].

rate, and determine the amount of surfactant or dye sorbed by the fibers (Q_t). The value of Q_t/Q_∞ is calculated, and from the appropriate graph (Figure 5–6) the corresponding value of $K = Dt/r^2$ is determined. Since t, r, and K are all known, D may be calculated from $D = Kr^2/t$. Obviously, replication of the experiment and determination of an average value for D provides a more reliable estimate. Davis and Taylor [86] have used this procedure for the determination of diffusion coefficients for orange II dye into nylon fiber, and Holmes [87] has used a modification of this procedure for evaluating diffusion coefficients for diffusion of dyes into human hair.

The Case of a Variable Diffusion Coefficient

The diffusion equations described in the previous section have been derived from Fick's second law for unidirectional diffusion with the assumption that the diffusion coefficient is constant throughout the reaction. Crank [80] has also derived equations for evaluating diffusion data for systems with a variable diffusion coefficient that can be used to test one's data.

King [88,89] found that the transport of either water or alcohol through keratin fibers is an example of reactions with a variable diffusion coefficient. For the wool-water vapor system [88] at 25°C , the diffusion coefficent is approximately 10^{-9} cm^2/sec as the fiber approaches dryness. However, at high regains, it is of the order of 10^{-7} cm^2/sec. Theoretically, the diffusion coefficient for this system could approach the limiting value of 2.43×10^{-5} cm^2/sec, the diffusion coefficient for water in water [90]. The variable diffusion coefficient in this case is caused by changes in internal fiber structure during the reaction that involve changes in water binding with increasing regain.

In general, the penetration of solvents (which promote swelling) into polymers may be described as processes with a variable diffusion coefficient. For more comprehensive treatment of this subject, see the books by Alexander et al. [91] and Crank [80].

The Influence of Temperature on the Diffusion into Keratin Fibers

The activation energy (E_{ACT}) describes the effects of temperature on reaction rates. For example, the rate of a reaction with a higher E_{ACT} will respond more readily to temperature changes than one with a lower E_{ACT}. The activation energy can also help to distinguish between diffusion through solution and diffusion through the fibers; for example, an E_{ACT} of 3 to 5 kcal/degree mole generally indicates that diffusion through solution is rate limiting, whereas an E_{ACT} of 10 to 30 kcal/degree mole generally indicates that diffusion through the fibers is rate limiting.

The E_{ACT} is an important parameter in the collision theory of reaction rates, and it approximates the energy of activation in the transition state theory of reaction rates [92]. As indicated, diffusion reactions for keratin fibers generally involve mutual diffusion coefficients, because they involve transport components other than temperature. The theoretical interpretations in terms of molecular motions that follow assume no complications from electrical gradients and other factors of forced and obstructed diffusion that are undoubtedly involved in these interactions.

According to the collision theory of reaction rates, the effect of temperature on the rate of a chemical reaction is defined by the Arrhenius equation:

$$K = Ae^{-E_{ACT}/RT}$$

where K = the specific reaction rate, A = the preexponential function (entropy-related term), R = the gas constant (1.987 cal/degree mole), and T = the absolute temperature (0°C = 273.16 K).

The following two equations have been derived from the Arrhenius equation and are convenient for determining the E_{ACT} experimentally.

$$\log K = \frac{-E_{ACT}}{2.303\ RT} + \log A$$

$$E_{ACT} = \frac{RT_2T_1}{T_2 - T_1} 2.303 \log \frac{K_2}{K_1}$$

The first equation suggests that E_{ACT} may be determined by plotting the logarithm of reaction rates against $1/T$ and multiplying the slope by $-2.303\ R$. The second equation suggests that E_{ACT} may be evaluated by determining the rates of reaction at two different temperatures and calculating this parameter from the corresponding expression above.

The E_{ACT} for diffusion of water into wool fiber decreases with increasing water content, from 7.5 kcal/degree mole at lower regains to about 4.8 kcal/degree mole at 16% water content [88]. The E_{ACT} at higher regains is essentially the same as for the diffusion of simple solute molecules (sulfonate dyes) in water [93]. This suggests a two-phase system at higher regains, with water molecules diffusing through the aqueous phase within the fibers [88].

Activation energies for diffusion of the simple dye orange II into both human hair and wool fiber have been reported. Gilbert [94] has shown that the rate of diffusion of orange II into keratin fibers obeys the Arrhenius equation between 0 and 80°C. He has also reported activation energies of 28 and 23 kcal/degree mole for diffusion of orange II into human hair and wool fiber, respectively. Robbins found similar activation energies (at low pH) for the diffusion of this same dyestuff into human hair and merino wool (29 and 24 kcal/degree mole, respectively). The higher activation energy for diffusion into human hair is probably related to its higher cross-link density.

In addition, Robbins and Scott [95] have found that the E_{ACT} for diffusion of orange II into merino wool is pH dependent, decreasing from 24 to 11 kcal/degree mole with increasing pH from 1 to 7. Robbins and Fernee [96], while studying the swelling of stratum corneum by anionic surfactant as a function of pH, suggested that ionic bonding dominates the reaction between keratin and anionic surfactant at pH 1, whereas hydrophobic bonding between surfactant and keratin is more important near neutral pH. Therefore, if this reaction between ionic surfactants or dyes and hair is predominately ionic in character at acidic pH, and the reaction near neutral pH involves a greater amount of hydrophobic character, the activation energy should show a corresponding change, as found.

The temperature-independent term of the Arrhenius equation (A, or the preexponential function) is generally considered to be analogous to the entropy of activation of transition state theory [78]. Robbins has found that this parameter varies from 10^4 to 10^{-1} cm^2/sec for the diffusion of orange II into merino wool at pH 1 and pH 6.7, respectively. Hudson [97] has reported a value of 10^{-2} for this parameter at an unspecified pH. If the analogy of the preexponential function and the entropy of activation holds for this mutual diffusion process, then the diffusion of anionic dye or surfactant into keratin fibers requires an entropy of activation that increases with decreasing pH. This suggests that there is less precise orientation in the activated state at low pH than at neutral pH for the diffusion of anionic dye or surfactant into keratins. Thus, the reaction of anionic surfactant with keratin that is dominated by hydrophobic bonding (near-neutral pH) requires a higher degree of molecular orientation than the ionic reaction at acidic pH.

Molecular Size and the Concept of Pore Size

Speakman [98,99] theorized that keratin fibers consist of a solid containing holes or pores. Although the cell membrane complex is not actually holes, with some imagination, one can visualize this region of entry into the fibers as not too far removed from Speakman's proposal. This concept suggests that the rate of diffusion into a fiber containing holes depends on the effective molecular radius of the diffusing species and on the size and frequency of holes in the solid. Apparently the size of the holes will depend on the swelling medium and reaction conditions employed.

Assuming this theory to be valid, Holmes [87] investigated the size of these holes in human hair via a dye diffusion study in 0.1 N hydrochloric acid, and suggests that they are approximately 15 Å in diameter. Wilmsmann [100], on the other hand, has attempted to determine the relationship between molecular size of cationic dyes and their penetration into human hair by microscopic observation of fiber cross sections that were previously dyed. Although his reaction conditions were limited (30 min at 36°C in strong alkali), Wilmsmann observed that none of the larger species of triphenyl methane dyes penetrated into the cortex, whereas the smaller aromatic diamines did. He concluded that there is a hindrance to the penetration of larger molecules, and the largest diamine that he examined, 4-amino diphenylamine, which has a corresponding molecular diameter of 6 Å, is near the critical molecular diameter for penetration.

The apparent discrepancy between the conclusions of Holmes and Wilmsmann probably stems from the use of a qualitative analysis for short reaction times by Wilmsmann versus a quantitative analysis for longer reaction times by Holmes, in addition to different hair and different experimental conditions in general. However, Wilmsmann's results do provide a feel for the reaction times and size requirements for the penetration of cationic ingredients through the cuticle, and Holmes' data suggests that any ingredient that is approximately spherical or larger in all dimensions than 15 Å may experience a slow rate of penetration into hair at low pH.

Cross-Link Density and Diffusion Rate

Table 5–14 describes the influence of cross-link density in different keratin fibers on diffusion rate. These data clearly show that the rate of diffusion into keratin fibers decreases with increasing cystine content and therefore with increasing cross-link density. One may conclude that reactions that decrease the cross-link density of hair (e.g., bleaching) will lead to hair that is more rapidly penetrated, and its penetrability will increase with increased bleaching. Decreasing cross-link density obviously increases the rate of transcellular diffusion.

The Binding of Ionic Groups to Hair

The interactions of ionic ingredients (i.e., acids, alkalies, and neutral salts) with keratin fibers are of major importance to shampoos, creme rinses, ionic conditioners, and the group of hair dyes referred to as rinses. In this section, these interactions are described:

TABLE 5–14. Cross-link density and diffusion rates.

Type of keratin fiber	Percent cystine calculated from percent sulfur[a]	Relative diffusion [101] coefficient at 60° using orange II dye
Human hair	14.0	1.0
80's Merino wool	11.3	1.9
6's Mohair	9.2	3.4
56's Down wool	8.8	5.0

[a] Calculated from percent of sulfur, assuming all sulfur exists as cystinyl residues, and a residue weight of 178 daltons.

Hydrogen ion interactions
Hydroxide ion interactions
Interactions of salts near neutrality with keratin fibers.

This section considers the hypothesis that ionic interactions with hair may be partly represented (at low or high pH) as hydrogen ion or hydroxide ion interactions with hair, even though most of us are concerned with the combination of surfactants or dyes with hair. This approach becomes more palatable when one considers that for every hydrogen ion or hydroxide ion that interacts with hair, an accompanying anion or cation must also interact to maintain electrical neutrality. The counterion that combines with the fibers is determined by its affinity for the hair and by its concentration relative to competing counterions. Hydrogen ion interactions are most important when only simple inorganic cations (e.g., sodium or potassium) are present. These ions have a low affinity for hair relative to hydrogen ion and therefore compete most effectively for sites on hair at higher pH values, that is, at low hydrogen ion concentrations.

The acid or hydrogen ion combinations are described in this manner: the combination of simple acids (hydrochloric and ethyl sulfuric) with hair; the influence of anions on the combination of hydrogen ions with hair; and the combination of low molecular weight organic acids with hair.

Hydroxide ion interactions are essentially a mirror image of the hydrogen ion interactions and are described in three analogous sections.

The interactions of salts near neutrality consider mechanisms of interactions near pH 7 that are somewhat different from those of low and high pH.

Hydrogen Ion Interactions with Keratin Fibers

The Maximum Acid-Combining Capacity

The maximum acid-combining capacity of keratin fibers, from reaction with simple acids such as hydrochloric, phosphoric, or ethyl sulfuric acids, is approximately 0.75 mmole/g for unaltered human hair and, about 0.82 mmole/g for wool fiber [102]. This value approximates the number of dibasic amino acid residues in the fibers [103], that is, the combined amounts of arginine, lysine, and histidine (Table 5–15). The primary sites for interaction with acid (protons) are probably the carboxylate groups of aspartic and glutamic acids (ionized by interaction with the dibasic amino acid residues) and the dibasic amino acid groups themselves.

This acid–base reaction involves protonation of a basic site on/in the fiber forming a positive charge on the fiber that attracts a negative ion to it. Steinhardt and Harris [102] have shown that the uptake of chloride ion by wool corresponds to the uptake of hydrogen ions during reaction with hydrochloric acid, and Robbins has shown the same to be true for human hair.

TABLE 5–15. Data on the acid-combining capacity of unaltered hair and wool.[a]

From 0.1 N HCl	Orange II combined from formic acid	Orange II combined from 0.1 N HCl	Arginine + Lysine + Histidine
Human hair			
0.77	0.67–0.77 [105]	—[b]	0.81 [104]
0.82 [121]			
0.87–0.91 [106]			
Wool fiber			
0.8–0.9 [112,126]	0.81 [105]	0.82–0.85 [107]	0.88 [104]
	0.83 [103]	0.96	

[a] Data expressed as mmole/g dry hair.
[b] Because of competing hydrolysis, etc., reliable values for equilibrium could not be obtained. For example, at 19 days at 50°C, uptake values for equilibrium could not be obtained because the hair was still picking up dye. Perhaps longer dyeing times and extrapolation would provide a reasonable estimate of equilibrium; however, the reaction was stopped after 19 days.

Maclaren [103] has taken advantage of this counterion effect and has developed a test for the acid-combining capacity of keratin fibers by measuring the uptake of the anion of orange II dye (*p*-hydroxy-1-naphthyl azobenzenesulfonic acid) from formic acid solution. Robbins et al. [104] have used this test to study the variation in the acid-combining capacity of hair among individuals, by cosmetic treatments, and from environmental effects.

Variation in the Acid-Combining Capacity of Unaltered Hair

Hair samples were collected from 20 female Caucasians aged 10 to 30 years who had never bleached, dyed, or permanent-waved their hair. These hair samples were analyzed by Maclaren's method for the acid-combining capacity. The average uptake was 0.70 mmole/g. Analysis of variance indicated significant differences among these hair samples beyond the $\alpha = .01$ level.

Variation in the Acid Combining Capacity of Altered Hair

Bleaching

Bleaching decreases the acid-combining capacity of both human hair [105,106] and wool fiber [107]. Analysis of hair samples bleached to different extents shows that the acid-combining capacity decreases with increased bleaching [105] (Table 5–16).

Amino acid analysis of these same hair samples showed no change in the basic amino acid residues. Therefore, the decrease in acid combination must be caused by the formation of cysteic acid in the fibers, which forms a strong ionic bond with the basic amino acid residues and inhibits their

TABLE 5–16. Acid-combining capacity of bleached hair.

Sample description	Acid-combining capacity (Mmole/g hair)[a]	Cysteic acid (Mmole/g hair)
Control (unbleached)	0.67	0.03
One bleach	0.60	—
Two bleaches	0.52	—
Three bleaches	0.48	—
Four bleaches	0.43	—
Frosted hair	0.30	0.66

[a] Determined by the method of Maclaren [103].

interaction with weaker acids, such as formic acid, thus decreasing the uptake of orange II dye.

Because the exocuticle and its A layer (see Chapters 1 and 2) are highly cross-linked with cystine [108] and are near the fiber surface, one would expect a large increase in cysteic acid in the cuticle and, in all probability, a decrease in the isoelectric point of hair. This should produce a decrease in acid dye combination and an increase in the combination of cationic substances at or near the fiber surface with increased bleaching or oxidative weathering.

Permanent Waving

Sagal [106] has suggested that the acid-combining capacity of hair increases with permanent waving. Robbins [105] could not find a change in the acid-combining capacity of human hair waved under "normal" conditions on live heads. Both of these studies involved determinations on whole fiber. A "surface" analysis method might be more sensitive to such a difference, if it actually exists.

Other Chemical Modifications to Hair

Modification to the number of acidic and basic groups have been made by Laden and Finkelstein [109], who added Bunte acid groups to hair, and by Robbins and Anzuino [110], who added polydimethylaminoethyl methacrylate groups to hair by in situ polymerization.

Weathering

Robbins [111] has shown that the acid-combining capacity of human hair decreases with weathering, although only to a small extent. This study involved a comparison of root and tip ends of five samples of long hair (longer than 18 in.) that was visually lighter in the tip ends than the root ends. The acid-combining capacity varied from approximately 3% to 13% less in the tip ends. The most severely affected sample was hydrolyzed and analyzed for amino acids and found to contain significantly less lysine and

TABLE 5–17. Acid-combining capacity of root and tip ends of human hair.

Sample description	Difference (tip minus root, %)
A	−9.9
B	−2.9
C	−12.9
D	−6.0
E	−2.9
	Average = −7.0%

	Root ends	Tip ends
Total basic amino acids[a]	0.75	0.69
Acid-combining capacity[b]	0.70	0.61
Cysteic acid[a]	0.02	0.04

[a] By hydrolysis and amino acid analysis [105]; data are mmole/g hair.
[b] By method of Maclaren [103].

histidine and a larger amount of cysteic acid in the hydrolyzates of the tip ends (see Table 5–17). This result is presumably from photochemical degradation.

Reaction Conditions and the Combination of Hydrogen Ions with Hair

Most of the subject matter in the following section has been studied thoroughly for wool fiber and confirmed in a few critical experiments with human hair.

Reaction Temperature

Steinhardt et al. [112] have studied the effect of temperature on the reaction of wool fiber with hydrochloric acid from 0°C to 50°C and found only small differences in the titration curve over the pH range where acid combines with wool. Heats of dissociation, from their titration data, are only 2,500 calories, in good agreement with those for the back-titration of carboxyl groups of simple acids and of soluble proteins.

pH and the Isoionic Point

The pH at which a protein or particle has an equivalent number of positive and negative charges as determined by proton exchange is the isoionic point. The pH at which a protein or a particle does not migrate in an electric field is called the isoelectric point. The isoionic point is a whole fiber property of hair and is reflected by the equilibrium acid-base properties of

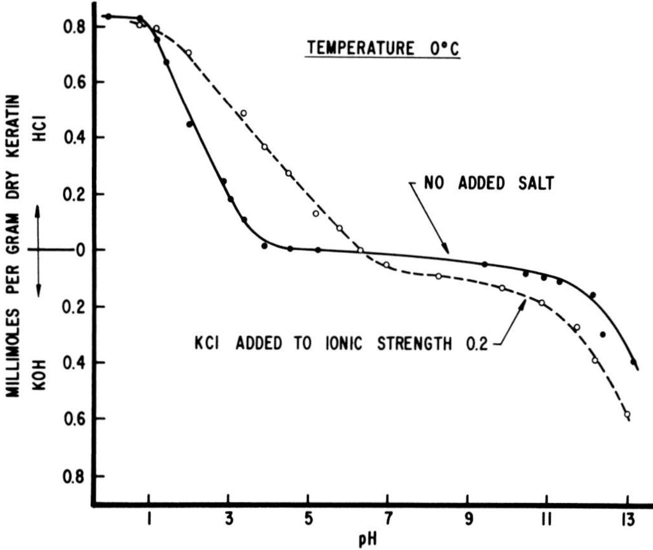

FIGURE 5–7. The influence of salt on the combination of simple acids and base with keratin fibers. (From data by Steinhardt and Harris [102].)

the total fiber; the isoelectric point is related to the acid-base properties of the fiber surface.

The isoionic point of human hair may be evaluated from titration data in the presence of salt (see Figure 5–7) or buffers. It may also be approximated by allowing thoroughly rinsed hair to equilibrate in deionized water and determining the pH of the resultant solution.

The isoionic point of human hair is close to that of wool fiber (generally near pH 6.0) and it varies among hair of different individuals. Freytag [113] has found isoionic points from pH 5.6 to 6.2 by following the pH changes of hair in buffer solutions, and an isoionic point of pH 5.8 ± 1.0 was found for unaltered hair from nine different individuals in a study by Robbins.

Wilkerson [114] found the isoelectric point of a single hair sample to be pH 3.67 by measuring the electrophoretic mobility of hair particles in buffer solutions. The isoionic point of wool fiber has been determined to be at pH 6.4 [102] and the isoelectric point between pH 3.4 and 4.5 [115,116].

Similar isoelectric points for hair and wool fiber are to be expected, because chemical compositions of the cuticle (see Chapter 2) are similar and both fibers show similar dye-staining characteristics. Cuticle from both fibers stains more readily with cationic dyes than with anionic dyes [94], whereas the cortex stains readily to anionic dyes [117].

Because bleaching increases the ratio of acidic to basic amino acids

[118], the isoionic point should decrease with increasing oxidation. One might also anticipate a similar decrease in the isoelectric point of hair with bleaching as the A layer of the cuticle cells is rich in cystine.

For longer term interactions, if the pH of the surrounding solution is less than the isoionic point of hair, the hair will pick up acid, above its isoionic point, it will attract hydroxide ions more readily. For short-term and surface interactions, the isoelectric point is more important than the isoionic point. The isoionic point becomes more important to whole-fiber treatments such as perms and bleaches.

In the absence of added salt in the pH region of 4 to 9, there is negligible combination of simple acids or alkalis such as hydrochloric acid or sodium hydroxide with wool or hair [102]. This phenomenon is not observed with soluble proteins and seems strange, because unbuffered solutions (near neutral pH) in the presence of hair free of acid or alkali drift toward the isoionic point of hair. The explanation is that in the presence of small solution-to-hair ratios (100:1 or less), the consumption of relatively small amounts of alkali or acid by the hair will provide a significant pH drift in the solution. In addition, in most cases, salts are present in the solution leading to greater interaction.

The Influence of Anions

Table 5–18 illustrates anion affinities of several acids [119] and shows that simple anions like chloride and ethyl sulfate have very low affinities for keratin, whereas surfactant anions, such as dodecyl sulfate or dodecyl sulfonate, and dye anions have relatively high affinities. In fact, the anion affinities of Table 5–18 show a correlation ($r = .94$ and $r^2 = .90$) with molecular weight, suggesting that 90% of the variance can be explained by molecular weight. Because most of these anions differ primarily by increasing size of either aliphatic or aromatic substituents, this type of

TABLE 5–18. Molecular weight and anion affinities of simple acids.[a]

Acid	Molecular weight	pH Midpoint	Total affinity, $\mu°$ (kcal)	Anion affinity[b] (kcal)
Hydrochloric	36.5	2.32 (0°C)	5.8	0.5
Ethyl sulfuric	126	2.33 (0°C)	5.8	0.5
Isoamyl sulfonic	152	2.58 (0°C)	6.4	1.1
Benzene sulfonic	158	2.63 (0°C)	6.6	1.3
Octyl sulfuric	210	3.47 (25°C)	9.1	3.8
Dodecyl sulfonic	250	4.02 (25°C)	11.1	5.8
Dodecyl sulfuric	266	4.08 (25°C)	11.0	5.7
Orange II	328	4.63 (25°C)	12.6	7.3

[a] Calculated from the pH midpoint titration data of Steinhardt et al. [119].
[b] Calculated assuming the hydrogen ion affinity to be 5.3 kcal [120].

affinity may be associated with van der Waals interactions. Therefore, the decreasing hydrophilic nature and increasing keratinophilic nature of these organic acids with molecular size cause the acid to partition from the aqueous phase to the keratin phase.

Anions of Low Affinity

The effect of increasing chloride ion concentration in hydrochloric acid solution is to produce a greater uptake of acid by the fibers at any given pH below the isoionic point. Steinhardt and Harris [102] have demonstrated this effect for wool fiber; Robbins has demonstrated it for human hair (Table 5–19; see Figure 5–7).

When one considers that essentially equivalent quantities of hydrogen and chloride ions combine with the fibers, it is apparent that the extent to which either one of these ions is taken up by the fibers will influence the other. The hydrogen ion has a much greater influence on the combination of chloride ion with the fibers than chloride has on hydrogen, because hydrogen ion has a greater affinity for keratin (see Table 5–18). However, because chloride ion does have some affinity for keratin, increasing its concentration in solution does increase the combination of chloride, and ultimately hydrogen, ions with hair or wool fiber.

Anions of High Affinity

A considerably greater amount of acid combines with hair or wool fiber, at any given pH, in the presence of anions of high affinity for keratin [121]. In fact, the extent of combination at low pH (pH 2.5 or lower) can be well in excess of the maximum combining capacity. This suggests that interaction between groups other than carboxylate or dibasic amino acid groups occurs. Interaction of the hydrophobic portions of the fibers with the hydrophobic group of the surfactant is involved, and protonation of amide groups has been suggested.

TABLE 5–19. Influence of salt on the combination of acid with keratin fibers [102].

	Wool fiber, 25°C[a]		Human hair, 25°C[a]	
pH	no salt	Ionic strength, 0.2	no salt	Ionic strength, 0.2
1.0	0.78	0.83	—	—
2.0	0.44	0.73	—	—
3.0	0.15	0.51	0.29	0.46
4.0	0.03	0.29	—	—

[a] Data are expressed in mmole/g dry keratin and are interpolations from graphs from foregoing references. Added salt is potassium chloride.

Competition of Cations with Hydrogen Ions

Cations of Low Affinity

In neutral dyeing or surfactant–hair interactions, competition of cations with hydrogen ions must play a role. When the concentration of hydrogen ions is low and cations of low affinity are present, the adsorption of anion is influenced by the concentration and affinity of cations for hair. If the cation affinity is high enough so that it is adsorbed, a counterion must accompany it to maintain electrical neutrality. In the presence of low-affinity cations, such as, sodium or potassium, hydrogen ions can be taken up until quite high pH values are reached [122]. However, competition between hydrogen ions and other cations will occur.

Cations of High Affinity

Long-chain quaternary ammonium compounds must have a high affinity for human hair, because they compete quite effectively with hydrogen ions for sites on hair at acid pH in many creme rinse formulations, and they are difficult to completely remove from hair with anionic surfactants.

Low Molecular Weight Organic Acids

The data of Table 5–20 suggest that the interactions of low molecular weight carboxylic acids with hair involve more than simply the back-titration of the carboxylate groups. Many of these acids are relatively weak (e.g., acetic, propionic, and butyric), and relatively high concentrations are required to achieve hydrogen ion concentrations approaching 0.1 *m*,

TABLE 5–20. The interaction of carboxylic acids with human hair.[a]

Acid used	pH	Concentration (%)	Swelling (%)
Water	7.0	100.00	32
Hydrochloric	1.0	0.36	34
Formic	1.1	25.0	47
Formic	0.4	50.0	62
Formic	—	98.0	110
Acetic	1.15	50.0	47
Propionic	2.6	5.0	33
Propionic	1.4	75.0	54
Butyric	2.6	5.0	34
Butyric	1.65	75.0	46
Monochloroacetic	1.1	50.0	47
Trifluoroacetic	1.9	0.5	34
Trifluoroacetic	—	25.0	50
Trifluoroacetic	—	75.0	110

[a] Data from Barnett [124] (at 24 h).

the concentration of hydrochloric acid required for its maximum combining capacity. However, these acids at hydrogen ion concentrations well below 0.1 N produce extensive swelling, suggesting that the undissociated acid itself combines with the fibers [123].

In the case of pure formic acid, extreme swelling results. However, the attraction for positive sites on the fibers must be greater for the anion of orange II dye than for formate ion, for Maclaren's acid-combining test [103] to be valid. For additional discussion concerning this type of interaction, see Barnett's thesis [124] and articles by Speakman and Stott [125,126].

Hydroxide Ion Interactions with Keratin Fibers

Maximum Alkali-Combining Capacity

The maximum alkali-combining capacity of keratin fibers from reaction with simple alkalis (e.g., potassium hydroxide) has been reported at 0.44 mmole/g for unaltered human hair [106] (no correction for decomposition) and at 0.40 mmole/g for wool fiber [102]. This reaction involves the back-titration of the conjugate acids of amino and guanidino groups of the fibers, forming negative sites that attract cations.

Variation in the Alkali-Combining Capacity

Keratin is more sensitive to alkaline hydrolysis than to acid hydrolysis, making this determination more difficult and complicated. Sagal [106] has shown a higher uptake of alkali in bleached hair and in permanent-waved hair than in cosmetically unaltered hair. This effect could be caused by a larger number of acidic sites; however, it is more likely due from an increased susceptibility of damaged hair to hydrolysis.

Influence of Cations on the Combination of Hydroxide Ions with Keratin Fibers

Quantitative cation affinities have not been determined for human hair; however, Steinhardt and Zeiser [123] have shown that for a series of quaternary ammonium halides, as with anions, the affinity for wool keratin increases with increasing molecular dimensions.

Organic ions of small size (<150 daltons) differ very little in affinity and are similar to inorganic alkali metal cations, but for ions larger than 150 daltons the affinity of organic cations increases rapidly. The high affinity of hexadecyltrimethylammonium and larger cations results from electrostatic attractions, van der Waals attractive forces, and the relatively low hydrophilic nature of the molecule. Scott et al. [127] have shown a similar phenomenon for human hair by comparing the sorption behavior of hexadecyl- and dodecyltrimethylammonium bromides. These scientists found that under similar conditions of adsorption and desorption, greater amounts of the larger hexadecyltrimethylammonium bromide combined with hair, attesting to its greater affinity.

Cations of low affinity, at high concentrations, increase the interaction of hydroxide ion with keratin. This has been described by Steinhardt and Zeiser [123] as an effect of salt on the base-binding behavior of keratin. Cations of high affinity produce a greater effect in increasing the interaction of hydroxide ion with keratin [123].

Low-Molecular Weight Organic Bases

Similar to the interactions of low molecular weight carboxylic acids with hair, the interactions of low molecular weight organic bases involve more than simply the back-titration of conjugate acids. Barnett [124] has described the interaction of mono-, di-, and triethanol amines at 25% concentration and higher with human hair (see Table 5–20). The reactions of these species with hair involve extensive swelling and ultimately lead to decomposition and disintegration of the hair.

Interactions of Salts near Neutrality with Keratin Fibers

The interactions of surfactants and ionic dyes with keratin fibers, near neutral pH (5–8), have not been studied as thoroughly as acid and basic dyeing. However, Vickerstaff [128] suggests that the mechanism for neutral dyeing is analogous to the action of surface-active agents at an air–water interface, where they orient with their hydrophobic tail extending into the air and the hydrophilic group in the water. Another analogy is the electrophoresis of proteins in sodium dodecyl sulfate, where the hydrophobic portion of the surfactant binds to the protein and the charged group projects toward the solvent or gel.

Thus, a mechanism for neutral interactions of surfactants with keratin fibers depicts the surfactant attaching to the fiber by its hydrophobic tail and the hydrophilic group (e.g., the sulfonate group) projecting toward the solution [129]. Evidence for a change in mechanism for the binding of surfactants to keratin as the pH of the system changes from acid to neutral has been provided by Robbins and Fernee [96] (see the discussion earlier in this chapter).

A "leading ion mechanism" has also been proposed by Peters [122] for interactions near neutrality. For this mechanism, the fiber surface bears a net negative charge because of its low isoelectric point (pH 3.7). Positively charged ions are attracted to the negatively charged surface, thus helping it to overcome the electrical barrier for anions. This view elevates the importance of the counterion (cations in particular) in neutral dyeing or surfactant binding to hair near neutral pH. The effect of salt addition on dye uptake is consistent with this mechanism, because the addition of electrolyte near neutral pH increases the amount of dye [130] or surfactant [131] that combines with keratin fibers. Since anion and cation affinities are independent of pH [63], surfactants and dyes with high affinities bind readily to keratin fibers even near neutrality. As mentioned before, a convenient rule of thumb is that anion or cation affinities increase with increasing molecular size of the organic moiety.

Most surfactant interactions with hair are above the critical micelle concentration, and aggregation introduces complexities to these mechanisms. However, because sorption of sodium lauryl sulfate continues to increase above the critical micelle concentration [132], higher concentrations of aggregate near the fiber surface may be capable of providing higher concentrations of monomer for diffusion into the fiber, for it is probably monomer rather than aggregate that diffuses into the fiber. Interestingly, nonionic surfactant has been shown to decrease the sorption of sodium lauryl sulfate, probably by decreasing the concentration of monomer available at the surface. Ethoxylation to sodium lauryl sulfate decreases the sorption too, although it is not clear whether this action is simply an effect on diffusion rate or on the anion affinity or both of these parameters.

Damage to Hair from Shampoos, Grooming, and Weathering

Hair damage is the breakdown or removal of structural components or parts of hair that either weaken it or make it more vulnerable to chemical or mechanical breakdown as occurs in shampooing and everyday grooming actions. Sunlight, pool water, and cosmetic products such as perms, bleaches, straighteners, and some hair dyes chemically alter hair and increase its propensity to further chemical and mechanical breakdown, as evidenced by an increased sensitivity to cuticle abrasion/erosion and fiber splitting [133].

Shampoos can damage the hair, primarily the cuticle, in different ways. They can damage hair by abrasion/erosion, both during the shampoo process itself, when hairs are rubbed against each other while lathering [134], or while towel drying or even when blow drying. Shampoos can also slowly dissolve or remove structural lipids and proteinaceous material from hair. Further, every time a person shampoos or conditions their hair, they either comb or brush it. Therefore, combing and brushing of hair and the resultant damage should be considered a part of shampoo and hair-conditioning damage, especially since some shampoos condition hair making it easier to comb and brush.

Okumura [134] suggested that a large amount of cuticle damage occurs in the lathering step during the actual shampooing of hair when fibers are rubbed against each other in the presence of detergents. Kelly and Robinson [135] have concluded that shampooing and towel drying of hair also damages hair. However, these scientists conclude that combing and brushing damages hair more than the lathering step of shampooing, and further that brushing is more damaging than combing. They have also shown that cuticle loss is greater from wet combing than from dry combing.

These shampooing/grooming actions all cause the cuticle to be more susceptible to further abrasion/erosion and the lifting of scales and other types of hair damage that are described in this section. These actions can also lead to increased diffusion of chemicals into hair and to additional damage by penetrating chemicals or products.

Shampooing, combing and brushing, and exposure to sunlight over time induce changes in hair that can be detected at the morphological level. These effects may be viewed as aging of hair (not the person) or of weathering damage. Weathering effects include damage to hair by environmental factors such as sunlight, air pollutants, wind, sea water, or even the chlorine in pool water. Several types of the following different actions produce rubbing of hairs against hairs or other objects that result in hair damage: combing and brushing, shampooing (during both the lathering and drying steps, including towel drying or blow-drying of hair), rubbing hairs during styling, such as curling, braiding, and tying or clamping hairs together frequently in the same spot in a bun or a knot, and rubbing hairs against other hairs while turning one's head during sleeping or lying down. All these rubbing actions, except the very last one, are a part of the hair-grooming process.

The process of cuticle chipping that results by rubbing objects such as grooming devices and even other hairs against hair fibers is a major factor in hair damage (Figure 5–8; also see the discussion in Chapter 1 on the different stages of cuticle wear over time and Figure 1–18). As indicated earlier, hair damage can be produced by either stretching or bending, by rubbing hairs, by chemical action, or even by penetration underneath the scales (intercellular route; see Figure 5–4). For example, the lifting of scales can be produced either by stretching or by penetration underneath the scales (see Figures 5–9 and 5–5, respectively). That this type of scale lifting can occur from combing hair tresses is illustrated by Figure 5–10. Removal of large sections or chunks of a single scale also results from rubbing action, particularly after scales have been raised (Figures 5–10 and 5–11).

Extraction or erosion can also remove lipid and even proteinaceous matter leaving the hair feeling dry and more susceptible to further damaging actions. Large segments or sections of scales can also be abraided, torn, or ripped from the hair (Figure 5–12). This type of damage is most likely to occur from back-rubbing such as by rubbing one hair against another as can occur in a tangle or a knot or even from back-combing or teasing.

The uneven removal of scale sections also occurs by continuously rubbing against the same area on a fiber as occurs in a twisted or noncircular fiber with a "high spot" (high region). This type of effect can occur even when the rubbing forces are extremely low, as for example by sliding a hair fiber under its own weight (only 0.58 mg) continuously (about 50 times)

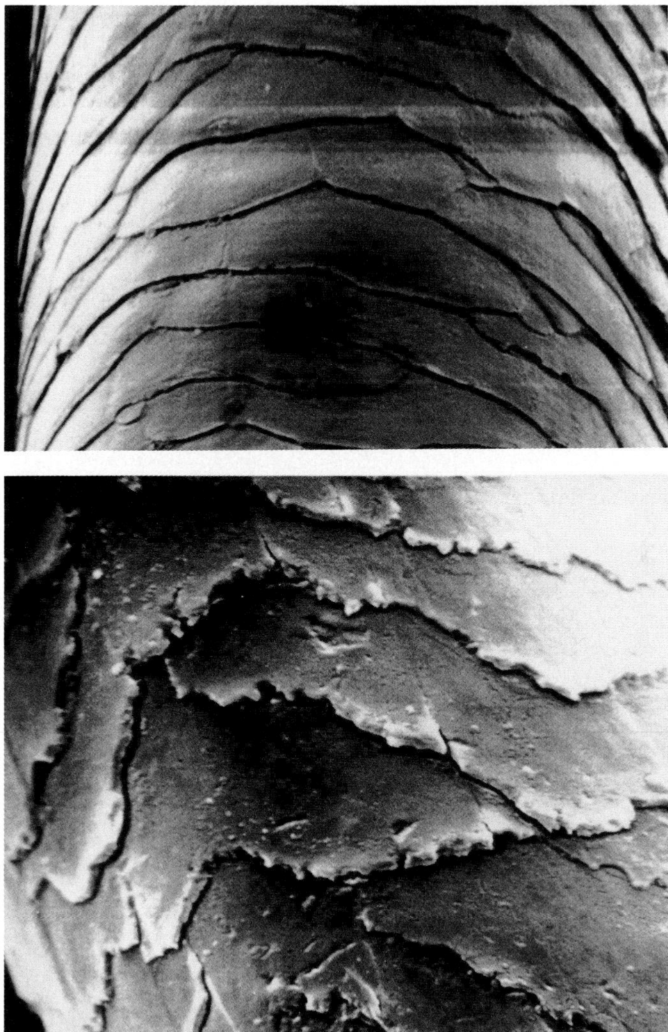

FIGURE 5–8. SEM illustrating damage from the chipping of scale edges. The top SEM illustrates the fiber surface close to scalp with virtually no damage; note the smooth scale edges and smooth scale faces. The bottom SEM illustrates the fiber surface about 6 in., down representing about 1 year of growth and wear. Note the rough scale edges where small chips or cuticle fragments have been broken off.

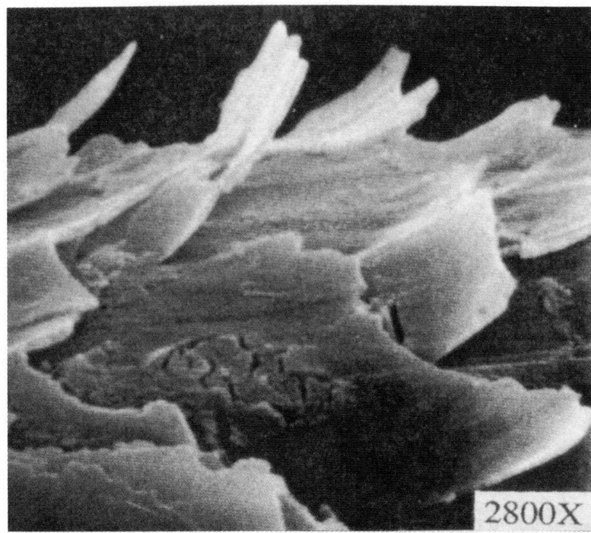

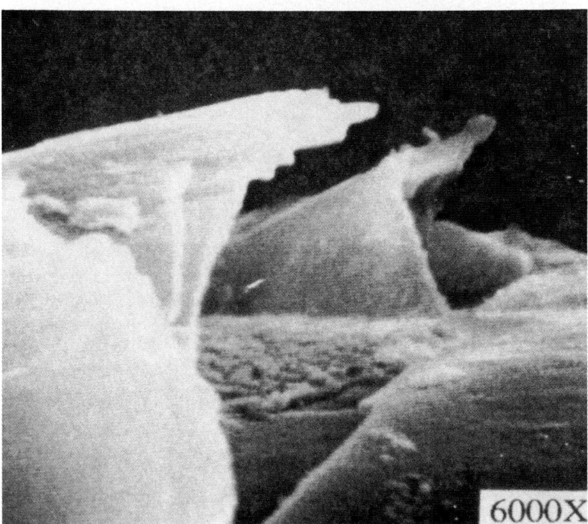

FIGURE 5–9. SEM illustrating lifted scales from stretching hair fibers less than 20% at 45% RH. (SEM provided by the courtesy of Dr. H.D. Weigmann of Textile Research Institute).

over two other parallel hairs (see Figures 5–13 and 5–14 and the axial wear patterns of the two hairs from this experiment). The fact that this type of wear can also be produced by combing hair is illustrated by Figure 5–15. Rubbing actions as in combing can even induce cortical lifting (Figure 5–16).

The foregoing types of damage can occur anywhere on the fiber, that is, near the root and midsections of hair and even near the tips, and most of

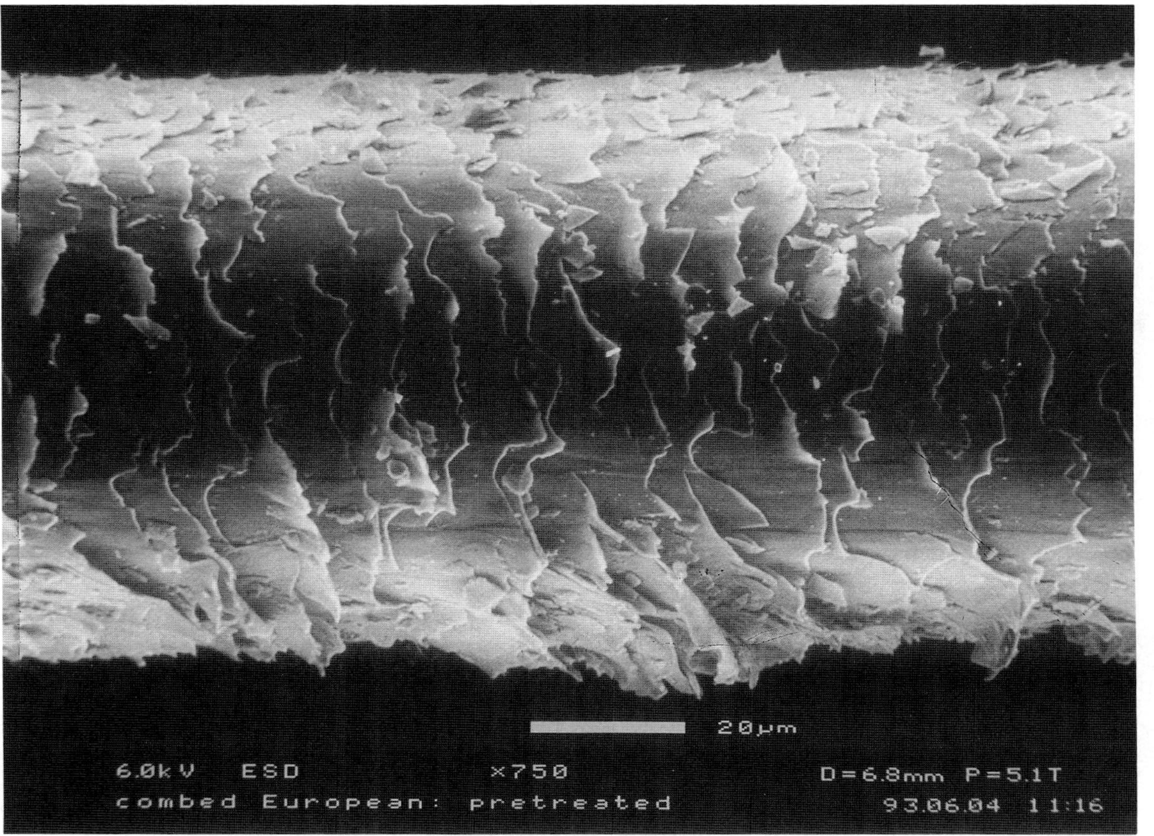

FIGURE 5–10. Scale lifting caused by vigorous combing of hair tresses.

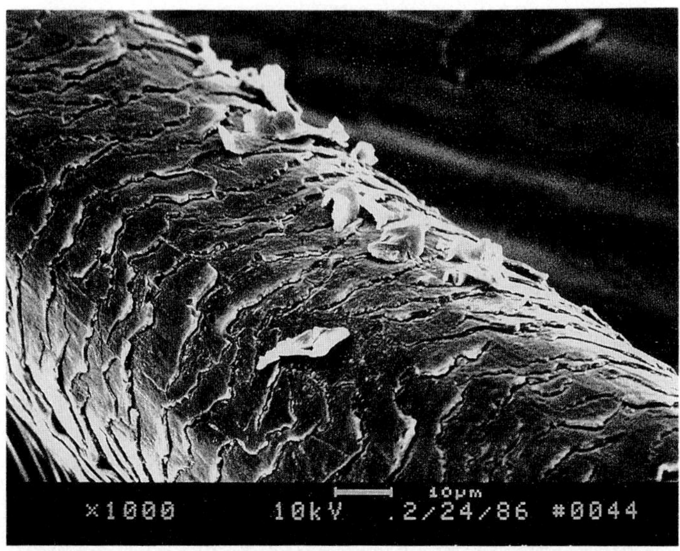

FIGURE 5–11. Tearing or breaking off of large sections of a single cuticle scale after scale lifting.

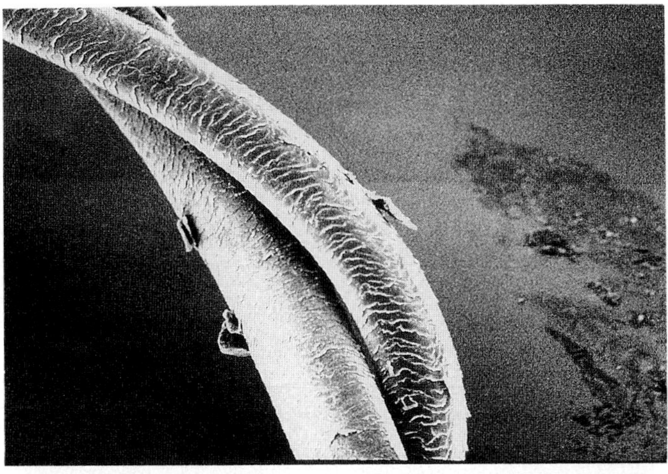

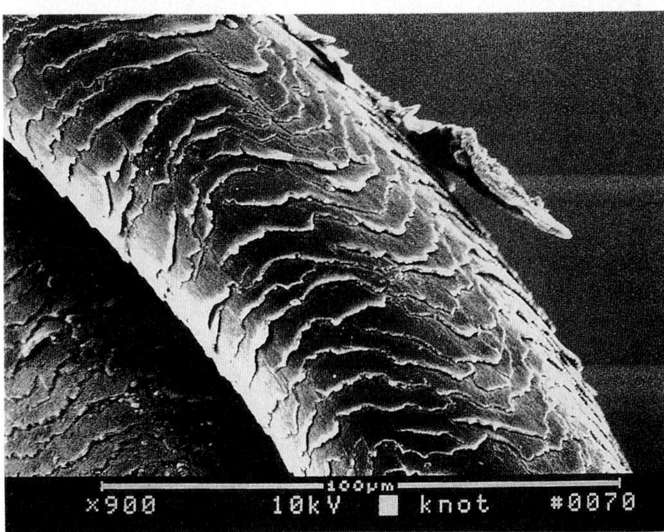

FIGURE 5–12. Tearing or ripping off of large segments of several scales caused by tying a single hair fiber in a loop and gradually pulling it to the state shown in top SEM.

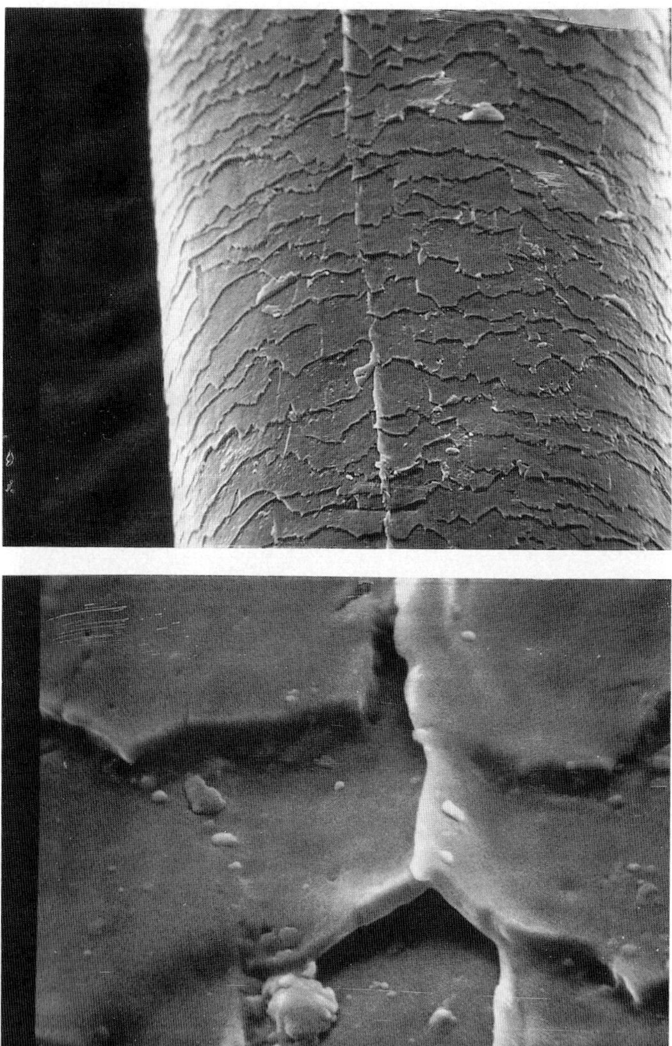

FIGURE 5–13. Cuticle wear caused by sliding one single fiber (20 cm long) in the form of a loop (weight, 0.58 mg) over two parallel fibers in a sliding friction experiment. The hair loop was allowed to slide approximately 25 times over the other two fibers in both a root-to-tip and a tip-to-root direction. This SEM illustrates the surface damage caused on one of the fibers. Note the cuticle wear line along the "high spot" of the fiber.

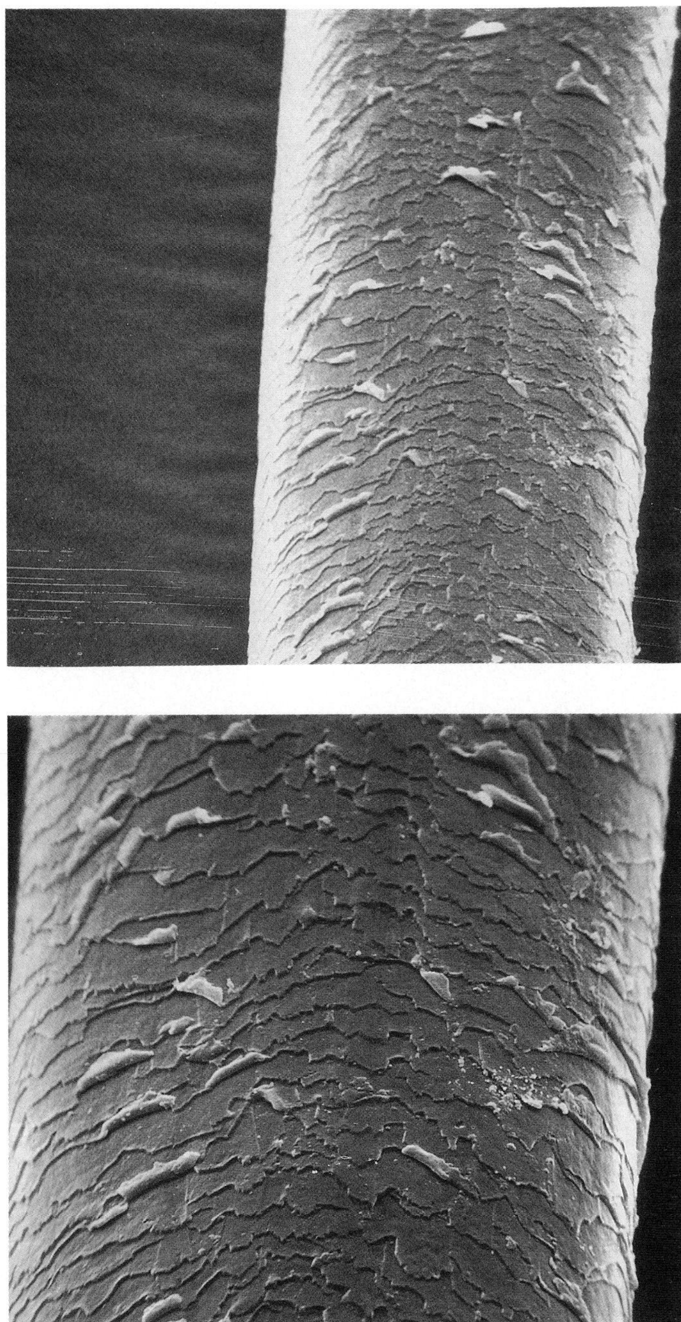

FIGURE 5–14. Cuticle wear from the other fiber from the experiment described in FIGURE 5–13. Note the lines of wear and the folded-back scale edges. This latter effect is probably from tip-to-root rubbing.

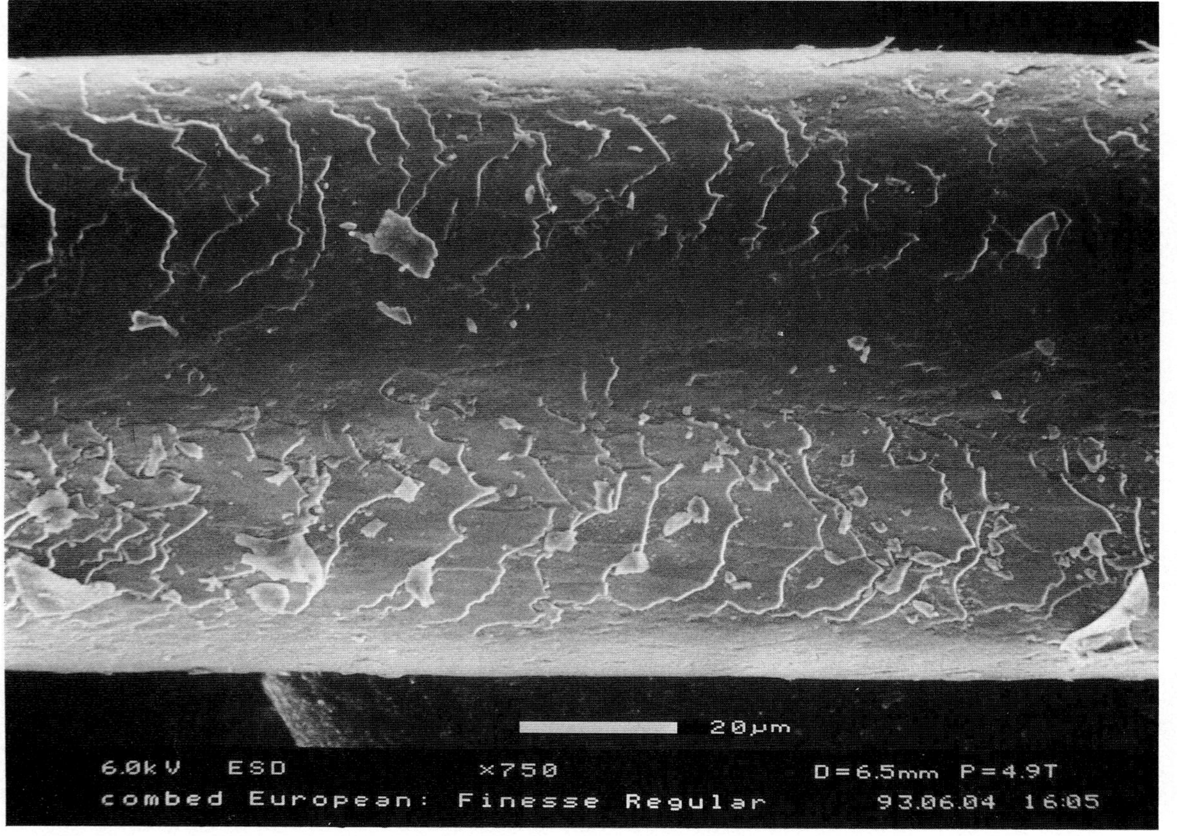

FIGURE 5–15. Axial wear along a hair fiber from tress combing. Note the pattern of wear along the axis rather than circumferential. This wear pattern is similar in type to that caused by the fiber loop experiment, illustrated in Figure 5–13. (Micrograph provided by the courtesy of Dr. E. Gretler.)

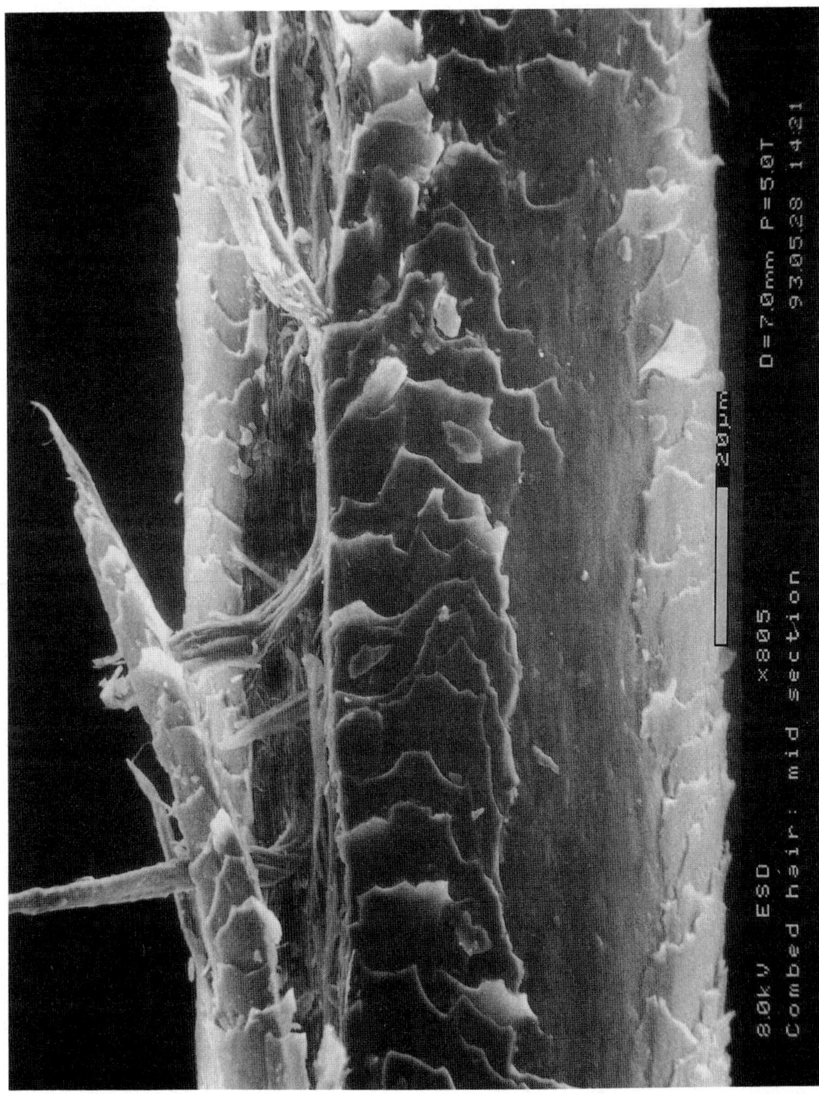

FIGURE 5–16. Cortical lifting caused by combing of tresses. (Micrograph provided by the courtesy of Dr. E. Gretler.)

these damaging actions with additional rubbing ultimately producing split hairs.

Garcia et al. [136] have developed a mathematical model to predict cuticle wear, assuming that wear occurs primarily by cuticle chipping, and these scientists conclude that cuticle erosion from grooming accelerates as the grooming operation moves closer to the tip end of the hair. This effect probably results from the fact that scale raising and removal of larger chunks of scales and even cortical lifting and other types of damaging action become greater as the grooming action moves closer to the tip ends.

In 1982, Kelly and Robinson [135] described the formation of split ends by the gradual erosion of cuticle scales during shampooing, drying, brushing, and combing of hair (see the section on fracturing of hair fibers in Chapter 8). Kambe et al. [137] concluded similarly, that the loss of or the gradual fragmentation of cuticle cell layers results in split ends. Robbins and Sandu [138] have taken the method of Swift and Bews [139] for the physical isolation of hair cuticle and modified it to a method to assess cuticle wear or damage to the cuticle.

Three important papers on the fracturing of human hair, published by Henderson et al. [140] and by Kamath and Weigmann [141,142], show that breaking or fracturing of hair occurs differently in the cuticle versus the cortex and that fracturing of hair fibers occurs in different patterns (see Figure 8–2). Further, these fracture patterns depend on the type of hair, the relative humidity (whether the hair is wet or dry), and whether or not the fiber is twisted or contains flaws (see the section on elastic and tensile deformations in Chapter 8). When the hair is wet, it tends to fracture most often in the smooth fracture pattern (see Figure 5–17). However, when the hair is dry, at less than 90% RH, the step fracture is the predominate fracture pattern (see Figure 5–17). Although fibrillation and splitting are not the predominate fracture patterns, fibrillation and splitting do tend to occur to some degree with more twisted or kinky fibers and when the relative humidity is lower, rather than when the fiber is wet. Thus, although fiber breakage does produce different end effects, with rubbing from grooming over time all the different fracture patterns eventually lead to split ends.

Extension of hairs to only 17% to 22% (at 45% RH) can induce failure in the endocuticule, which is the weakest region of the cuticle, resulting in the separation of the surface scales from the underlying layers producing an uplifting of scales (see Figures 5–9 and 5–10). Extensions to 30% to 32% can produce multiple circumferential fracturing with separation of cuticle sections from the cortex (Figure 5–18) [140].

The dissolution or the removal of structural lipids or proteinaceous matter from hair, probably from the cell membrane complex or epicuticle, by shampoos, surfactant solutions, or other cosmetic treatments, has been demonstrated by several different scientists. For example, Marshall and Ley [143] demonstrated the extraction of proteinaceous components

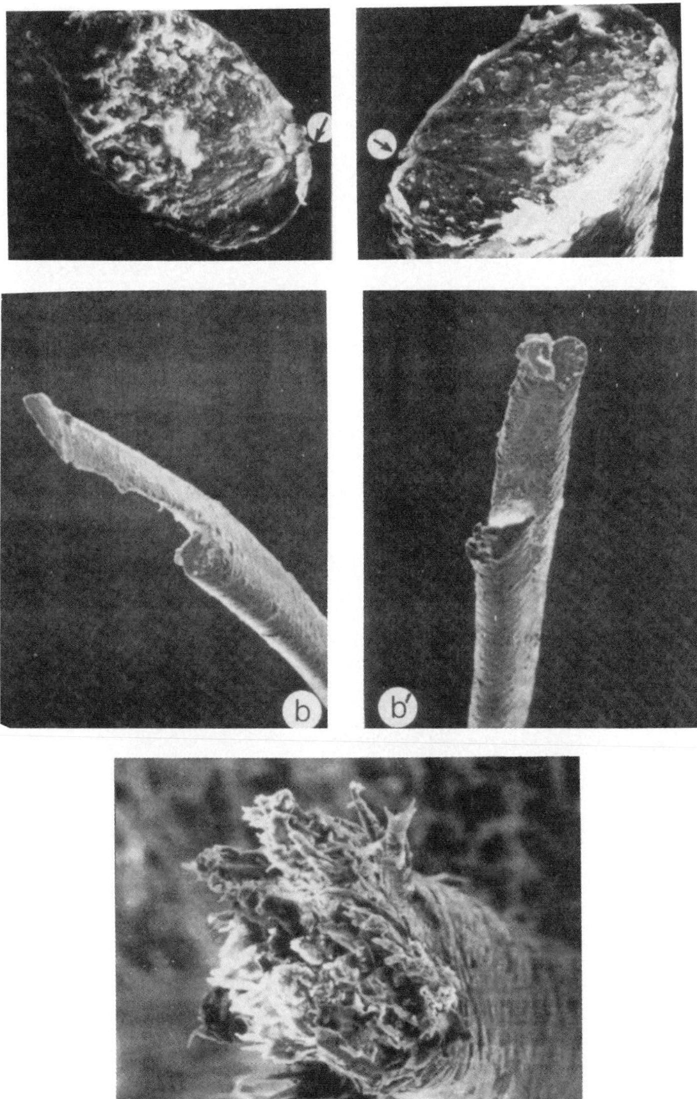

FIGURE 5–17. SEMs of three common fracture patterns for human hair. (SEMs from article by Weigmann et al. [142]. Published with permission of the *Journal of the Society of Cosmetic Chemists.*)

FIGURE 5–18. Light micrographs of hair fibers stretched to break in water. Note the circumferential cracks in the cuticle caused by extensive stretching. (From Henderson et al. [140]. Published with permission of the *Journal of the Society of Cosmetic Chemists.*)

from the cuticle of wool fiber by surfactant solutions of sodium dodecyl sulfate, cetyltrimethylammonium bromide, or triton X-100. Zahn et al. [144,145] and Mahrle [146] independently showed that part of the lipid components of the cell membrane complex were removed from hair by permanent waving.

Zahn [145] also described findings that show that intercellular lipids can be extracted from hair during repeated washing with detergents. Gould and Sneath [147] examined root and tip end sections of scalp hair by transmission electron microscopy (TEM) and observed holes or vacancies in the thin cross sections. These holes were greater (more frequent and larger) in tip ends than in root ends. These scientists attribute these holes to damaging effects by shampooing, or the breakdown and removal of components of the nonkeratinous portions of the hair by shampoos leaving the intercellular regions more susceptible to rupture.

To summarize, damage to the nonkeratinous regions of hair can result from stretching or extension to hair that causes fracturing between the scales at the nonkeratinous junctions, and this damage is more likely to occur in tip sections or in weathered or damaged hair than in the root ends of hair. Further, research in our laboratories shows that the action of detergents can lead to an analogous type of damage to the nonkeratinous regions by chemically and physically breaking down the nonkeratinous matter between the scales. This type of damage is more likely to occur on hair that has been damaged by perms and weathering than on root sections of hair.

Thus, it is becoming increasingly clear that shampooing and rubbing actions such as occur in grooming and other such actions as were described over time actually do damage hair by abrasion, erosion, and dissolution. In addition, stretching of hairs when a snag is encountered when combing or brushing also produces rupturing in the nonkeratinous regions of the cuticle, leading to scale lifting. This will produce even further damage by rendering those areas more susceptible to cuticle loss and to the penetration of chemicals into the hair, because the nonkeratinous regions are the primary regions of entry for penetration into the hair fiber. Further, the weakening of the nonkeratinous regions either by stretching or by penetrating chemicals will ultimately lead to the degradation of the cuticle and to an even faster rate of penetration of damaging treatments into the hair.

Thus, without question, normal cleaning and grooming practices that involve rubbing, stretching, and washing hair with simple shampoos or even with soap ultimately contribute to cuticle and even to cortical damage by abrasive, erosive, and stretching actions and by the dissolution of components from the nonkeratinous regions of the fiber. These damaging effects are ultimately detected by consumers as split ends, brittle hair, dry and dull ends, and by an increased sensitivity of their hair to damaging actions during grooming and to other damaging cosmetic treatments. In addition, sunlight [148] and chemical processing treatments such as bleaches, permanent waves, straighteners, and some hair dyes, or even chlorinated water, can accelerate rubbing, stretching and shampoo damaging actions by making the hair even more susceptible to all other damaging actions.

Dandruff and Scalp Flaking

Dandruff results from a scalp malfunction and is not directly related to the chemistry and physics of the human scalp hair. However, antidandruff products must be compatible with other hair products, and they have become an increasingly important hair care category during the past three decades. The subject of dandruff and antidandruff active ingredients therefore, merits some mention in a book dedicated to the chemistry and physics of human scalp hair, and thus this section offers an entry into the literature relevant to antidandruff products. Dandruff (seborrhea sicca, pityriasis sicca, or sicca capitis) has been defined by Ackerman and Kligman as "chronic noninflammatory scaling of the scalp" [149], as observed clinically. This clinical definition allows differentiation between dandruff and other scaly scalp diseases such as psoriasis, atopic dermatitis, etc. Others have demonstrated histologically that inflammation does exist in the upper dermis in dandruff [150]. However, this does not negate or reduce the utility of Kligman's clinical definition of dandruff.

The stratum corneum in the dandruff scalp is thinner [151] than in the normal scalp, and the epidermal turnover rate is also increased [149,152]. It has been suggested that this rapid transfer of cells to the scalp surface inhibits complete keratinization. Therefore, the stratum corneum is less coherent, cracks develop, and large flakes result.

Dandruff is age related [153]; it is rarely seen before puberty, but is common with the onset of puberty. It peaks in the early twenties and declines in middle and old age. Dandruff appears to be seasonal, being most severe in the winter months (October through December) and milder in the summer [153]. Dandruff occurs equally among males and females [154], but the cause(s) of dandruff, that is, the cause of the increased rate of epidermal turnover in the dandruff scalp, have not been well defined.

One potential concern in this area is the assumption that dandruff is a single "disease" with a single cause. One theory suggests the fungus *Pityrosporum ovale* as the cause of dandruff [154,155]. However, the Advisory Review Panel has concluded that *P. ovale* is not the cause of dandruff [153].

Van Abbe and Dean [155] have suggested that dandruff is an adaptive response to a threshold irritation. The irritation could result from scalp microflora or their metabolic products or from other sources. This conclusion is consistent with the observation of Heilengotter and Braun-Falco [150] that inflammation can be detected histologically in dandruff.

TABLE 5–21. Active ingredients for dandruff treatment.

Ingredient	Concentration specified (%)	Use specified
Coal tar preparations	0.5–5.0	Shampoos
Salicylic acid	1.8–3.0	Body and scalp products
Selenium sulfide	1	Topical use
Sulfur	2.0–5.0	Topical use
Zinc pyrithione	1.0–2.0	Shampoos
Zinc pyrithione	0.1–0.25	Hair-grooming products

The OTC monograph recommended three classes of potential antidandruff ingredients [153]:

Category I: Active ingredients considered safe and effective for use for dandruff, seborrheic dermatitis and psoriasis.
Category II: Ingredients not recognized as safe and effective or misbranded.
Category III: More data are required.

Actually, at this time only two categories are recognized: category I, as defined above, and category II. No other ingredients are recognized as safe and/or effective in this more recent classification.

Five ingredients are currently recognized as safe and effective for use against dandruff in the United States (Table 5–21). The OTC monograph recommends each ingredient at specific concentration levels for specified purposes (products and applications), as indicated in Table 5–21.

Other ingredients either reported or shown to be effective against dandruff and described either in the OTC monograph, the published literature, or the patent literature include alkyl isoquinolinium bromide, allantoin, benzethonium chloride, magnesium omadine, climbazole (1-Imidazopyl-1-(p-chlorophenoxy)-3,3-dimethyl butan-2-one), octopirox (1-hydroxy-4-methyl-6-(2,4,4 trimethyl pentyl)-2 (1H) pyridine ethanolamine and ketoconazole. These latter ingredients have not been described in category I by the OTC monograph.

Several methods have been described to evaluate dandruff, such as brushing off the hair and/or the scalp with various devices and weighing the scurf [152,156]. However, the most popular approach involves partitioning the scalp into several areas, rating each area for dandruff severity, and analyzing the combined data statistically [157,158]. The scalp partitioning method using appropriate statistical procedures provides a powerful tool to evaluate dandruff severity and the efficacy of antidandruff products.

Safety Considerations for Shampoos and Conditioners

Shampoos and conditioners are among the safest consumer products sold today when used for their intended purpose and in the manner described on the package label. Cautionary eye warning labels appear on most medicated products and on some cosmetic brands, attesting to the fact that eye irritation can occur if some products accidentally spill into the eyes. Warning against internal consumption also appears on many shampoo labels and on a few creme rinses or hair conditioners, although many conditioners contain no cautionary warnings, because they are mild and of such low toxicity.

Bergfeld [158] has reviewed the most frequent adverse effects of hair products from patients at the Cleveland Clinic Dermatology Department over a 10-year period. Bergfeld found relatively few adverse effects from shampoos, and the majority of these are caused by sensitization rather than to irritation or hair breakage. Further, Bergfeld attributes these few adverse effects either to preservatives or medicated ingredients of these products rather than to the active ingredients.

Ishihara [159] surveyed five large hospitals in Japan for contact dermatitis in 1970. Only 0.2% of cases of the total number of outpatients at all dermatologic clinics were from adverse reactions to all hair preparations; only 0.008% of these adverse reactions were to shampoos, and these few cases involved contact dermatitis. From these results, Ishihara concluded that most cases of contact dermatitis from shampoos are not serious enough to be treated in a hospital.

References

1. Grote, M.B.; et. al. U.S. patent 4,741,855.
2. Eccleston, G.M. J. Soc. Cosmet. Chem. 41:1 (1990).
3. Flick, E.W. Cosmetic and Toiletry Formulations, 2nd Ed., Noyes Publications, Park Ridge NJ (1989).
4. Hunting, A.L.L. Encyclopedia of Shampoo Ingredients. Micelle Press, Cranford, NJ (1983).
5. Hunting, A.L.L. Encyclopedia of Conditioning Rinse Ingredients. Micelle Press, Cranford, NJ (1987).
6. Faucher, J.A.; Goddard, E. J. Colloid Interface Sci. 55:313 (1976).
7. Sykes, A.R.; Hammes, P.A. Drug Cosmet. Ind. 62 (1980).
8. Idson, B.; Lee, W. Cosmet. Toiletries 98:41 (1983).
9. Koch, J., et al. J. Soc. Cosmet. Chem. 33:317 (1982)
10. Schebece, F. Personal communication.
11. Ramachandran Bhat, G.; et al. J. Soc. Cosmet. Chem. 30:1 (1979).
12. Milosevic, M.; et al. Arkh. Hig. Rad. Toksikol. 31(3):209 (1980).
13. Breuer, M. J. Soc. Cosmet. Chem. 32:437 (1981).
14. Kissa, E. Text. Res. J. 51:508 (1981).

15. Knott, C. A.; et al. Int. J. Cosmet. Sci. 5:77 (1983).
16. Shaw, D.A. Int. J. Cosmet. Sci. 1:317 (1979).
17. Thompson, D.; et al. J. Soc. Cosmet. Chem. 36:271 (1988).
18. Ludec, M.; et al. Proc. 10th IFSCC Cong., p. 693 (Australia, 1978).
19. Stamm, R.; et al. J. Soc. Cosmet. Chem. 28:571 (1977).
20. Scott, G.V.; Robbins, C.R. J. Soc. Cosmet. Chem. 31:179 (1980).
21. Robbins, C.R.; Bahl, M.K. J. Soc. Cosmet. Chem. 35:379 (1984).
22. Dobinson, G.C.; Petter, P.J. J. Soc. Cosmet. Chem. 27:3 (1976).
23. Robbins, C.R.; Crawford, R. J. Soc. Cosmet. Chem. 35:369 (1984).
24. Robbins, C.R. In Chemical and Physical Behavior of Human Hair, p. 107. Van Nostrand Reinhold, New York (1979).
25. Schuster, S.; Thody, A. J. Invest. Dermatol. 62:172 (1974).
26. Clarke, J.; Robbins, C.; Schroff, B. J. Soc. Cosmet. Chem. 40:309 (1989).
27. Clarke, J.; Robbins, C.; Schroff, B. J. Soc. Cosmet. Chem. 41:335 (1990).
28. Robbins,C.; Reich,C.; Clarke, J. J. Soc. Cosmet. Chem. 40:205 (1989).
29. Gould, J.G.; Sneath, R. J. Soc. Cosmet. Chem. 36:53 (1985).
30. Hilterhaus-Bong, S.; Zahn, H. Int. J. Cosmet. Sci. 11:167 (1989).
31. Jacob, C. Personal communication.
32. Curry, K.; Golding, S. J. Soc. Cosmet. Chem. 22:681 (1971).
33. Gloor, M. Dermatol. Monatssch. 160:730 (1974).
34. Eberhardt, H. J. Soc. Cosmet. Chem. 27:235 (1976).
35. Gloor, M. In Cosmetic Science, Vol. 1, Breuer, M., ed., p. 218. Academic Press, New York (1978).
36. Minor, F.; et al. Text. Res. J. 29:931 (1959).
37. Robbins, C.; Reich, C. 4th Int. Hair Sci. Symp., Syburg, W. Germany German Wool Res. Int. Publ. Abstracts (November 1984).
38. Crawford, R.; Robbins, C.R. J. Soc. Cosmet. Chem. 31:273 (1980).
39. Hughes, L. Text. Chem. Colourist 10:20/88 (1978).
40. Robbins, C.R.; et al. Text. Res. J. 38:1197 (1968).
41. Dawber, R.P.R.; Calnan, C.D. Clin. Exp. Dermatol. 1:155 (1976).
42. Faucher, J.A.; et al. Text. Res. J. 47:616 (1977).
43. Hannah, R.B.; et al. Text. Res. J. 48:57 (1978).
44. Woodard, J. J. Soc. Cosmet. Chem. 23:593 (1972).
45. Gloor, M., et al. Kosmetologie 3:193 (1975).
46. Bereston, E.S. JAMA 156:1246 (1954).
47. Neu, G.E. J. Soc. Cosmet. Chem. 11:390 (1960).
48. Hart, J.R.; DeGeorge, M.T. J. Soc. Cosmet. Chem. 31:223 (1980).
49. Hart, J.R.; DeGeorge, M.T. Drug Cosmet. Ind. 134:46 (1984).
50. Hart, J.R.; Levy, E.F. Soap Cosmet. Chem. Spec. 53(8):31 (1977).
51. Ross, J.; Miles, G.D. Oil Soap 18:99 (1941).
52. Domingo Campos, F.J.; Druguet Toutina, R.M. Cosmet. Toiletries 98(9): 121 (1983).
53. Powers, D. In Cosmetics, Science and Technology, Sagarin, E., ed., Chap. 17. Interscience, New York (1957).
54. Gerstein, T. Cosmet. Perfumery 90:35 (1975).
55. Radar, C.; Tolgyesi, W. Cosmet. Perfumery 90:29 (1975).
56. Robbins, C.; Reich, C.; Patel, A. J. Soc. Cosmet. Chem. (in press).
57. Wilkerson, V. J. Biol. Chem. 112:329 (1935–1936).
58. Leeder, J.D.; Rippon, J.A. J. Soc. Dyers Col. 99:64 (1983).

59. Hall, R.O. J. Soc. Dyers Col. 53:341 (1937).
60. Leeder, J.D.; et al. Tokyo Wool Res. Conf. (1985).
61. Jurdana, L.E.; Leaver, I.H. Text. Res. J. 62(8):463 (1992).
62. Vickerstaff, T. The Physical Chemistry of Dyeing, Chap. 4. Interscience, New York (1954).
63. Lemin, D.; Vickerstaff, T. J. Soc. Dyers. Col. 63;405 (1947).
64. Han, S.K.; et al. J. Soc. Cosmet. Chem. 36:1 (1985).
65. Vickerstaff, T. In The Physical Chemistry of Dyeing, p. 95. Interscience, New York (1954).
66. Vickerstaff, T. In The Physical Chemistry of Dyeing, pp. 356–376. Interscience, New York (1954).
67. Gilbert, G.; Rideal, E. Proc. R. Soc. A182:355 (1944).
68. Speakman, J.; Peters, L. J. Soc. Dyers Col. 65:63 (1949).
69. Delmenico, J.; Peters, R. Text. Res. J. 34:207 (1964).
70. Breuer, M. J. Text. Inst. 58:176 (1967).
71. Peters, L. J. Text. Inst. 58:179 (1967).
72. Oloffson, B. J. Soc. Dyers Col. 67:57 (1951).
73. Oloffson, B. J. Soc. Dyers Col. 68:506 (1952).
74. Erhardt, H. Colgate Palmolive Internal Res. Rep. No. 1868 (1961).
75. Williams, J.; Cody, L. Chem. Rev. 14:171 (1933).
76. Crank, J. In The Mathematics of Diffusion, Chap. 11. Clarendon Press, Oxford, UK (1967).
77. Alexander, P.; et al. In Wool, Its Chemistry and Physics, pp. 146–148. Chapman and Hall, London (1963).
78. Jost, W. In Diffusion in Solids, Liquids and Gases, Chap. 1. Academic Press, New York (1952).
79. Crank, J. In The Mathematics of Diffusion, Chap. 1. Clarendon Press, Oxford, UK (1967).
80. Crank, J. In The Mathematics of Diffusion, Chap. 12. Clarendon Press, Oxford, UK (1967).
81. Crank, J. In The Mathematics of Diffusion, Chap. 5. Clarendon Press, Oxford, UK (1967).
82. Weigmann, H.D. J. Polymer Sci. A-1(6):2237 (1968).
83. Hill, A. Proc. R. Sci. 104 B:39 (1929).
84. Alexander, P.; Hudson, R. Text. Res. J. 20:481 (1950).
85. Vickerstaff, T. In The Physical Chemistry of Dyeing, Chap. 5. Interscience, New York (1954).
86. Davis, G.; Taylor, H. Text. Res. J. 35:405 (1965).
87. Holmes, A. J. Soc. Cosmet. Chem. 15:595 (1968).
88. King, G. Trans. Faraday Soc. 41:479 (1945).
89. King, G. Nature 154:575 (1944).
90. Parrington, J.; et al. Nature 169:583 (1952).
91. Alexander, P.; Hudson, R.F. In Wool, Its Chemistry and Physics, pp. 136–146. Chapman and Hall, London (1963).
92. Ingold, C. In Structure and Mechanism in Organic Chemistry, 2nd Ed., p. 50. Cornell University Press, Ithaca, New York (1969).
93. Valko, E. J. Soc. Dyers Col. 55:173 (1939).
94. Gilbert, G. Proc. R. Soc. (Lond.) A183:167 (1944).

95. Robbins, C.R.; Scott, G.V. Text. Res. J. 40:951 (1970).
96. Robbins, C.R.; Fernee, K.M. J. Soc. Cosmet. Chem. 34:21 (1983).
97. Hudson, R.F. Discuss. Faraday Soc. 16:14 (1954).
98. Speakman, J. Proc. R. Soc. (Lond.) A132:167 (1931).
99. Speakman, J.; Elliot, G. Symposium on Fibrous Proteins, p. 116. Soc. Dyers Col., Leeds, UK (1946).
100. Wilmsmann, H. J. Soc. Cosmet. Chem. 12:490 (1961).
101. Speakman, J.; Smith, S. J. Soc. Dyers Col. 52:121 (1936).
102. Steinhardt, J.; Harris, M. J. Res. Natl. Bur. Stand. 24:335 (1940).
103. Maclaren, J. Arch. Biochem. Biophys. 86:175 (1960).
104. Robbins, C.R.; Kelly, C. Text. Res. J. 40:891 (1970).
105. Robbins, C.R.; et al. Text. Res. J. 38:1130 (1968).
106. Sagal, J. Text. Res. J. 35:672 (1965).
107. Smith, A.; Harris, M. J. Res. Natl. Bur. Stand. 19:81 (1937).
108. Swift, J.; Bews, B. J. Soc. Cosmet. Chem. 27:289 (1976).
109. Laden, K.; Finkelstein, P. Am. Perfumer Cosmet. 81:39 (1966).
110. Robbins, C.R.; Anzuino, G. J. Soc. Cosmet. Chem. 22:579 (1971).
111. Robbins, C. Text. Res. J. 37:337 (1967).
112. Steinhardt, J.; et al. J. Res. Natl. Bur. Stand. 25:519 (1940).
113. Freytag, H. J. Soc. Cosmet. Chem. 15:265 (1964).
114. Wilkerson, V. J. Biol. Chem. 112:329 (1935–1936).
115. Sookne, A.; Harris, M. J. Res. Natl. Bur. Stand. 23:471 (1939).
116. Harris, M.; Sookne, A. J. Res. Natl. Bur. Stand. 26:289 (1941).
117. Vickerstaff, T. The Physical Chemistry of Dyeing, p. 350. Interscience, New York (1954).
118. Robbins, C.R.; Kelly, C. J. Soc. Cosmet. Chem. 20:555 (1969).
119. Steinhardt, J.; et al. J. Res. Natl. Bur. Stand. 28:201 (1942).
120. Vickerstaff, R. The Physical Chemistry of Dyeing, p. 373. Interscience, New York (1954).
121. Speakman, J.; Elliot, G. Symposium on Fibrous Proteins, p. 116. Soc. Dyers. Col., Leeds, UK (1946).
122. Peters, R.H. Ciba Rev. 2 (1964).
123. Steinhardt, J.; Zeiser, E. J. Biol. Chem. 183:789 (1950).
124. Barnett, G. M.S. thesis, Polytechnic Institute of Brooklyn, Brooklyn, New York (1952).
125. Speakman, J.; Stott, C. Trans. Faraday Soc. 30:539 (1934).
126. Speakman, J.; Stott, C. Trans. Faraday Soc. 31:1425 (1935).
127. Scott, G.V.; et al. J. Soc. Cosmet. Chem. 20:135 (1969).
128. Vickerstaff, T. The Physical Chemistry of Dyeing, p. 413. Interscience, New York (1954).
129. Rosen, M.J. J. Am. Oil Chem. Soc. 49:293 (1971).
130. Vickerstaff, T. The Physical Chemistry of Dyeing, p. 397. Interscience, New York (1954).
131. Faucher, J.; Goddard, E. J. Soc. Cosmet. Chem. 29:323 (1978).
132. Goddard, E.; Hannah, R.B. J. Colloid Interface Sci. 55:73 (1976).
133. Swift, J.A.; Bews, A.C. J. Soc. Cosmet. Chem. 23:695 (1972).
134. Okumura, T. 4th Int. Hair Sci. Symp., Syburg, W. Germany (Nov. 1984).
135. Kelly, S.C.; Robinson, V.N.E. J. Soc. Cosmet. Chem. 33:203 (1982).

136. Garcia, M.L.; et al. J. Soc. Cosmet. Chem. 29:155 (1977).
137. Kambe, T., et. al. Proc. 6th Int. Hair Sci. Symp. DWI, Luneberg, Germany, German Wool Res. Int. Publ. Abstracts, (1988).
138. Sandhu, S.; Robbins, C.R. J. Soc. Cosmet. Chem. 44:163 (1993).
139. Swift, J.A.; Bews, B. J. Soc. Cosmetic Chem. 25:13 (1974).
140. Henderson, G.H.; et. al. J. Soc. Cosmet. Chem. 29:449 (1978).
141. Kamath, Y.; Weigmann, D. J. Appl. Poly. Sci. 27:3809 (1982).
142. Kamath, Y.; Weigmann, D. J. Soc. Cosmet. Chem. 35:21 (1984).
143. Marshall, R.C.; K.F. Ley Text. Res. J. 56:772 (1986).
144. Hilterhaus-Bong, S.; Zahn, H. Int. J. Cosmet. Sci. 11:167 (1989).
145. Kaplan, I.J.; Schwan, A.; Zahn, H. Cosmet. Toiletries 97:22 (1982).
146. Mahrle, G. et al. In Hair Research, Orfanos, C.E.; Montagna, W.; Stuttgen, G., eds., pp. 524–528. Springer-Verlag, New York (1981).
147. Gould, J.G.; Sneath, R.L. J. Soc. Cosmet. Chem. 36:53 (1985).
148. Tolgyesi, E. Cosmet. Toiletries 98:29 (1983).
149. Ackerman, A.B.; Kligman, A. J. Soc. Cosmet. Chem. 20:81 (1969).
150. Heilengotter, G.; Braun-Falco, O. In Hair Research, Orfanos, C.E.; Montagna, W.; Stuttgen. G., eds., p. 568. Springer-Verlag, Berlin (1981).
151. Kligman, A. J. Soc. Cosmet. Chem. 27:111 (1976).
152. Laden, K.; Finkelstein, P. J. Soc. Cosmet. Chem. 19:669 (1968).
153. Federal Register 47FR54646 (Dec. 3, 1982).
154. Klauder, J.V. J. Soc. Cosmet. Chem. 7:443 (1956).
155. Van Abbe, N.J.; Dean, J. J. Soc. Cosmet. Chem. 18:439 (1967).
156. Van der Wyke, R.W.; Raia, F.C. J. Soc. Cosmet. Chem. 15:761 (1964).
157. Van Abbe, N.J. J. Soc. Cosmet. Chem. 15:609 (1964).
158. Bergfeld, W.F. In Hair Research, Orfanos, C.E.; Montagna, W.; Stuttgen G., eds., p. 507. Springer-Verlag, Berlin (1981).
159. Ishihara, M. In Hair Research, Orfanos, C.E.; Montagna, W.; Stuttgen, G., eds., p. 536. Springer-Verlag, Berlin (1981).

6

Dyeing Human Hair

There are three ways to modify the color of hair: it may be made lighter by bleaching (see Chapter 4); artificial colors may be added to the hair; or a combination of both of these methods may be employed. Adding color is the subject of this chapter.

Hair coloring has been carried out for more than 2,000 years [1,2] using various vegetable, mineral, and animal substances as coloring agents. Most of the dyes that have been considered for human hair may be described as either oxidation dyes, ionic dyes, metallic dyes, or reactive dyes.

The classification used for discussion in this chapter consists of four groups:

permanent or oxidation dyes
semipermanent dyes
temporary dyes or color rinses
other dyes.

Beyond updating, the primary addition to this chapter from earlier editions is a section on hair dye formulations for permanent dyes, semipermanent dyes, color rinses, and lead acetate-sulfur hair dyes. Oxidation dyes are often referred to as permanent hair dyes and are the most important of the commercial hair dyes. Permanent hair dyes generally consist of *p*-diamines and *p*-aminophenols that are oxidized by hydrogen peroxide to active intermediates [3]. These active intermediates then react in the hair with color couplers to provide shampoo-resistant hair dyes.

Semipermanent products consist of nitroaromatic amines or aromatic amino nitroanthroquinone dyes [4,5] that diffuse into and bind to the hair but do not bind firmly. Because these dyes are not firmly bound, they diffuse out of the hair after a few (4–6) shampoos.

Temporary dyes or color rinses are acid dyes [6–8] similar to those used in wool dyeing. These dyes do not bind firmly either, and they may be removed by a single shampooing. Wool dyeing is very different from hair dyeing primarily because of temperature. In fact, low-temperature dyeing in wool is 80°C to 85°C. Solvent-assist dyeing permits dyeing wool at 60°C to 70°C, which is still too high for human hair.

Other dyes developed and described in the literature for hair include basic dyes [9], metallic dyes, reactive dyes, and vegetable dyes [2], which are discussed in the last part of this chapter. The basic physical chemistry of the interactions of ionic dyes with hair is related to that of ionic surfactants and is described in Chapter 5, including definitions for diffusion coefficients, ion affinities, and experimental procedures to determine these important parameters.

Oxidation Dyes or Permanent Hair Dyes

Compositions and Dyeing Conditions

Oxidation hair dyes consist of dye precursors, which form active intermediates; dye couplers, which condense with the active intermediates; and an oxidizing agent (hydrogen peroxide). These reactions are usually carried out at alkaline pH, generally from 8 to 10 [2,10–13]. By adjusting the proportions of peroxide, precursors, and couplers, the hair may be made lighter or darker in one process.

Oxidation dye precursors are derivatives of aniline (Table 6–1). Precursors are difunctional ortho- or para-diamines or aminophenols that are capable of oxidizing to diiminium (IX) or quinoniminium (X) ions, proposed by Corbett [3] as the active intermediates of this process. This article [3] by Corbett is an excellent review of the chemistry of oxidative dyeing.

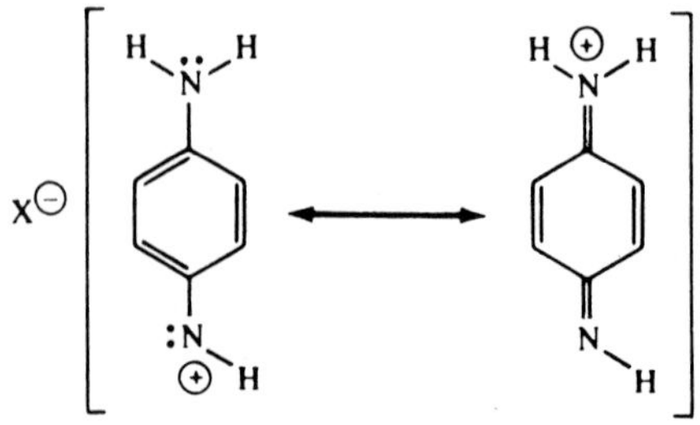

Diiminium ion of p-phenylenediamine (IX)

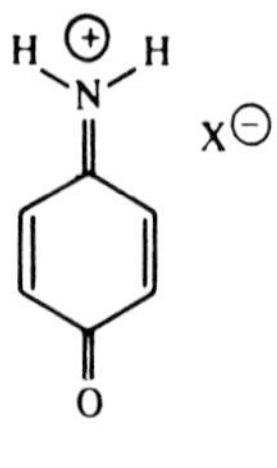

Quinoniminium ion of p-aminophenol (X)

TABLE 6–1. Some oxidation dye precursors.[a]

p-Phenylenediamine (I)	*Toluene-2,5-diamine* (II)
2-Methoxy-p-phenylenediamine (III)	*2-Chloro-p-phenylenediamine* (IV)
Toluene-3,4-diamine (V)	*N,N-Bis(2-hydroxyethyl)-p-phenylenediamine* (VI)
o-Aminophenol (VII)	*p-Aminophenol* (VIII)

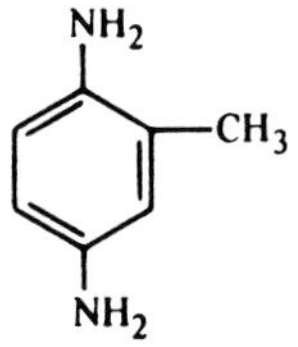

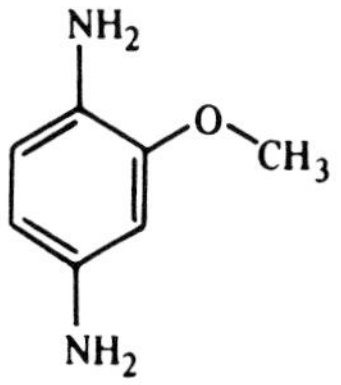

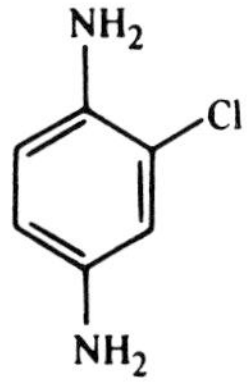

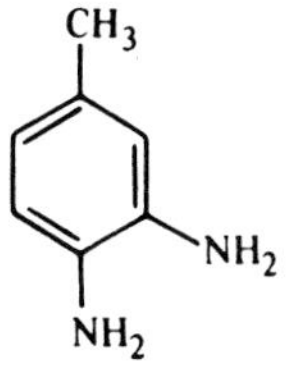

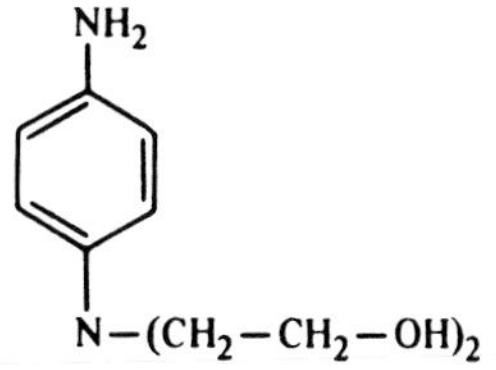

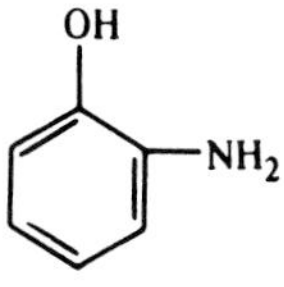

[a] The ingredients are described in references 2, 13–18, and on product ingredient labels.

Oxidation dye couplers are electron-rich aromatic species. They are commonly substituted resorcinols or meta-phenylenediamines, usually containing a vacant position para to the amine or phenolic group (Table 6–2). Oxidation dye precursors, when oxidized in the absence of couplers, form colored compounds, usually gray or brown-black shades. On the other hand, couplers themselves usually produce little or no color, but in the pres-

TABLE 6–2. Some oxidation dye couplers.[a]

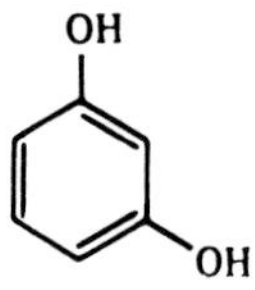

Resorcinol (XI)	*1-Naphthol* (XII)	*Pyrogallol* (XIII)
4-Chlororesorcinol (XIV)	*4-Methoxy-m-phenylenediamine* (VX)	*m-Phenylenediamine* (XVI)
2-Methyl-5-aminophenol (XVII)	*Hydroquinone* (XVIII)	

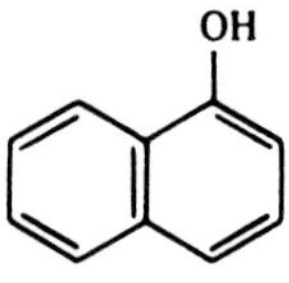

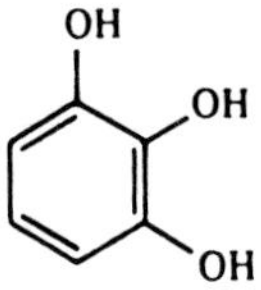

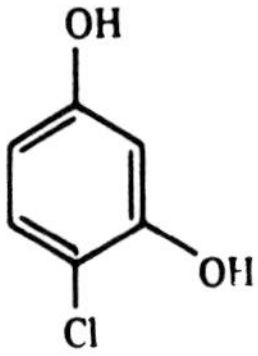

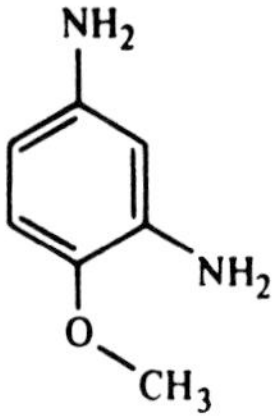

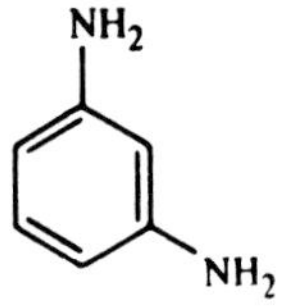

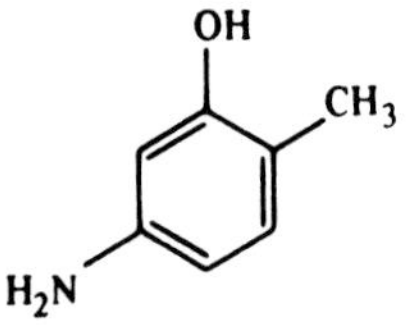

[a] The ingredients are described in references 2, 13–18, and on product ingredient labels.

ence of precursors and oxidizing agent, they modify the color formed by the precursor.

Some oxidation dye formulations contain two or three ingredients that act as either dye precursors or couplers but others contain five to seven or more such ingredients. In this latter case, several reactions are involved, and multiple dye products are formed.

As indicated earlier, hydrogen peroxide is usually the oxidizing agent of choice for oxidative hair dyeing. However, peracids [3] and autoxidation [18] or air oxidation of highly electron-rich dye precursors have also been considered. The nature of the autoxidation process must be analogous to the self-condensation reactions described for dye precursors in the next section. Internal cyclization of the dinuclear indo dye frequently occurs, too [18].

A few of the trisubstituted benzene derivatives that have been described for autoxidation dyeing are shown in the following structures.

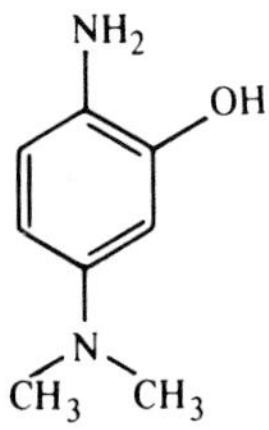

NH2, OH, N, CH3, CH3

3-Hydroxy-N,N-dimethyl-p-phenylenediamine

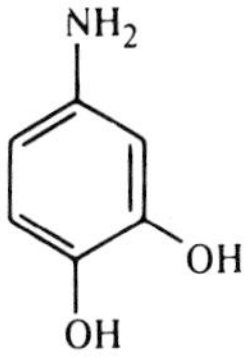

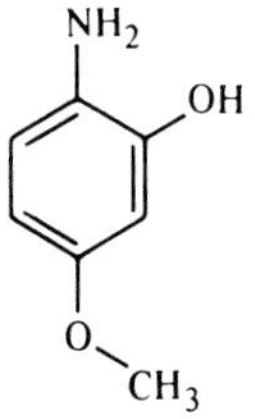

NH2, OH, OH   NH2, OH, O, CH3

4-Aminocatechol   *3-Hydroxy-4-amino-anisole*

Analogous naphthalenes [18,19] and quinolines have also been described, along with some of the dye products of autoxidation reactions [18].

Modern oxidation dyes sometimes contain coloring agents in addition to dye precursors and couplers; for example, direct dyes like disperse blue 1 and nitro-phenylenediamines are sometimes included.

One might predict that these ingredients could enter into oxidation dye reactions, but because the strong electron-withdrawing nitro or anthraquinone groups are present, these groups should decrease the rates of oxidation and coupling of these species below that of oxidation dye reactants. Thus, these dyes probably function primarily as direct color modifiers.

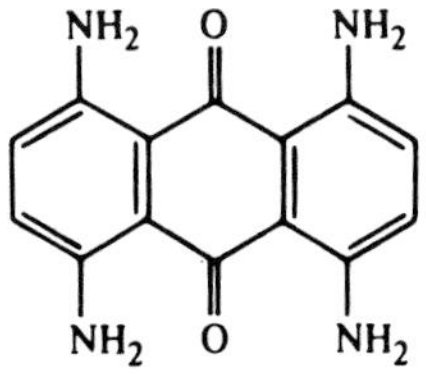

NH2 O NH2, NH2 O NH2

Disperse blue 1

NH2, NO2, NH2   NH2, NH2, NO2

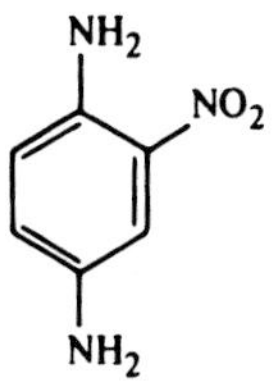

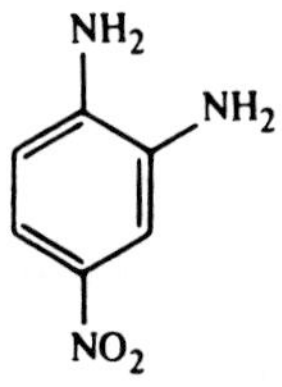

2-Nitro-p-phenylenediamine (*red*)   *4-Nitro-0-phenylenediamine* (*yellow-orange*)

Tables and data describing the colors formed by reaction of many of the precursors and couplers shown in Tables 6–1 and 6–2, as well as related ingredients, have been compiled by Wall [2], Tucker [8], and Corbett [3,18].

Summary of the Reactions of Oxidation Dyes

Table 6–3 summarizes a scheme for formation of oxidation dyes and provides some examples of the types of dyes that have been isolated from these reactions. This scheme shows that a dye precursor (e.g., *p*-phenylenediamine) is oxidized to its corresponding diiminium ion (IX), and this active intermediate condenses with an electron-rich dye coupler, forming a dinuclear product that is oxidized to an indo dye.

This reaction may stop at the dinuclear dye stage, or additional condensation-oxidation reactions may occur, forming trinuclear or even poly-

TABLE 6–3. Scheme and examples of indo dyes formed in oxidative hair dyeing.

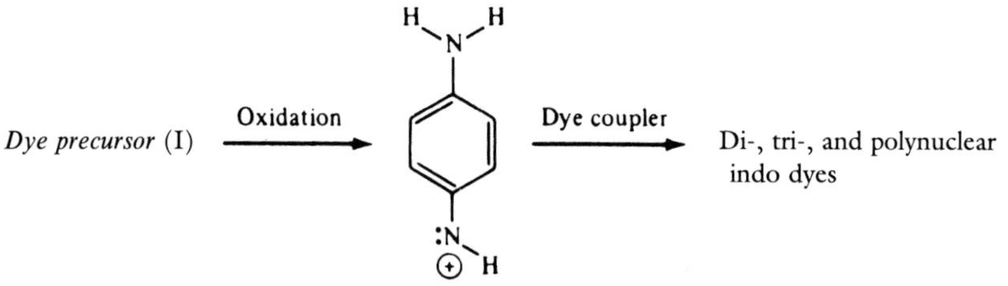

Dye precursor (I) —Oxidation→ [H, H, N, :N, ⊕, H] —Dye coupler→ Di-, tri-, and polynuclear indo dyes

Diiminium ion[a]

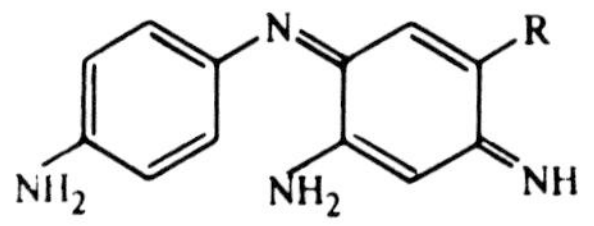

[N, R, NH₂, NH₂, NH]

Dinuclear indo dye from (I) *and 4-substituted m-phenylenediamines* [3] (XIX)

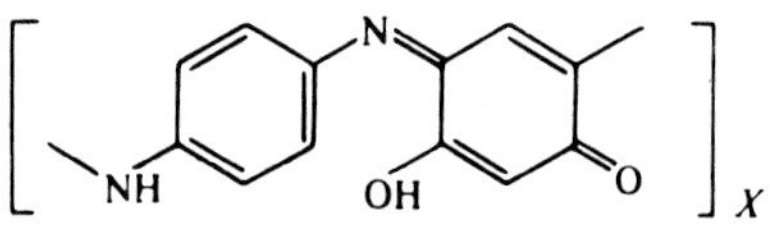

[N, NH, OH, O, x]

Polynuclear indo dye from (I) *and resorcinol* [20] (XX)

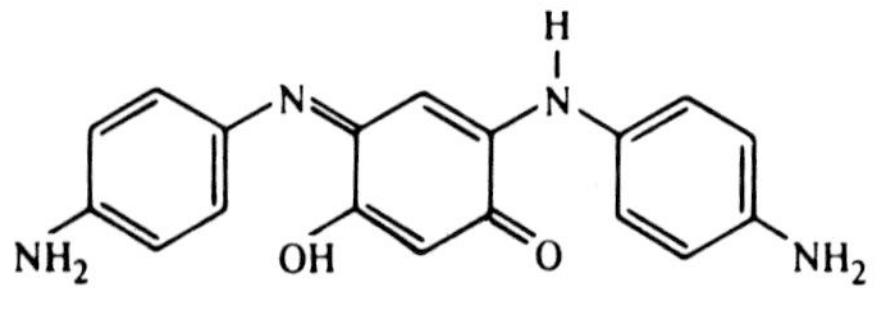

[H, N, N, NH₂, OH, O, NH₂]

Trinuclear dye from (I) *and resorcinol* [21] (XXI)

[a] Of the several possible resonance forms of (IX), this one is used in the next few sections because it enables one to clearly visualize the mechanisms proposed to explain the products shown. The mechanisms presented are intended more as tools for structure prediction than as proven descriptions of the molecular actions involved.

nuclear dyes and pigments [3,20]. More detailed mechanisms describing the formation of these and other indo dye products are presented in the next section.

Because most oxidation dye products contain several ingredients capable of acting as either dye couplers or precursors, mixtures of di-, tri-, and polynuclear indo dyes are formed in these reactions. In addition, it is conceivable that nucleophilic groups in hair might even add to the indo dyes, covalently bonding dye molecules to hair. Penetration of the dye precursors and the couplers can occur, but penetration must be limited to the outer regions of the hair, because the condensation reactions that occur are relatively fast compared with diffusion and the larger condensation products (at least in the hair) are resistant to shampooing.

Possible Mechanisms for Oxidation Dye Reactions

The Active Intermediates in Oxidation Dye Reactions

The diimine (XXII) has been described as a vital intermediate in oxidative hair dyeing [22]. More recently, Corbett [3] has described the protonated diimine (IX) as the reactive species that attacks dye coupling agents, ultimately forming indo dyes.

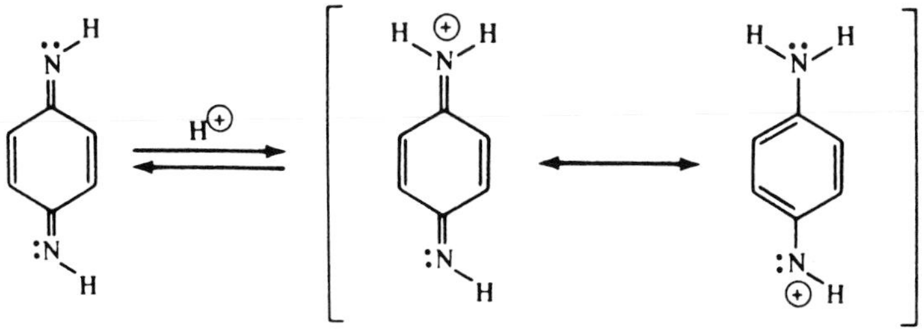

Diimine of (I)
(XXII)

Diiminium ion of (XXII)
(IX)

Certainly diiminium ions are more electrophilic and therefore more capable of serving as the active species in these reactions than are diimines. As such, diiminium ions are described as the active intermediates in the mechanisms considered in the subsequent discussions. By analogy, quinoniminium ions, such as (X), would be the active intermediates formed from ortho- and para-aminophenols.

If one assumes that diimines are formed by two one-electron transfer reactions [23,24] and the loss of two protons, and that the entire sequence occurs stepwise, a diiminium ion is formed before dimine.

Therefore, although diimines may form in these interactions, they are not necessary intermediates for forming the di-, tri-, and polynuclear indo dyes that have been shown to form. For example, compound (VI), an

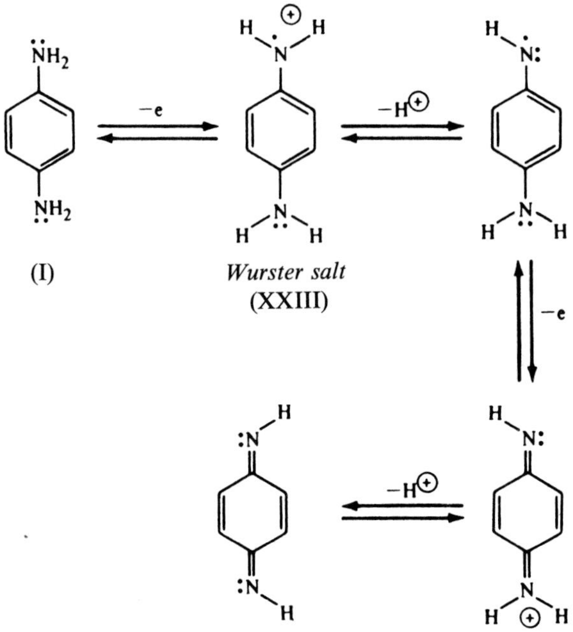

(I) Wurster salt (XXIII)

Diimine Diiminium ion (IX)

N,N-dialkyl-substituted *p*-phenylenediamine, is used in several popular hair dyes. This species should be capable of forming a diiminium ion (XXIV), although it cannot form a corresponding diimine.

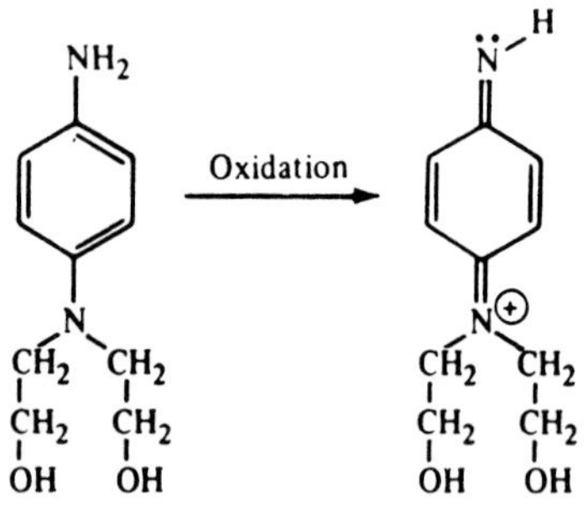

(VI) (XXIV)

One might speculate that (VI) functions only as a dye coupler; however, related *N,N*-dialkyl *p*-phenylenediamines have been used with common dye couplers (in the absence of unsubstituted *p*-phenylenediamines), suggesting that this type of species does act as a dye precursor in oxidative hair dyeing [8,25,26].

Mechanisms may also be written involving Wurster salts (XXIII), which provide the di-, tri-, and polynuclear indo dyes formed in these reactions.

Lee and Adams [27] have generated the Wurster salt of *p*-phenylenediamine by electrochemical oxidation in buffered media. Above pH 6, the radical stability decreases rapidly, indicating the low stability of these species under hair-dyeing conditions.

For the discussion that follows, a five-step reaction mechanism explains the formation of di-, tri-, and polynuclear indo dyes that have been isolated from oxidation dye reactions.

Step 1. Formation of the diiminium ion from the dye precursor.
Step 2. The diiminium ion attacks a coupler (generally para to an amino or phenolic group), forming a dinuclear species.

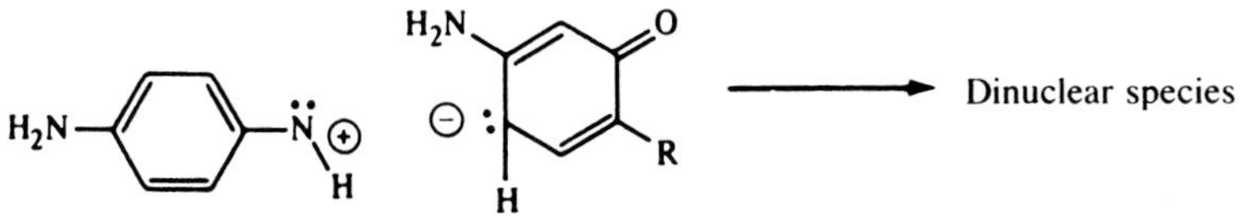

H_2N Dinuclear species R H

m-Aminophenolate ion

Step 3. Oxidation of the dinuclear species to a dinuclear indo dye. If the 4 position of the indo dye is blocked (bears a substituent other than hydrogen), the reaction tends to stop at this step.

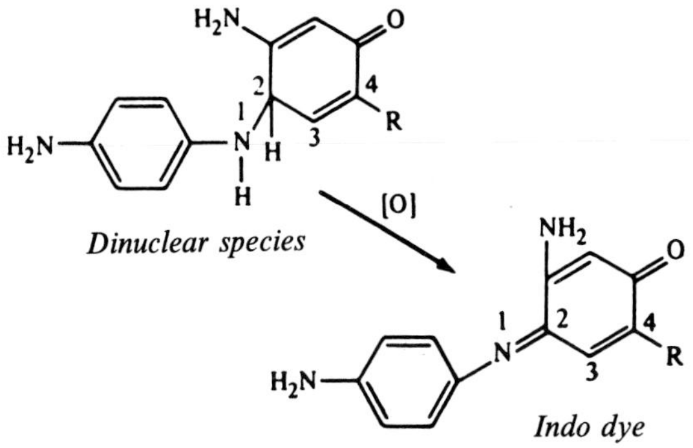

Dinuclear species [O] *Indo dye*

Step 4. Dye precursor or another molecule of indo dye may add by 1,4 addition across the indo dye, forming a trinuclear or polynuclear species (see Table 6–3).
Step 5. Oxidation of trinuclear or polynuclear species to higher indo dyes. Steps 4 and 5 may be repeated, forming higher polymeric dyes.

Some Products Formed in Oxidation Dye Reactions

Bandrowski's Base

It had been proposed [14,28] that *p*-phenylenediamine diffuses into the fibers and is oxidized to diimine (XXII), which in turn condenses with *p*-phenylenediamine to form Bandrowski's base, a brown-black indo dye.

However, a great deal of discussion and work concerning the actual chemical structure [29–31] and importance of Bandrowski's base to oxidative dyeing have taken place [29,30]. Altman and Rieger [29] and Dolinsky et al. [30] have independently provided evidence that the structure shown (XXV) represents the correct tautomer, in contrast to the structure proposed earlier by Bandrowski [31]. Altman and Rieger [29] also suggest that Bandrowski's base is probably the end product of an undesirable side reaction in hair dyeing and is not the main colorant of hair dyed with *p*-phenylenediamine. Corbett [3] confirmed this conclusion by showing that modern dye couplers are several orders of magnitude more reactive to diiminium ion than is *p*-phenylenediamine, which precludes the formation of significant quantities of Bandrowski's base in modern oxidation dyes.

The formation of Bandrowski's base may be described by a mechanism consistent with the general scheme described in the previous section.

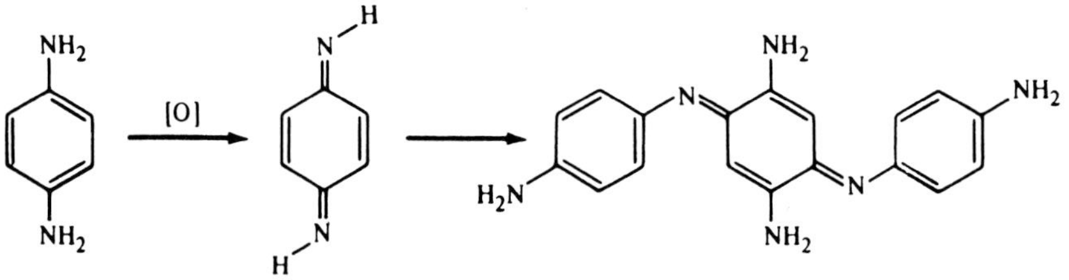

NH_2 [O] N H N H NH_2 NH_2 N N H_2N NH_2 NH_2

Diimine (XXII) *Bandrowski's base* (XXV)

Polyindophenols from Resorcinol and *p*-Phenylenediamine

Resorcinol is one of the components of most oxidation dyes, and as such is probably the most commonly used oxidation dye coupler. Brody and Burns [20] have shown that *p*-phenylenediamine, in the absence of hair and phenols, is oxidized to Bandrowski's base. However, when resorcinol is present, polyindophenols (XX) are formed, and the formation of Bandrowski's base is effectively prevented [20]. These brown polymeric polyindophenol pigments have been identified by elemental analysis, acetyl values, and hydrolysis to *p*-phenylenediamine. Low molecular weight di- and trinuclear species were not detected by Brody and Burns, but Shah et al. [21] have isolated a green pigment from hair dyed with mixtures of *p*-phenylenediamine and resorcinol. This pigment was identified as the trinuclear indophenol (XXI).

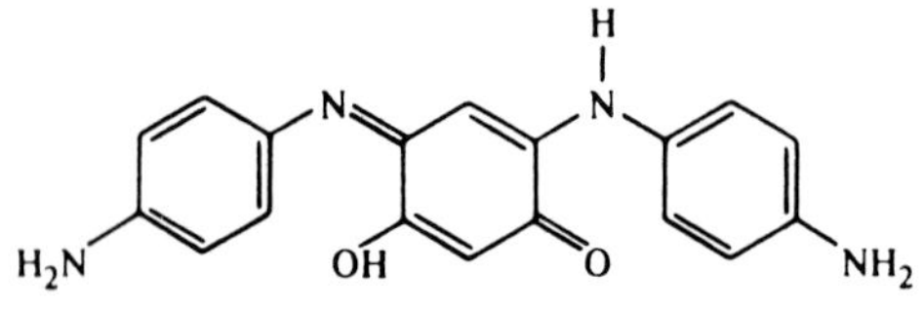

H N N H_2N OH O NH_2

Trinuclear green pigment from p-phenylenediamine and resorcinol [21] (XXI)

The following scheme describes the formation of the trinuclear green pigment and polyindophenols from (I) and resorcinol and is consistent with the general mechanism described earlier. This scheme suggests that the diiminium ion (IX) is the actual active species, and its formation has already been described. In step 2, this electrophilic species attacks a resorcinol anion, para to the phenolic group, forming (XXVI), which is oxidized to in indophenol step 3.

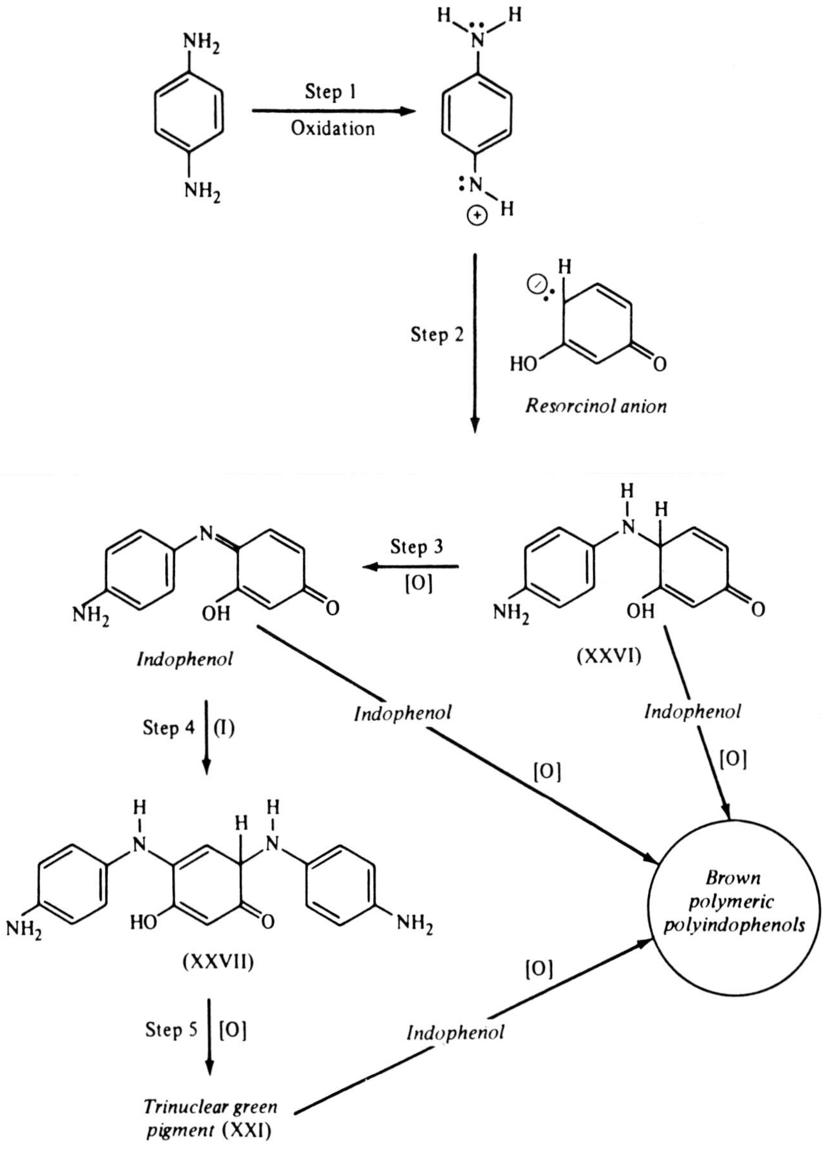

Step 1
Oxidation

Step 2

Resorcinol anion

Step 3
[O]

Indophenol

(XXVI)

Step 4 (I)

Indophenol

Indophenol

[O]

[O]

(XXVII)

Brown polymeric polyindophenols

[O]

Step 5 [O]

Indophenol

Trinuclear green pigment (XXI)

In step 4, *p*-phenylenediamine adds to the indophenol in a 1,4 manner [24,32], producing the trinuclear species (XXVII), which is then oxidized to the trinuclear green pigment (XXI). Several routes exist for formation of polymeric indophenols, and all are analogous to steps 4 and 5. Repetition of these steps will result in the formation of higher polymers. Because most oxidation dyes contain both *p*-phenylenediamine and resorcinol, the formation of these tri- and polynuclear pigments is important in the oxidative dyeing of human hair.

Indamines from *m*-Phenylenediamines

In addition to resorcinol, 4-methoxy *m*-phenylenediamine (XV) is an important dye coupler, and is representative of the class of *m*-phenylenediamine coupling agents.

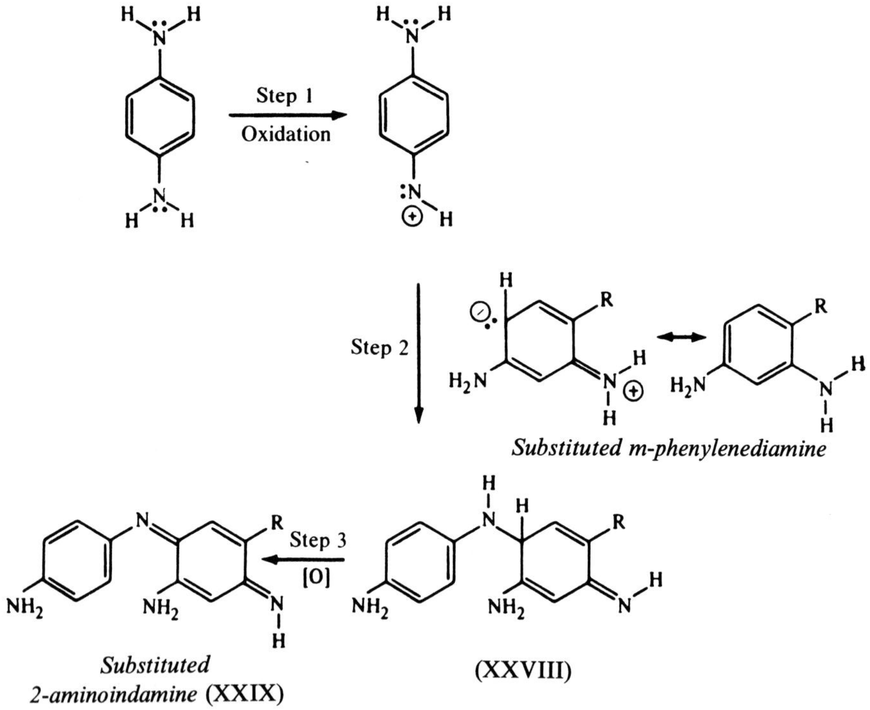

Step 1
Oxidation

Step 2

Substituted m-phenylenediamine

Step 3
[O]

Substituted 2-aminoindamine (XXIX)

(XXVIII)

This type of dye coupler reacts with dye precursors to give blue-violet dyes, which have been shown to be 2-aminoindamines [33]. Mechanistically, the formation of 2-aminoindamines is related to the formation of indophenols and fits the general scheme described earlier.

When the R group of structure (XXIX) is methoxy, it represents the structural formula for the 2-aminoindamine of *p*-phenylenediamine and 4-methoxy *m*-phenylenediamine, which is a relatively stable dye. Other, less frequently used *m*-phenylenediamines form unstable blue dyes that cyclize internally at high humidity, forming red 2,8-diaminophenazines [3].

The parent compound *m*-phenylenediamine (XVI) reacts with (I) to produce a trimer (XXX) analogous to the trinuclear pigment that has been isolated from resorcinol and *p*-phenylenediamine (XXI) [3].

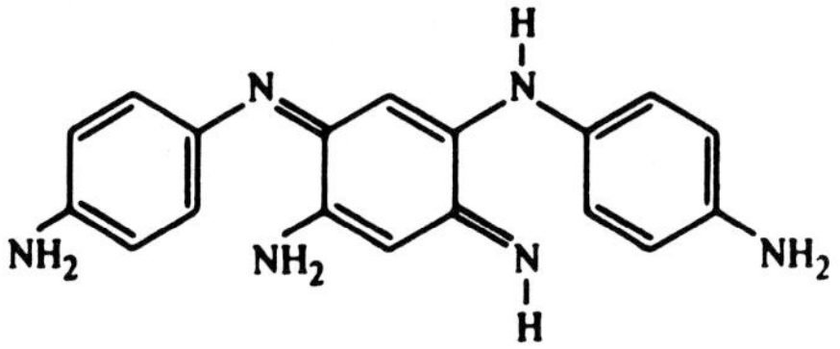

Trimer of p-phenylenediamine and (XVI)
(XXX)

Indo Dyes from *m*-Aminophenols

The reactions of *m*-aminophenols with *p*-phenylenediamine are similar to those described for *m*-diphenols and *m*-diamines and are summarized by the reaction scheme shown here.

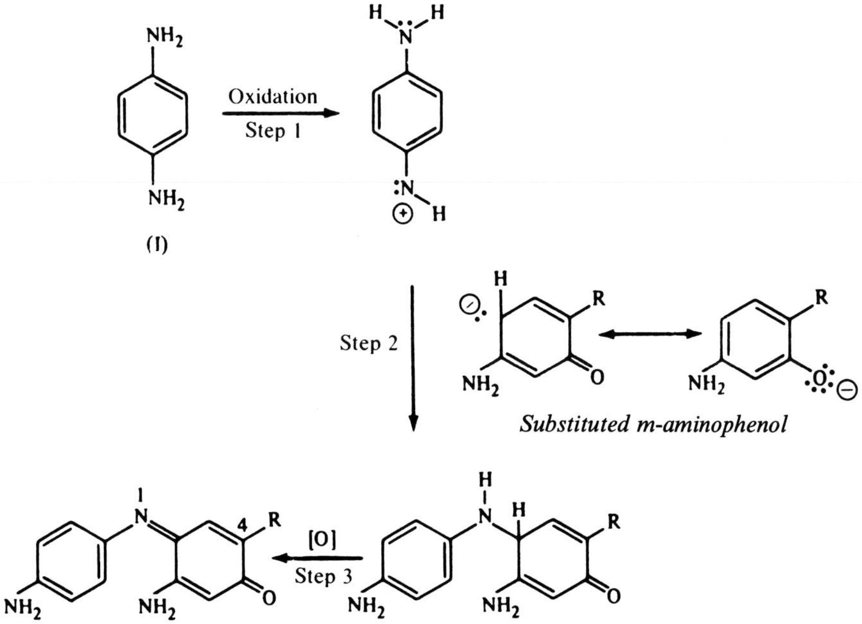

Substituted m-aminophenol

2-aminoindoaniline (XXXII) (XXXI)

Once again (I) is oxidized to the electrophilic diiminium ion (IX), which attacks the aminophenol para to the phenolic group, forming (XXXI). This species is then oxidized to the dimeric indo dye (XXXII) [3], analogous to indophenol. If the R substituent of (XXXII) is alkyl (anything but hydrogen), the reaction generally stops at the dimer stage. However, if R is hydrogen, it adds another molecule of (I) at the 4 position and is oxidized to a trimeric indo dye [3], analogous to the trimeric pigment of (I) and resorcinol; see structure (XXI). If the position para to the phenolic group is blocked, then the diiminium ion attacks para to the amino group, giving 2-hydroxyindamines [3]; see structure (XXXIII).

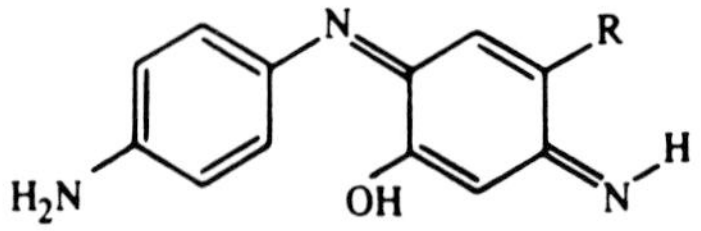

2-Hydroxyindamine (XXXIII)

Speculation about Dye Products Formed with Other Commonly Used Coupling Agents

1-Naphthol is a commonly used dye coupler, and its reactions in oxidative dyeing are probably similar to those described in the previous sections; that is, the active diiminium ion probably attacks para to the naphthol group, forming (XXXIV). Oxidation of (XXXIV) should provide the indonaphthol (XXXV).

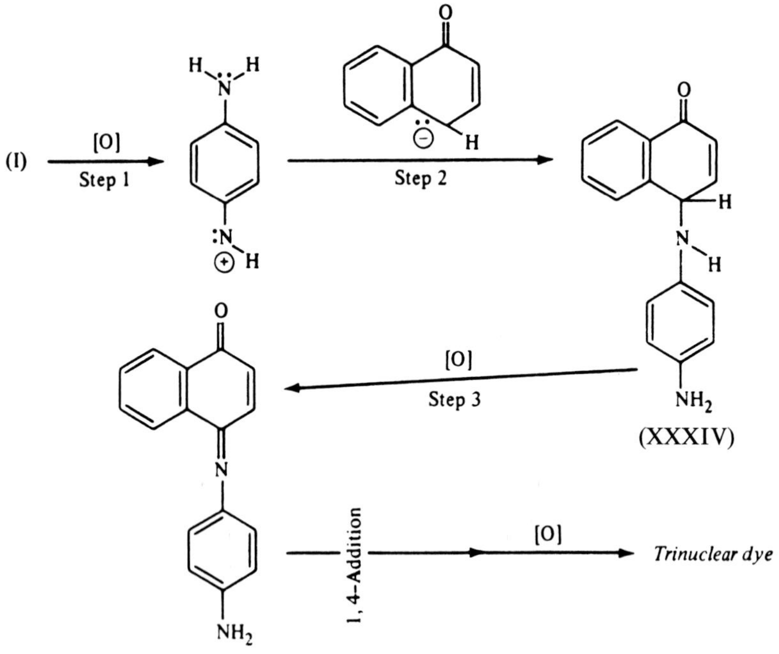

(I) —[O], Step 1→ … —Step 2→ … —[O], Step 3→ …

(XXXIV)

(XXXV) —1,4-Addition→ —[O]→ *Trinuclear dye*

Reaction may stop at this stage, or 1,4-addition and oxidation may occur, producing a trinuclear dye analogous to that from resorcinol and *p*-phenylenediamine (XXI).

Schematic reactions similar to those for resorcinol and diiminium ions can be written for pyrogallol and diiminium ions, suggesting di-, tri-, and polynuclear indophenolic dyes from the reaction of this species with dye precursors.

4-Chlororesorcinol, in the presence of diiminium ions, probably tends to stop at the dimeric stage, forming (XXXVI), because the positioning of the chloro group would tend to inhibit 1,4-addition.

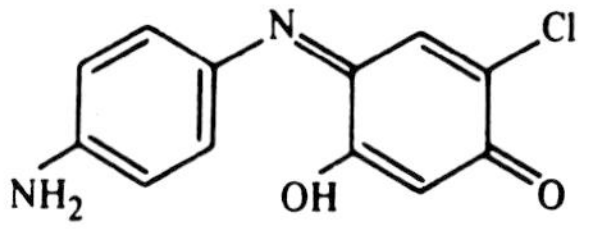

NH_2 OH O Cl N

(XXXVI)

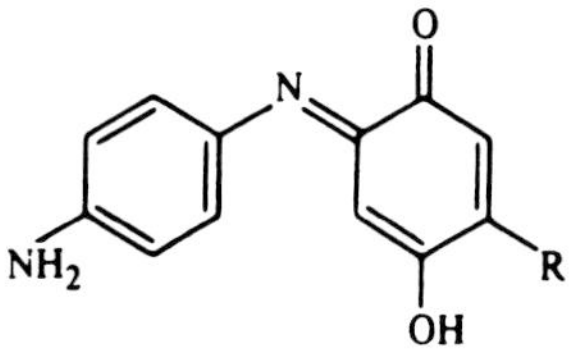

O N NH_2 OH R

(XXXVII)

OH N NH_2 N OH NH_2

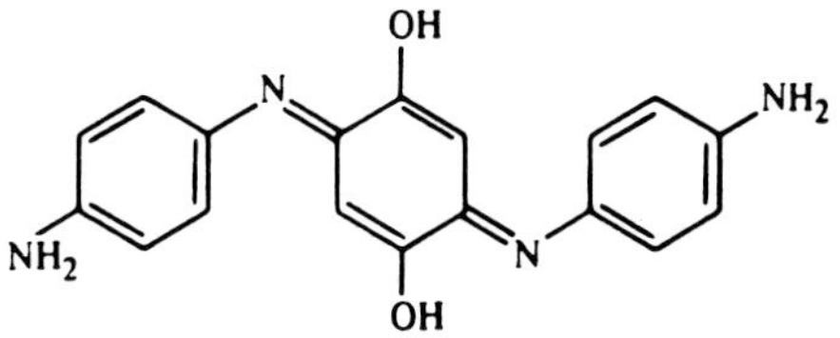

(XXXVIII)

Hydroquinones probably form dinuclear dyes analogous to (XXXVII) and then stop at this stage, whereas the parent compound (unsubstituted hydroquinone) probably forms analogous dinuclear species and even trinuclear dyes (XXXVIII) that are analogous to Bandrowski's base.

Fading of Highly Colored Indo Dyes

Although this phenomenon has not been studied at length, Corbett [3] proposes that fading most likely involves the addition of aromatic moities to dinuclear indo dyes, together with hydrolytic degradation to *p*-diamines and hydroxybenzoquinones, which further interact.

The Formulation of Permanent Hair Dyes

Permanent hair dyes should be formulated with two different compositions or parts. The first composition is a precursor-coupler base containing surfactants (to help dissolve the precursors and couplers, to assist in spreading the dye evenly over the hair, and to help thicken the product so it does not run during use), alkalinity (to facilitate the oxidation reaction), a low concentration of a reducing agent (to inhibit oxidation of the precursors by air), the precursors and couplers, and water. The second composition is an oxidizing base containing oxidizing agent, stabilizer (for the peroxide), and sometimes surfactant (for thickening during use) (Table 6–4).

To formulate the foregoing hair precursor-coupler bases, add the sulfonate and neodol to the water with stirring at room temperature. Add the

TABLE 6–4. Precursor-coupler base.[a]

Ingredient	Percent of ingredient for desired hair color			
	Dark brown	Light brown	Red	Black
Dodecyl benzene Sulfonate (50%)	14	14	14	14
Cocodiethanolamide	9	9	9	9
Neodol 91-2.5	6	6	6	6
Ammonium hydroxide	6	6	6	6
Sodium sulfite	0.3	0.3	0.3	0.3
p-Phenylenediamine	0.4	—	—	0.4
o-Aminophenol	0.3	0.4	—	0.2
p-Aminophenol	—	0.4	0.4	—
4-Methyl-5-aminophenol	—	—	0.4	—
m-Aminophenol	—	—	—	0.2
Water	64	63.9	63.9	63.9

[a] The colors listed can be achieved when starting with naturally light blonde hair. Of course, the actual shade achieved with depend on the starting hair color, the condition of the hair, and the time permitted for reaction.

TABLE 6–5. Oxidizer base.

Ingredient	Percentage
Hydrogen peroxide (30%)	50
Dodecyl benzene sulfonate (50%)	33
Phosphoric acid	1
Water	16

sodium sulfite and then the dye precursors and couplers, stirring until they dissolve completely. Then add the alkalinity, followed by the amide, whereupon the product will thicken considerably. The oxidizer base is detailed in Table 6–5:

To formulate the oxidizer base, dissolve the sulfonate in water with stirring at room temperature. Add the phosphoric acid to the peroxide separately and then add this mixture slowly to the detergent in water with stirring.

Usage Instructions

The usage of this type of product consists of three steps:

the allergy test,
the strand test
application to the hair on the head.

The Allergy Test. Wash a small area inside the arm at the elbow. Apply a few drops of the dye solution with a cotton swab to that area. Leave this area of skin uncovered for 24 hours. If any itching, redness, burning, or any other allergic symptom is noticed, do not use the product. However, if there is no allergic reaction, apply the product.

The Strand Test. Snip a small strand of hair about $\frac{1}{4}$ in. wide, cutting it close to the scalp. Apply tape to the cut ends to hold the hairs together during treatment. Mix a small amount ($1\frac{3}{4}$ tsp.) of precursor-coupler base and oxidizer base, apply the dye solution to the strand of hair, wait 20 min, then wipe the dye from the strand so as to check the color. If the desired shade is not dark enough, apply more dye and recheck. Continue until the desired depth of shade is achieved.

Now the consumer is ready to use the product. Mix 100 g of the precursor-coupler base with 75 g of the oxidizer base and spread it through the hair, allowing the reaction to proceed for 10–20 minutes before rinsing thoroughly with water.

Synopsis of Oxidative Dyeing of Human Hair

The literature on the interactions of oxidation dyes shows that dye precursors and dye couplers are involved. Dye precursors are oxidized to active intermediates, probably diiminium or quinoniminium species. These active intermediates react with resorcinol to form polyindophenols and trinuclear dyes, with *m*-diamines to form indamines, with *m*-aminophenols to form indamines; and with naphthol and hydroquinone to form indophenols. Bandrowski's base is a trinuclear species, the product of a side reaction that does not occur to a significant extent in oxidative dyeing.

NH_2 / NH_2 (I) → Oxidation → (I)

Resorcinols → Indo and polyindophenols
m-Diamines → Indamines
m-Aminophenols → Indoanilines or indamines
Naphthol or hydroquinone → Indophenols
→ Bandrowski's base

Some oxidation dyes contain 3 to 7 components capable of acting as either dye precursors or dye couplers, and most contain *p*-phenylenediamine and resorcinol; in these cases, several di-, tri-, and polynuclear colored species of this general type are formed in the reactions.

Semipermanent Hair Dyes

The term "semipermanent hair dye" refers to those products that dye the hair to a color that persists four to six shampoos and do not use hydrogen peroxide to develop the hair color [5]. For this type of product, preformed hair dyes are used (Table 6–6). To achieve the desired shade, each product contains a combination of as many as 12 dyes [34], such as those described in Table 6–6. Other ingredients in these products are solvents (primarily water and glycols or glycol derivatives), surfactant(s), amide, fragrance, and acid or alkali for pH adjustment.

Semipermanent hair dyes are generally applied to freshly shampooed hair and allowed to remain on the hair for approximately 20 min. The hair is then rinsed with water. Often a "conditioner," packaged with the product, is added and the hair rinsed again and then dried.

The hair dyes of Table 6–6 generally consist of neutral aromatic amine, nitro aromatic amine, or anthraquinone derivatives. They are all highly polar ingredients and can be classified as mono-, di-, or trinuclear (ring) dyes. Wong [35] has studied the kinetics of dye removal from hair for this type of hair dye. Wong concludes that under all conditions the larger, trinuclear dyes rinse more slowly from hair than the smaller, mononuclear hair dyes.

For both bleached and unbleached hair, Wong [35] showed that dye rinse-out for small dye molecules (corresponding to the mononuclear species) of semipermanent hair dyes is diffusion controlled with relatively weak dye–hair interactions. Han et al. [36] confirmed this finding in a study of the diffusion of HC Red 3 (Table 6–6) into and out of hair. For a discussion of techniques to determine diffusion coefficients, see Chapter 5. These workers found essentially no hysteresis in the adsorption versus desorption kinetics, suggesting weak binding between this mononuclear dye and hair. Han et al. [36] indicate that HC Red 3 has a pK_a of 3.7; therefore, this dye is not positively charged, but it is neutral under hair-dyeing conditions, providing a rationale for the weak binding found between this dye and hair.

Wong [35] also found that the larger trinuclear dyes, analogous to those of Table 6–6, have a greater affinity for hair than the simple mononuclear dyes undoubtedly arising from a larger number of polar and van der Waals binding sites between dye and hair.

The principles of this study by Wong are being utilized today in semipermanent hair dye products in the following manner. It is well known that diffusion into and removal of dye are faster in weathered tip ends than in the root ends of hair [5]. Thus, blends of dyes are used not only to obtain the right blend at root and tip but to provide a more even wash-fastness in both root and tip ends of hair. For example, blends of single ring dyes diffuse more readily into and are retained more readily in root ends, whereas blends of dinuclear and trinuclear dyes are retained more readily in tip ends

TABLE 6–6. Examples of some dyes used in semipermanent products.

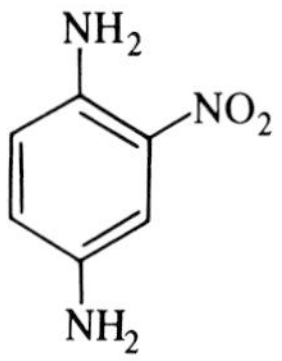

2-Nitro-p-phenylenediamine (red)

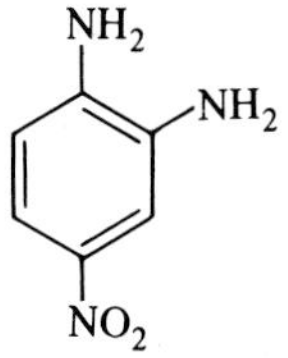

4-Nitro-o-phenylenediamine (yellow-orange)

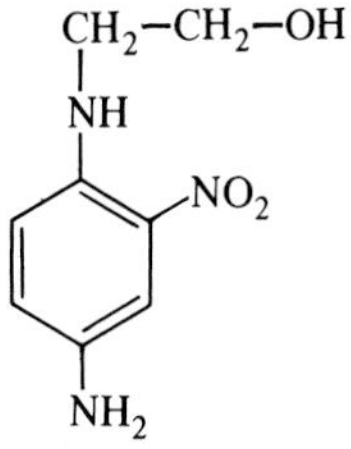

HC red No. 3

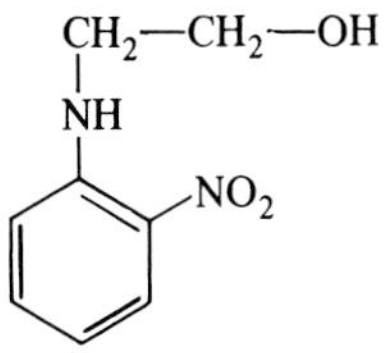

HC yellow No. 2

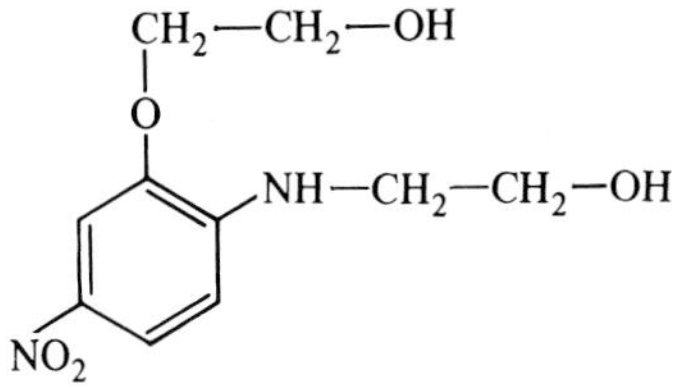

HC yellow No. 4

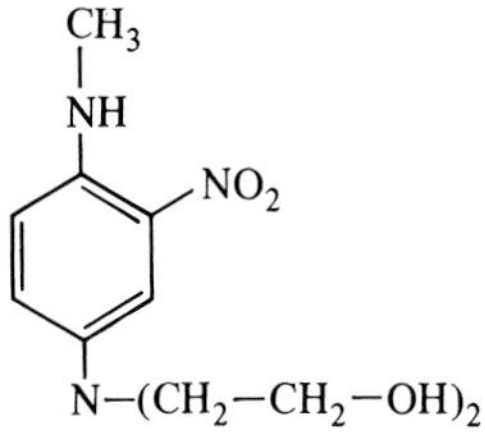

HC blue No. 1

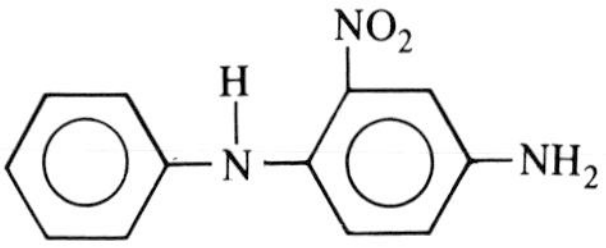

HC red No. 1

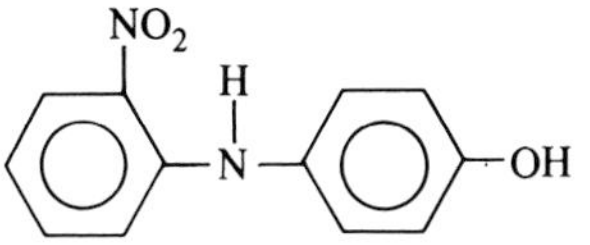

HC orange No. 1

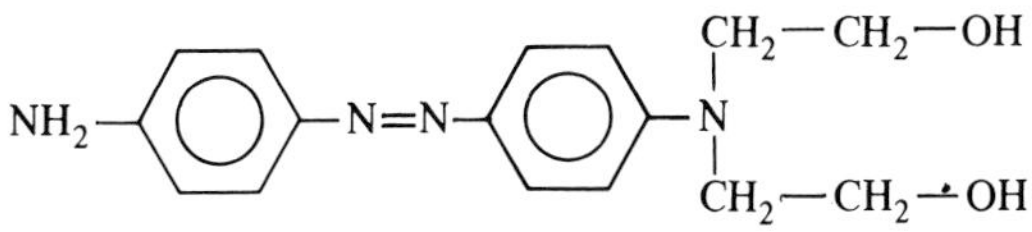

Disperse black 9

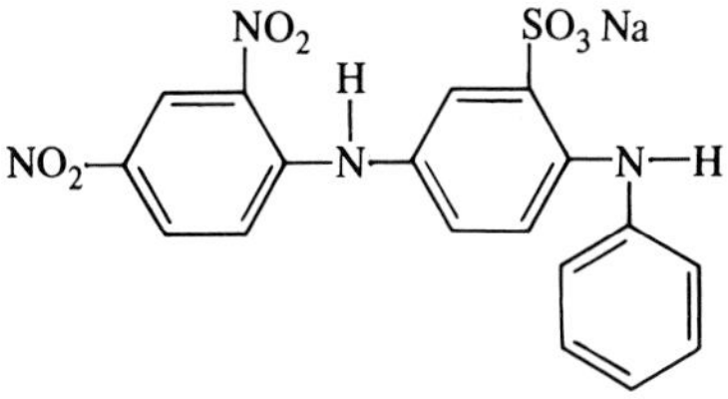

Acid orange 3

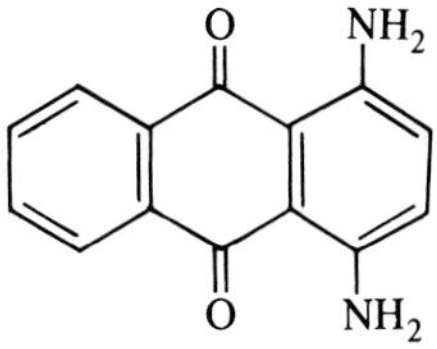

Disperse violet 1

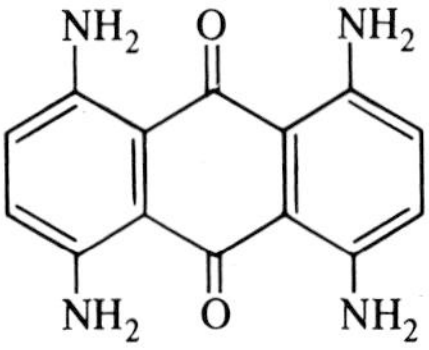

Disperse blue 1

[37]. Thus the proper blending of mononuclear and dinuclear with trinuclear dyes will provide a more even wash-fastness to both root and tip areas of hair.

Han et al. [36] further found that the diffusion coefficient of the mononuclear HC Red 3 increased with pH, with dye concentration (1.0–5.0 g/liter), and with increasing temperature (25°–60°C). However, changing the dyebath solvent from water to 50 vol % aqueous ethanol decreased dye uptake, but the diffusion coefficients remained similar in magnitude. The pH effect can be explained by increased swelling of the hair, and the dye concentration and temperature effects are consistent with expectation for a diffusion-controlled interaction. The solvent effect occurs because the dye is more soluble in the ethanol–water system than in water alone, thereby increasing the affinity of dye for the solvent phase relative to the keratin and causing more of the dye to partition into the aqueous–ethanol phase and less into the hair.

Blankenburg and Philippen [38], using a scanning photometer microscope, studied the reaction of this same dye with hair. At short dyeing times, these authors demonstrated maximum absorption of dye near the fiber exterior, whereas at longer dyeing times the dyestuff concentration increased in the center of the fiber. These results are also consistent with a diffusion-controlled process.

Wong [35] also found that for the smaller mononuclear hair dyes, bleaching increased the rate of dye rinse-out; however, for the larger dyes, the rinse-out rate decreased with a small amount of bleaching, and went through a minimum until the rinse-out rate began to increase with bleaching. Wong suggested that with a small amount of bleaching the larger dyes are able to reach more hindered positions in the hair substrate and thus bind more firmly, and are therefore more difficult to remove with rinsing. However, with additional bleaching, the hair becomes even more penetrable, and the rinse-out rate of the larger dyes increases but never approaches the rinse-out rates for smaller dyes.

Corbett [37] has related dye color and light stability of some semipermanent hair dyes to dye structure. With regard to light-fastness, Corbett demonstrated that for monosubstituted nitrobenzene dyes, the ortho-substituted dyes are the most stable to light and the para-substituted the least light stable. For nitrobenzene dyes containing 2 electron-donating substituents, those with 2,5 substitution are the most light stable, and those with 2,4 and 3,4 substitution are less stable to light.

Formulation of Semipermanent Hair Dyes

To formulate the dye products in Table 6–7, first dissolve the sulfate in water, then add the sulfonate, the neodol and the amide while stirring. Add the dyes and stirr until they are completely dissolved.

TABLE 6–7. Semipermanent hair dyes.[a]

Ingredients	Percent of ingredients for desired hair color		
	Light brown	Dark brown	Red-auburn
Cocodiethanolamide	10	10	10
Sodium dodecyl benzene sulfonate (52%)	4	4	4
Neodol 91-2.5	6	6	6
Sodium lauryl sulfate	2.5	2.5	2.5
2-nitro-*p*-phenylenediamine	0.4	0.4	0.4
HC Red No. 3	0.2	0.2	—
HC Yellow No. 2	0.2	0.2	0.2
HC Blue No. 2	—	0.1	—
Water	76.7	76.6	76.9

[a] These colors can be achieved when starting with naturally light blonde hair. Of course the actual shade achieved will depend on the starting hair color, the condition of the hair, and the time permited for reaction.

Usage Instructions for Semipermanent Hair Dyes

There are three basic steps to the usage instructions for this type of product: first, the allergy test; second, the strand test (both are described under the formulation of permanent hair dyes); and third, the actual use of the product on the hair. To apply a semipermanent hair dye, first shampoo the hair, rinse, and towel dry. Thoroughly saturate the hair with the product, using plastic gloves and being careful to minimize contact of the dye solution with the scalp. Allow the product to react with the hair for about 30 min or the time indicated by the strand test. A creme rinse/conditioner may be added to the hair, which is then lathered and rinsed. Or, the hair may be rinsed thoroughly with water, but not shampooed. It is set and styled as usual.

In summary, semipermanent hair dyes are products that persist through four to six shampooings. They are mixtures of preformed dyes, generally with 10 to 12 dyes mixed to achieve the desired shade. These dyes are generally mononuclear, dinuclear, or trinuclear species and are usually aromatic amines, amino nitrobenzenes, or anthraquinone derivatives. These dyes generally diffuse into hair and are retained by weak polar and van der Waals attractive forces, so the affinity of the dyestuff increases with increasing molecular size. Peroxide is not used to develop the color, and therefore no major chemical changes occur to the fibers during this type of dyeing.

Temporary Hair Dyes or Color Rinses

The objective of temporary hair dyes is to provide color that can be shampooed out of the hair with a single shampooing. Table 6–8 depicts structures for a few hair-coloring ingredients used in today's hair color rinses. A large number of hair color ingredients previously used in color rinses are described in the article by Wall on hair dyes [39]. Each color rinse product consists of a mixture of color additives, either among those described in Table 6–9 or similar FD&C or D&C colors. Generally two to five color ingredients are mixed to achieve the desired shade [39], because a single ingredient will not provide the desired color to the hair. Two dyes are

TABLE 6–8. Some hair color ingredients used in color rinse products.[a]

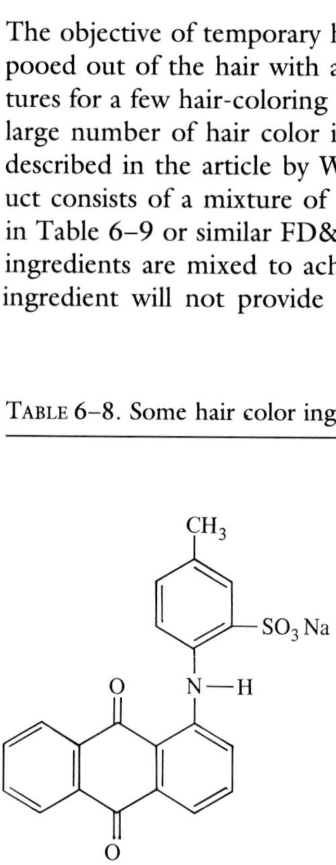

Acid violet 43
External D&C violet No. 2
(*name for certified batches*)

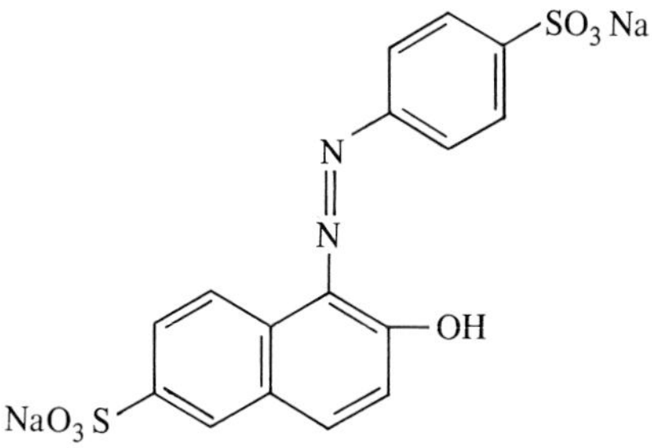

F D&C yellow No. 6

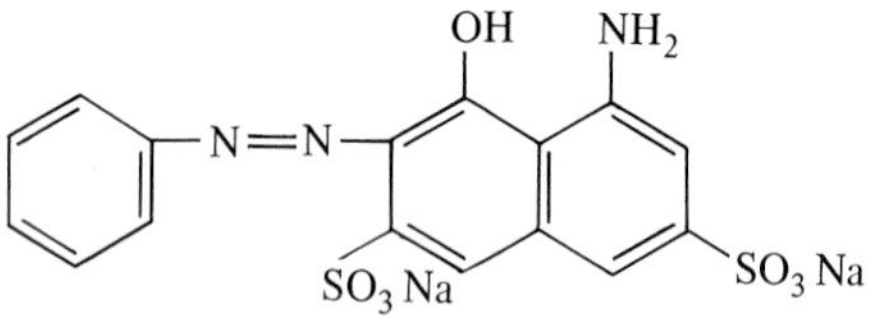

D&C red No. 33

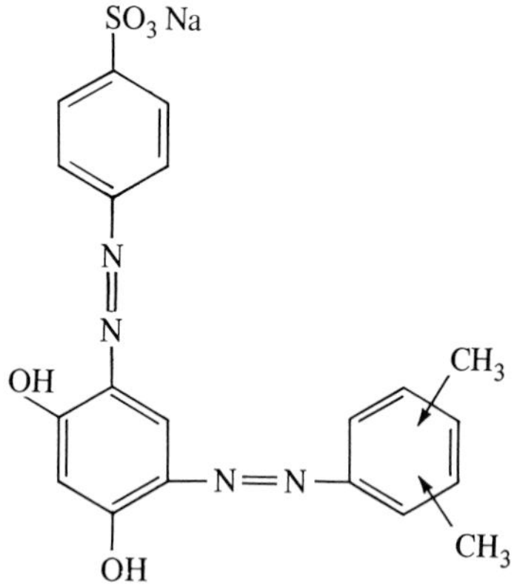

Acid orange 24
D&C brown No. 1
(*name for certified batches*)

[a] Some other dyes used in this type of product are direct black 51, direct red 80, acid black 2, D&C yellow No. 10, and other FD&C and D&C dyes.

TABLE 6–9. Prototype hair rinse formulations.[a]

Ingredients	Percent of ingredients for desired hair color		
	Brown	Red	White
Nonoxynol-9	1.0	1.0	1.0
Hydroxyethylcellulose (HHR)	0.7	0.7	0.7
Cetyltrimmonium chloride	0.6	0.6	0.6
Neodol 91-2.5	0.5	0.5	0.5
Citric acid trihydrate	0.5	—	0.5
Trisodium phosphate	—	0.3	—
Direct black 51	0.05	0.01	—
Acid violet 43	0.04	0.03	—
Direct red 80	0.03	0.05	—
Acid orange 24	0.04	0.02	—
External D&C Violet 2	—	—	0.03
D&C Red 33	—	—	0.01
FD&C Yellow 6	—	—	0.005
D&C Yellow 10	—	—	0.005
Water	96.54	96.79	96.65

sometimes used to provide tints for gray hair, and four to five dyes are generally mixed to achieve reds, browns, or black.

These products are usually applied to freshly shampooed hair and combed through. An alternative is to spray the product onto the hair and comb it through to achieve even distribution of the dyes. The hair is then set and dried without rinsing to minimize penetration of dyes into the fibers. The dyes used in color rinses are generally larger molecular species than those used in semipermanent hair dye products. Color rinse dyes are generally anionic or acid dyes and are selected to provide maximum water solubility and minimum penetration so they can be shampooed out of the hair.

Temporary dyes or color rinses are acid dyes [6–8] similar to those used in wool dyeing. These dyes do not bind firmly either and they may be removed by a single shampooing. Wool dyeing is very different from hair dyeing primarily because of temperature. In fact, low-temperature dyeing in wool is 80°C to 85°C. Solvent-assisted dyeing permits dyeing wool at 60°C to 70°C, which is still too high for human hair. For a more complete discussion of the dyeing of wool fiber, see the book *Wool Dyeing* edited by D.M. Lewis [40].

Formulation of Color Rinses

Temporary hair dye products frequently contain thickeners, a surfactant (cationic and or nonionic), sometimes a hair-setting or thickening polymer, and a buffer that is generally an acid such as tartaric, acetic, or citric to provide an acid medium for application of the dyes to the hair.

To make the hair rinses described in Table 6–9, first hydrate the hydroxyethylcellulose (0.7 g) with 49 g of water by stirring, then add the neodol, the nonoxynol, and the cetrimmonium chloride individually while stirring. Add the remaining water followed by the buffer, and then slowly dissolve the dyes in the product with stirring.

Use of Color Rinses

First shampoo the hair with a good cleaning shampoo. Rinse the hair thoroughly and towel dry. Thoroughly saturate the hair with the product, using plastic gloves if desired, being careful to minimize contact with the scalp. Dry the hair and style as desired.

Other Dyes for Hair

Metallic Dyes

Salts of several metals, including lead, silver, bismuth, cobalt, copper, iron, and mercury, have been used in the past for dyeing hair [41]. Among these metallic dyes, lead dyes are the only ones in commercial use today (Table 6–10). Lead dyes contain lead acetate and sulfur [42] which react with hair to darken slowly, presumably forming lead-sulfur complexes in the cuticle layers. These products are popular among men because of the slow gradual buildup of color requiring many applications to achieve the desired darkening; however, the shades are limited, and the dye can react if treated subsequently with bleaches, permanent waves, and even with certain other hair dyes [15].

Formulation of a Lead Acetate-Sulfur Hair Dye

Dissolve the lead acetate in water with stirring. Disperse the sulfur in propylene glycol and add this dispersion to the lead acetate in water while stirring vigorously. Continue to stir while adding the fragrance and preservative dissolved in alcohol and the surfactant.

Another form of metallic dye is called premetalized dyes. These compounds are metal complexes of anionic dyes with chromium, cobalt, or some other metal, generally in a ratio of 1:2 (metal:dye); see structure (XXXIX) [43–45]. The chemistry of premetalized dyes and their interactions with wool fiber have been described by Schetty [44]. Some tradenames are Irgalan, Cibalan, Lanasyn, Carbolan, and Isolan. Premetalized dyes have been patented and described for use with hair [46,47]. Premet-

TABLE 6–10. Lead acetate-sulfur hair dye.

Ingredient	Percentage
Sulfur (finely divided)	1.0
Lead acetate	1.0
Nonoxynol-9	0.8
Denatonium benzoate	0.5
Fragrance	0.2
Propylene glycol	20.0
Ethanol	3.0
Water	73.5

alized dyes are often classified as acid dyes rather than as metallic dyes, even though from substantivity and bonding considerations they are probably more like metallic dyes. The use of premetalized dye in the presence of an alcohol (benzyl or amyl) is called solvent-assist dyeing [43], an interesting technique that produces a larger uptake of dye relative to a pure aqueous solvent.

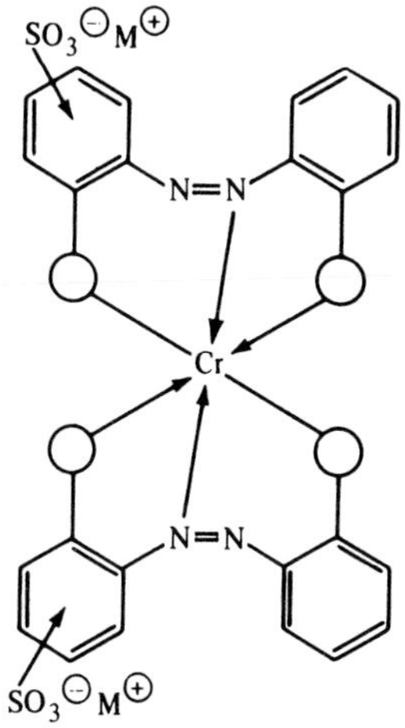

$SO_3^{\ominus}M^{\oplus}$ N=N Cr N=N $SO_3^{\ominus}M^{\oplus}$

(XXXIX)

Vegetable Dyes

Natural organic substances from plants are the earliest known hair dyes, but of the many substances of this type that have been tried over the years, only henna and camomile are currently used to a significant extent commercially [2]. The active coloring material of henna (Lawsone) is 2-hydroxy-1,4 naphthaquinone. The structure of this active dye ingredient was established by Tommasi [47] and Cox [48].

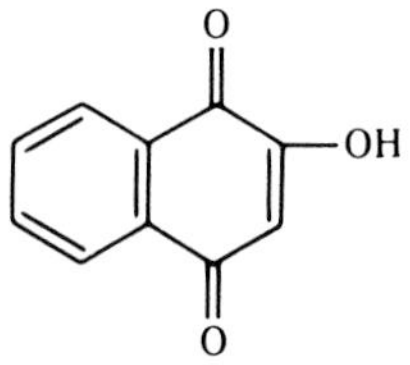

2-Hydroxy-1,4-naphthaquinone (*henna*)

This substance is the principal dyeing agent of the shrub henna, known as Egyptian privet, and has been used for dyeing hair and staining nails since ancient times in Egypt [1]. Henna is found in the leaves of the plant and is usually extracted with aqueous sodium bicarbonate. Henna produces yellow to reddish shades in proteins, is normally used in combination with camomile or with another dye substance to improve the color, and is normally applied in acidic media.

The active coloring substance from camomile flowers is a polyhydroxy flavone, 4′,5,7-trihydroxy flavone [2]. This substance is also found in parsley; and its common name is apiginin [49].

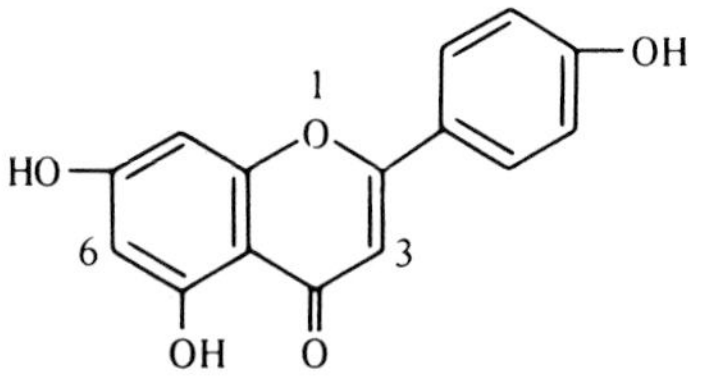

4′5,7-Trihydroxy flavone (*camomile*)

Other polyhydroxy flavones exist in plants [47]. These flavones are usually combined as glucosides or rhamnosides and are generally yellow. These materials have been used for dyeing cloth and hair, and they are usually mordanted with chromium, tin, or iron salts for dyeing cloth. This mordanting process changes the color as well as the binding character of the dye.

These two dyestuffs, henna and camomile, are chemically related by the α-, β-unsaturated grouping. This group is capable of undergoing 1,4-addition reactions with free amino or other nucleophilic groups of the side chain residues of hair proteins. The review by Wall [2] describes vegetable dyes in greater detail, including their history, sources, active ingredients, and even dyeing conditions.

Fiber-Reactive Dyes

A fiber-reactive dye binds to the hair or wool through covalent bonds, usually involving either nucleophilic displacement by amino, guanidino, hydroxyl, or sulfhydryl groups of the keratin on a dyestuff containing either a labile halogen or sulfate grouping, or nucleophilic addition onto a dyestuff

TABLE 6–11. Some examples of fiber-reactive dyes.

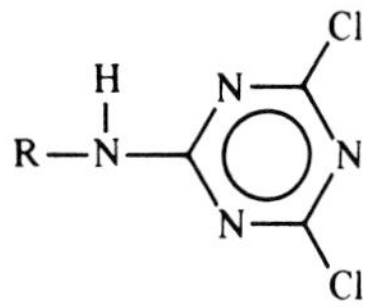

Dichlorotriazine type

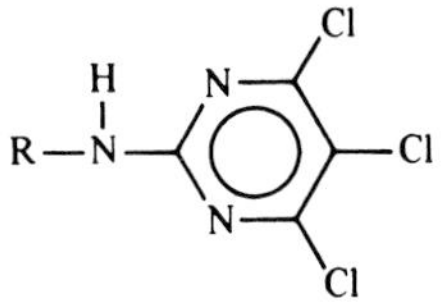

Trichloropyrimidine type

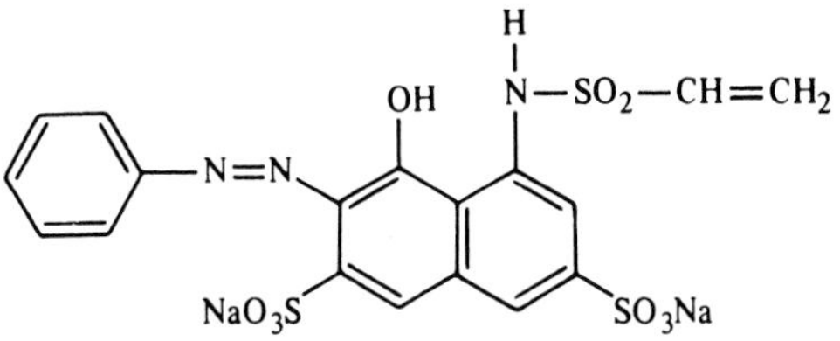

An activated olefin type

containing an activated olefinic linkage. This type of dyestuff has not been used commercially for dyeing human hair; nevertheless, these dyes offer a novel approach to permanent dyeing. A few examples of this type of dye are described in Table 6–11. Fiber-reactive dyes will react with unreduced keratin; however, prior reduction provides additional mercaptan groups for reaction with the dye and therefore enhances dyeing. For additional information, see the articles by Brown [12], Stead [43], and by Shansky [50] and references therein.

Safety Considerations for Hair Dyes

The general toxicology of hair dyes has been reviewed by Corbett [5,51], and the mutagenicity and carcinogenicity of hair dyes has been reviewed by Kirkland [52]. In addition, clinical observations of adverse reactions of hair dyes have been reported by Bergfeld [53], Ishihara [54], and Bourgeois-Spinasse [55]. The following discussion is a synopsis of these toxicological reviews and reports.

The primary toxicological concerns of hair dyes, primarily oxidation hair dyes, are with contact dermatitis and long-term "potential" systemic effects [5]. Of all hair products, the most sensitizing are the paradiamine oxidation dyes [55], and the most sensitizing ingredients of these products are *p*-phenylenediamine, *p*-toluenediamine sulfate (II, Table 6–1), and *o*-chloro-*p*-phenylenediamine (IV, Table 6–1), although other related aromatic amines have also been shown to have some sensitizing potential [54].

Because *p*-phenylenediamine is the major component of oxidation hair dyes and oxidation dyes are the most widely used of all hair dyes, *p*-phenylenediamine is the sensitizer of prime concern. Although Corbett [5] points out that while *p*-phenylenediamine is a strong allergin, the incidence of allergic reactions by oxidation dyes is of low frequency and the reaction when observed is generally mild. Corbett suggests that this low incidence of sensitization results from the rapid reaction and decreasing concentration of this reactive ingredient during actual hair dyeing [5].

The sensitization symptoms from oxidation dyes are a discrete dermatitis at the periphery of the scalp and on the edges of the ears, an itching scalp, and occasionally eruptions that occur on the face, especially around the eyelids [54,55]. Eruptions on the trunk and limbs are rare [54]. Usually symptoms appear several hours after the dyeing process. Treatment of this allergic reaction consists of oxidation of residual paradiamine with peroxide in saline solution and application of corticoid creams or lotions [55]. Allergic reaction to hair color rinses is rare [50]; however, a few incidents of allergic reaction to semipermanent hair dyes have been reported [51].

Misra et al. [56] has provided data to suggest that the skin toxicity of *p*-phenylene diamine might relate to the interaction with lipophilic biomolecules and the subsequent biotransformation products.

Gagliardi et al. [57] examined the alleged claim that exposure to vapors of *p*-phenylenediamine in hairdressing salons might cause lung allergies. Their results show that this is not the case and that it is unreasonable to consider that hairdressers are at risk to *p*-phenylenediamine-induced asthma.

Long-term toxicological risk from semipermanent hair dyes is low [51]. However, 2-nitro-*p*-phenylenediamine and 4-nitro-*o*-phenylenediamine, currently used in Europe but not in the United States in semipermanent hair dyes, have been tested extensively for potential carcinogenicity. The latter dye was shown to be noncarcinogenic in animal feeding studies, while the former diamine caused adenomas in rats, but only at the highest feeding level. At moderate and low feeding levels, this dye did not produce adenomas [51].

Several hair dye ingredients, primarily permanent hair dye components, have been tested for potential carcinogenicity [5,52]. The impetus for this testing arose from the finding that several aromatic diamines provide a positive reaction as mutagens [5,51] in the Ames bacterial screening assay against the bacterium ***Salmonella typhimurium*** [56]. Corbett summarized the details of the mutagenicity screening of oxidation dye components, concluding that the Ames test for aromatic diamines does not correlate as well with results of carcinogenicity in animal feeding studies, as for other chemical types [5].

Both Corbett [5] and Kirkland [52] have summarized the results of several common hair dye ingredients in in vitro mutation tests, in dominant lethal animal testing, and in epidemiologic studies. One of these testing programs consists of a CTFA study involving more than 35 oxidation dye components and 34 textile dyes used in temporary hair color products. From this work and other studies, the FDA has determined (in October 1979) that all hair dye products containing 2,4-diamino anisole must bear a cancer warning label [57–59]. Industry has responded by removing this ingredient from oxidation hair dyes.

Corbett concludes that "all animal studies completed to date have shown hair dyes to be safe for their intended use," and Kirkland [52] concludes that "epidemiological and human monitoring studies have not detected any such risk (carcinogenicity) in exposed human populations." However, Kirkland suggests that more controlled epidemiological studies and more extensive monitoring of exposed populations are in order.

References

1. Thompson, R.H. Naturally Occurring Quinones, p. 56. Academic Press, New York (1957).
2. Wall, F.E. In Cosmetics Science and Technology, Sagarin E., ed., Chap. 21. Interscience, New York (1957).
3. Corbett, J. J. Soc. Cosmet. Chem. 24:103 (1973).
4. Milos, Z.B.; Brunner, W.H. U.S. patent 3,168,441 (1965).
5. Corbett, J. Cosmet. Toiletries 91:21 (1976).
6. Societe Monsavon, L'Oreal. Br. patent 758,743 (1956).
7. Lyons, J.R. U.S. patent 3,480,377 (1969).
8. Tucker, H. J. Soc. Cosmet. Chem. 22:379 (1971).
9. Therachemie Chemisch Therapeutische G.M.B.H. Br. patent 909,700 (1962).
10. Cook, M. Drug Cosmet. Ind. 99:52 (1966).
11. Kass, G. Am. Perfumer Aromat. 68(1):25 (1956).
12. Brown, J. J. Soc. Cosmet. Chem. 18:225 (1967).
13. Cox, H. Analyst 65:393 (1940).
14. Heald, R. Am. Perfumer and Cosmet. 78:40 (1963).
15. Alexander, P. Am. Perfumer Cosmet. 82:31 (1967).
16. Lange, F. Fette Seifen Anstrichm. 67:222 (1965).
17. CTFA Cosmetic Ingredient Dictionary, Estrin, N., ed. Cosmetic, Toiletry and Fragrance Association, Washington, DC (1973).
18. Corbett, J. J. Soc. Dyers Col. 84:556 (1968).
19. Therachemie. Br. patent 1,023,327 (1966).
20. Brody, F.; Burns, M. J. Soc. Cosmet. Chem. 19:361 (1968).
21. Shah, M.J.; Tolgyesi, W.; Britt, A.D. J. Soc. Cosmet. Chem. 23:853 (1972).
22. Corbett, J. J. Soc. Cosmet. Chem. 20:253 (1968).
23. Taylor, T.; Baker, W. In Sidgwick's Organic Chemistry of Nitrogen, pp. 97–102. Clarendon Press, Oxford, UK (1945).
24. Fieser, L.; Fieser, M. In Advanced Organic Chemistry, pp. 853-858. Reinhold, New York (1961).
25. Brody, F.; Pohl, S. U.S. patent Appl. 186,475 (Oct. 1971).
26. Husemeyer, H. J. Soc. Cosmet. Chem. 25:131 (1974).
27. Lee, H.; Adams, R. Anal. Chem. 34:1587 (1962).

28. Kass, G. Am. Perfumer Aromat. 68:34 (1956).
29. Altman, M.; Rieger, M. J. Soc. Cosmet. Chem. 19:141 (1968).
30. Dolinsky, M., et al. J. Soc. Cosmet. Chem. 19:411 (1968).
31. Bandrowski, E. Monatsschr. Chem. 10:123 (1889).
32. Rodd, E. In Chemistry of Carbon Compounds, Vol. III, p. 700. Elsevier, New York (1956).
33. Corbett, J. J. Chem. Soc. B:827 (1969).
34. Brown, K. J. Soc. Cosmet. Chem. 33:375 (1982).
35. Wong, M. J. Soc. Cosmet. Chem. 23:165 (1962).
36. Han, S.K.; Kamath, Y.K.; Weigmann, H.-D. J. Soc. Cosmet. Chem. 36:1 (1985).
37. Corbett, J. J. Soc. Cosmet. Chem. 35:297 (1984).
38. Blankenburg, G.; Philippen, H. J. Soc. Cosmet. Chem. 37:59 (1986).
39. Wall, F.E. In Cosmetics Science and Technology, Sagarin, E, ed., pp. 486–488. Interscience, New York (1957).
40. Lewis, D.M. Wool Dyeing. Society of Dyers & Colourists, UK (1992).
41. Rostenberg, A.; Kass, G. Soap Perfumery Cosmet. 39(1):45 (1966).
42. Cook, M. Drug Cosmetic Ind. 101:40 (1967).
43. Stead, C. Chem. Br. 1:361 (1965).
44. Schetty, G. J. Soc. Dyers Col. 71:705 (1955).
45. Peters, L.; Stevens, C. Br. patent 826,479 (1960).
46. Stead, C. Am. Perfumer Cosmet. 79:31 (1964).
47. Tommasi, G. Gazz. Chim. Ital. 50(I):263 (1920).
48. Cox, H.E. Analyst 63:397 (1938).
49. Morton, A.A. The Chemistry of Heterocyclic Compounds, pp. 169–174. McGraw-Hill, New York (1946).
50. Shansky, A. Am. Perfumer Cosmet. 81:23 (1966).
51. Corbett, J. In Hair Research, Orfanos, E.; Montagna, W.; Stuttgen, G., eds., pp. 529–535. Springer-Verlag, Berlin (1981).
52. Kirkland, D.J. Int. J. Cosmet. Sci. 5:51 (1983).
53. Bergfeld, W.F. In Hair Research, Orfanos, C.E.; Montagna, W.; Stuttgen, G., eds., pp. 507–511. Springer-Verlag, Berlin (1981).
54. Ishihara, M. In Hair Research, Orfanos, C.E.; Montagna, W.; Stuttgen, G., eds., pp. 536–542. Springer-Verlag, Berlin (1981).
55. Bourgeois-Spinasse, J. In Hair Research, Orfanos, C.E.; Montagna, W.; Stuttgen, G., eds., pp. 543–547. Springer-Verlag, Berlin (1981).
56. Misra, V.; Gupta, V. & Viswanatnan, P.N. Int. J. Cosmet. Sci. 12:209 (1990).
57. Gagliardi, L.; Ambroso; Mavro J. & Discalzi, G. Int. J. Cosmet. Sci. 14:19 (1992).
58. Fox, J.L. Chem. Eng. News 55:34–46 (1977).
59. Grief, M.; Wenninger, H. Cosmet. Technol. 2:43 (1980).

7

Polymers and Polymer Chemistry in Hair Products

Polymers have become increasingly important components of cosmetics over the past few decades. The more important uses of polymers are as primary ingredients or adjuncts in shampoos, conditioning products, styling products (lotions and gels), mousses, and hair sprays. Polymers have been used to condition hair [1,2] and to improve the substantivity of other ingredients to hair [3,4]; to improve combing [1], manageability [1,2], body [5], and curl retention [2,6,7]; to thicken formulations [8,9]; and to improve emulsion stability [8].

During the past 5 years, the use of polymers in 2-in-1 shampoos has become increasingly important commercially, although relatively little in terms of new polymers of commercial significance has occurred in the area of hair fixatives. Nevertheless, because of environmental concerns, a great deal of research has taken place to develop polymers that release water more readily and polymers that are more compatible with water, with the aim of developing spray products that can be formulated with lower levels of volatile organic compounds (VOC). The driving force at this time has become the standards proposed by the California Air Resources Board (CARB), which has proposed that the maximum allowable VOC emissions for hairsprays be 80% by January 1, 1993, and 55% by January 1, 1998.

Other CARB regulations relevant to hair care are hair mousses, 16% VOC, by January 1, 1994; and hair styling gels, 6% VOC, by January 1,

1994. These regulations specify a 1-year period after the effective dates to sell existing products that do not comply with these regulations in the San Francisco Bay Area Air Quality Management District. Hair sprays are an exception; hair sprays that are not in compliance must be off the shelves by the effective dates.

The following areas are of special relevance to application of polymers in hair products:

the binding interactions of polymers to hair
the chemical nature of polymers used in hair products
in situ polymerization reaction mechanisms
rheological or flow properties of polymer solutions
film formation and adhesional properties of polymers.

The major emphasis in this chapter is on the first three of these subjects: the chemical and/or binding interactions of polymers to hair; the chemical nature of hair sprays, setting products, and mousses; and in situ polymerization reactions in hair. A new section presenting an introduction into the formulation of hair fixative products has been added to this chapter. Although the rheological properties of polymer solutions are especially important to formula viscosity and to the sensory perceptions of cosmetics, they are not emphasized here. It suffices to say that cellulosic ethers [8,9] are probably the most important thickening agents in hair products and that ethoxylated esters and carboxy vinyl polymers are also important.

For hair sprays, polymer setting products, and polymeric conditioners, film properties as well as the ability of the polymer solution to spread over the fiber surface are important to product performance. The spreading characteristics of the polymer are governed by its solution viscosity and the wettability of the fiber surface against the polymer solution. For optimum spreading, low solution viscosity is important, and both polar and dispersion interactions have been shown to be important to the ease of spreading over hair fiber surfaces [10]. The surface of cosmetically unaltered hair is generally considered to be hydrophobic, whereas the cortex is more hydrophilic.

Kamath et al. [10] have shown that both bleaching and reduction of hair increase the wettability of hair, making its surface more hydrophilic. For a more thorough treatment of wettability and the spreading of liquids on solids, see the references for this chapter [10–13].

The Binding of Preformed Polymers to Hair

Chapter 5 describes the work of Steinhardt and Harris and the affinities of organic acids [14] and quaternary ammonium hydroxide compounds [15] to keratin. It illustrates the importance of increasing molecular size and even nonprimary bonds to the substantivity of ingredients to hair. In the case of polymeric ingredients, these same principles are operative, and even more important. Chapter 5 also describes the important interactions leading to deposition of conditioning ingredients from both anionic and cationic surfactant systems, including a theory that considers a continuum between a charge-driven adsorption process and a hydrophobically driven process. This same theory is not repeated in the current discussion; however, it is highly relevant to the following discussion, which focuses on factors that are important to the deposition of polymeric ingredients onto human hair.

Chemical Bonding and Substantivity

It is convenient to consider three extreme types of bonds between polymer and hair:

Primary valence bonds (ionic and covalent bonds)
Polar interactions (primarily hydrogen bonds)
Dispersion forces (van der Waals attractions).

It should be noted that bond classifications of this type are not rigorous, and the transition from one type to another is gradual. Therefore, intermediate bond types do exist [16], although for simplicity in the following discussion a rigorous classification of bond type is presented.

Primary valence bonds include ionic and covalent bonds and are the strongest binding forces. They generally have bond energies of approximately 50 to 200 kcal/mole [17]. Ionic bonds are extremely important to the interactions of polymeric cationic ingredients and hair, whereas covalent bonds are probably involved between polymer and hair in certain in situ polymerization reactions or in the reaction of oxidation dyes with hair.

Hydrogen bonds are the most important polar interactions and are the next strongest binding forces, with bond energies generally of the order of 4 to 10 kcal/mole [17]. These bonds are important to the binding of polymers containing polyalcohol or polyamide units, including polypeptides and proteins.

Dispersion forces or van der Waals attractions have bond energies generally of the order of 1 kcal/mole [18]. van der Waals attractions are relatively weak and are dipolar in nature. Because electrons are in constant motion, at any instant in time the electron distribution is probably distorted, creating a small dipole. This momentary dipole can affect the electron distribution in an adjacent molecule, and if contact is just right, attraction is induced [19]. These attractive forces are short range and act only between the surfaces of molecules. Therefore, the total strength of van der Waals bonding increases with molecular surface area (i.e., with increasing molecular size), and in polymers it can approach the strength of primary valence bonds.

TABLE 7–1. Polar and dispersion forces in polymers.

Polymer	Structural unit	Approximate molar cohesions for structural units shown[a] (kcal/mole)
Polyethylene	$(-CH_2CH_2CH_2CH_2-)$	1.0
Poly isobutylene	$(-CH_2C(CH_3)_2-CH_2C(CH_3)_2-)$	1.2
Polystyrene	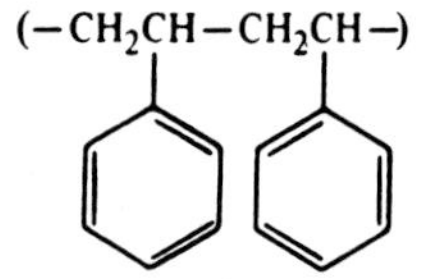 $(-CH_2CH(C_6H_5)-CH_2CH(C_6H_5)-)$	4.0
Polyvinyl alcohol	$(-CH_2CH(OH)-CH_2CH(OH)-)$	4.8
Polyamides	$(-CH_2CH(C{=}O\,NH_2)-CH_2CH(C{=}O\,NH_2)-)$	5.8

[a] Molar cohesions listed are actually for a chain length of 5 angstrom units (Å). The structural units shown above are approximately 4.6 Å long, assuming a constant carbon-carbon bond length of 1.54 Å [21].

Molecular Size and Substantivity

Mark [20,21] has described the forces involved in multiple polar and dispersion binding in polymers by means of molar cohesions (Table 7–1). These data show that even in polymers of low molecular weight, 10,000 daltons, the cohesive energy approaches that of primary chemical bonds. Therefore, by analogy, one may predict the importance of molecular size to substantivity of ingredients to hair.

Obviously, multiple sites for attachment of even stronger bonds (i.e., polar and especially primary valence bonds) are even more important to substantivity. Multiple covalent attachment sites could conceivably occur via in situ polymerization reactions and with bifunctional cross-linking agents (see Chapter 3). However for pragmatic reasons, this is not nearly as important as multiple ionic attachments to hair as, for example, with cationic polymers.

The appropriate spacing of groups in the polymer so that maximum bonding can occur, especially to ionic and polar groups on the keratin structure, is also important to substantivity. As a first approximation, maximum frequency of primary bonds and maximum molecular size will provide maximum substantivity [22].

Isoelectric Point of Hair and Polymer Substantivity

Although polymers may penetrate to a limited extent into human hair [23], the key interactions between hair and most polymers occur at or near the fiber surface. Because ionic bonds are the most important primary valence bonds for binding to hair under low temperature conditions in an aqueous or aqueous alcohol system, the net charge at the fiber surface is critical to polymer hair interactions.

Wilkerson has shown that unaltered human hair has an isoelectric point near pH 3.67 [24]; therefore, the surface of hair bears a net negative charge at all pH values above this. Because most cosmetic hair treatments are formulated above this pH, cationics are attracted to hair more readily than anionics, and polycationics are far more substantive to hair than polyanionics.

Desorption and Breaking of Multiple Bonds

Faucher et al. [25] have shown that the desorption of a polymeric cationic cellulose (polymer JR) (Table 7–2) from hair is slower than would be expected from a simple diffusional release predicted by the square root of time law [26]. They suggest that desorption of a polymer occurs only after all sites of attachment are broken. Statistically, the process of breaking all attachments simultaneously is of low probability. Therefore, one would expect high substantivity and a slow rate of release with increasing molecular size and increasing primary valence binding sites.

Penetration of Polymers into Hair

Low molecular weight polypeptides [27] ($\overline{M}_n = 1000$) and polyethyleneimine ($\overline{M}_n = 600$) [28] have been shown to diffuse into hair. Somewhat larger polypeptides ($\overline{M}_n = 10,000$) [27] and polymer JR (polyquaternium 10; formerly quaternium 19), with an average molecular weight of 250,000 [23], have also been reported to penetrate into hair. The polymer JR study involved bleached hair. These data suggest that penetration is limited to about 10% of the hair after 7 days and about an order of magnitude less in unaltered hair [25]. Sorption of a higher molecular weight JR polymer (average molecular weight of 600,000) by bleached hair is similar to the smaller polymer by unaltered hair.

It appears that some limited penetration into human hair can occur by the lower molecular weight species of low molecular weight polymers (<10,000 daltons). Larger polymers, to 500,000 daltons, may even contain species that can diffuse into the cuticle and perhaps further. Intercellu-

TABLE 7–2. Approximate structural formulas for three cationic polymers used in hair care products.

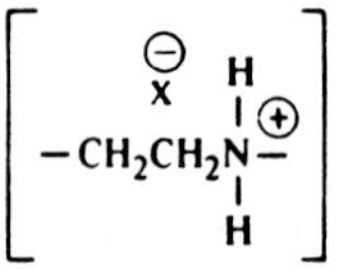

Polyethyleneimine (PEI) with a charge density of 176, assuming 25% protonation at pH = 8 [41]

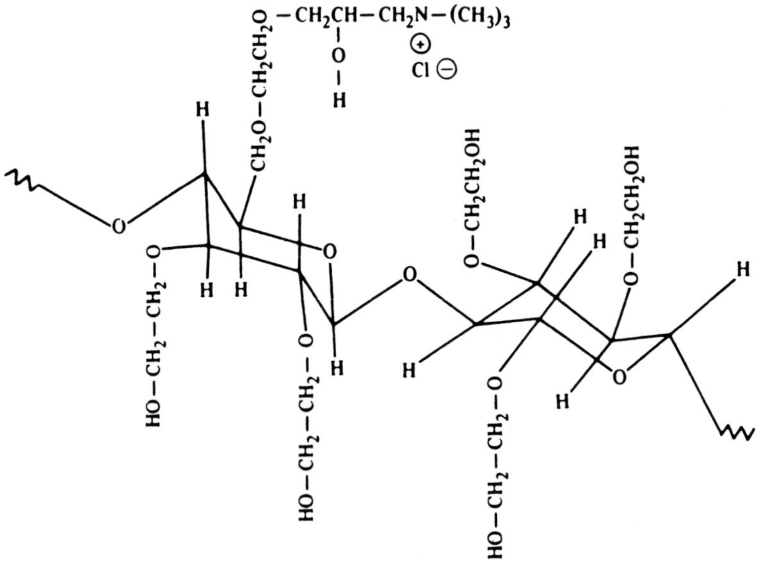

Quaternized hydroxyethyl cellulose with a charge density of 689 (polymer JR) (polyquaternium-10)

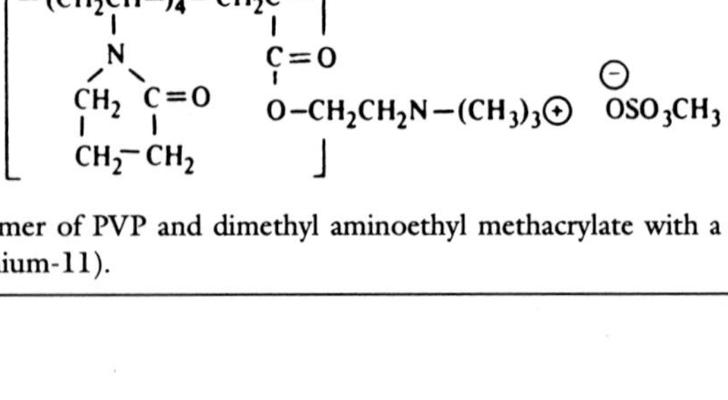

Quaternized copolymer of PVP and dimethyl aminoethyl methacrylate with a charge density of 616 (polyquaternium-11).

lar diffusion or diffusion via the low-sulfur regions is probably the preferred route for these large molecules (see Chapter 5; Figure 5–4). If the hair is degraded sufficiently or if the degree of polymerization for the polymer provides a very broad distribution, intracellular diffusion is also likely. However, it is highly unlikely that large polymers penetrate to a significant extent into human hair. Neutral or anionic polymers have not been studied for penetration effects; however, as a first approximation, one might draw similar conclusions with regard to the size and extent of penetration of these polymers also.

Cationic Polymers and Their Interactions with Hair

Cationic polymers as a group are one of the more important types of polymers used in hair products. Because of their high degree of substantivity to hair, they are useful in shampoos and conditioners. Their major asset, high substantivity, is also a potential problem, because they can be so substantive that they are difficult to remove from hair with ordinary shampoos.

Cationic ingredients in general are highly substantive to hair because of its low isoelectric point, which is approximately pH 3.67 [24] in cosmetically unaltered hair, and even lower in bleached hair. Therefore, at any pH above the isoelectric, the surface of hair bears a net negative charge and positively charged (cationic) ingredients are attracted to it.

Even monofunctional cationics are substantive to hair; that is, they resist removal by water rinsing. For example, stearyl benzyl dimethyl ammonium chloride and cetyl trimethyl ammonium chloride are major active ingredients in creme rinse products because they are substantive to water rinsing of hair, and they condition the hair fiber surface.

Dye staining tests [29,30] show that substantivity of monofunctional cationics to water rinsing does not occur unless the hydrocarbon portion is approximately 8 to 10 carbon atoms; that is, unless there are sufficient van der Waals attractive forces in addition to the electrostatic bond to bind the molecule to the keratin in the presence of the aqueous phase. As the molecular size of the cationic structure is increased, even greater sorption and substantivity result [30]. This enhanced substantivity is partly from an increase in dispersion binding and partly because the structure becomes less hydrophilic and partitions from the aqueous phase to the keratin phase because of entropy and energy considerations.

Approximate structural formulas for three cationic polymers that have been used in hair care applications are described in Table 7–2.

Interactions of Quaternized Cellulosic Polymers with Human Hair: Polymer JR

Polymer JR has been used in several different commercial hair products, including many different conditioning shampoos, as a conditioning ingredient. Polymer JR has a relatively low charge density 670 [31] and a high density of polar groups. Charge density is the residue molecular weight per unit of positive charge. This type of polymer has been studied in three different molecular weight versions (250,000, 400,000, and 600,000) in several excellent publications by Faucher, Goddard, and Hannah [23,25,31–33].

The formula in Table 7–2 approximates the structure of this polymer and is based on information in the CTFA cosmetic ingredient dictionary [34] and the charge density value of approximately 670 [33]. Note that the positions and numbers of ethoxamer units may vary for this structure, as does the position of the hydroxypropyl quaternary grouping.

Adsorption and Absorption to Hair

Faucher and Goddard [23] have studied the uptake of polymer JR onto bleached and unaltered hair. Their data suggest limited penetration into bleached hair, and possibly also some penetration into unaltered hair, for the lower molecular weight versions. At 0.1% polymer concentration, approximately 35 mg polymer per gram hair was sorbed onto bleached hair after 8 days and 8 mg polymer per gram hair after about 1 h (polymer molecular weight, 250,000).

Effect of Molecular Weight

Three different molecular weight versions (250,000, 400,000, 600,000) were studied with respect to sorption onto bleached hair [23]. The lowest molecular weight species was sorbed fastest and to the greatest extent. The sorption curve for the highest molecular weight species shows a rapid uptake followed by leveling, indicating saturation of the hair fiber surface and limited, if any, penetration.

Effect of Charge

Polymer JR uptake was compared to an analogous uncharged hydroxyethyl cellulose polymer. The uptake of the charged polymer was 50 times that of the uncharged polymer [23].

Effect of Concentration

The uptake of polymer JR increased sixfold with concentration from 0.01% to 1.0% [23]. However, apparent diffusion coefficients from initial slopes indicate a slower diffusion rate with increasing concentration. Faucher and Goddard [23] explain this anomaly by suggesting that a more compact polymer deposits on the hair at the higher concentrations. This may be somewhat analogous to the effects of pH on the activation energy for diffusion of orange II dye into keratin fibers [35]. In this latter situation, the activation energy for diffusion of dye into the fibers increases with decreasing pH where a higher concentration of dye enters the fibers. Apparently, the steeper concentration gradient with decreasing pH increases the energy required for each dye molecule to enter the fibers.

Effect of pH

The influence of pH on polymer JR sorption was studied in unbuffered media [23] at a starting pH of 4, 7, and 10. The largest uptake was at pH 7, with about 15% less polymer sorbed at pH 4 (which can be attributed to a decreasing net negative charge on the fiber surface). However, there is about 30% less polymer pickup at pH 10 than at pH 7, which would not be predicted on the basis of electrostatics or swelling of the hair. Drifting pH caused complications, and the observed differences, although small, await a satisfactory explanation.

Effects of Salt

Added salt produces a larger effect on the uptake of polymer JR than does pH [25] and may in fact help to explain the pH effect. The addition of 0.1% sodium chloride decreased pickup by almost two-thirds. This may be attributed to shielding of sorption sites on the hair, that is, competitive inhibition. Although the affinity of sodium ion for hair should be much less than for polymer JR, at this concentration sodium ion has more than 20 times the cationic charge concentration of polymer JR.

Other salts, such as lanthanum and calcium, had an even greater effect in decreasing polymer pickup. Trivalent ions (lanthanum, aluminum, and iron) had the largest effect, followed by divalent ions (calcium and ferrous iron). Monovalent ions showed the least effect. Faucher et al. [25] suggested the analogy to hair of a strong acid ion exchange resin and postulated that the decrease in polymer uptake by inorganic cations is caused by competitive inhibition.

Effect of Hair Damage

Most of the studies with polymer JR employed bleached hair. Bleached hair has a higher concentration of negative sites at and near the fiber surface to attract and bind cations, and is more porous than chemically unaltered hair. As one might predict, uptake of polymer JR onto unaltered hair was an order of magnitude lower than for bleached hair [25].

Desorption of Polymer JR from Hair

Desorption of polymer JR from hair by distilled water is very slow, and less than 15% was removed in 30 min [25,32]. Sodium dodecyl sulfate (SDS at 0.1 M) solution, analogous to a shampoo, was much more effective, removing more than 50% of the polymer in 1 min and nearly 70% in 30 min. However, a small amount of strongly bound polymer was still attached to the hair after SDS treatment [32]. Attempts to remove this strongly bound polymer by multiple treatments with SDS were not examined.

Salts were also found to be effective in removing a portion of the polymer, and trivalent salts were more effective than divalent, which were more effective than monovalent. However, even after 1 week in 0.1 M lanthanum nitrate solution (La^{3+}), approximately 40% of the polymer was still bound to the hair [25]. Most of these results were on bleached hair, but desorption experiments on chemically unaltered hair indicate similar behavior.

Effect of Surfactants on the Sorption of Polymer JR

All surfactants that have been examined, whether neutral, anionic, or cationic, decrease the uptake of polymer JR onto hair [23,33]. Pareth-15-9, a nonionic surfactant, exhibited the smallest effect in decreasing the uptake of polymer JR. Faucher and Goddard [23] attribute this to the relatively low affinity of this surfactant for both keratin and the polymer.

$$R{-}O{-}(CH_2{-}CH_2{-}O)_9{-}H$$

Pareth-15-9 ($R = C_{11}$ to C_{15})

Cocoamphoglycinate had a slightly greater effect in decreasing the uptake of JR.

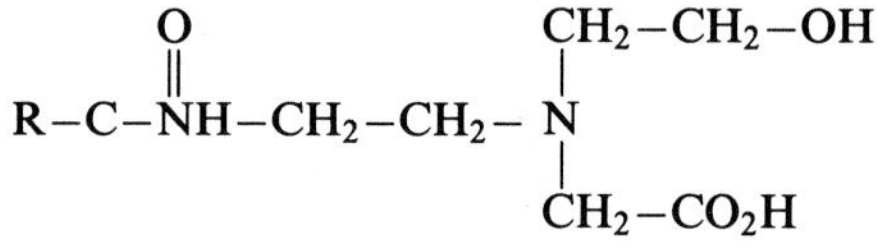

$$R{-}\overset{\overset{\displaystyle O}{\|}}{C}{-}NH{-}CH_2{-}CH_2{-}\underset{\underset{\displaystyle CH_2{-}CO_2H}{|}}{\overset{\overset{\displaystyle CH_2{-}CH_2{-}OH}{|}}{N}}$$

Cocoamphoglycinate

Pickup was greater in the presence of potassium laurate than cocoamphoglycinate. Goddard et al. [33] found a relatively thick, nonuniform deposit in the presence of laurate that they attributed to precipitated calcium laurate (soap) with polymer.

Anionic and cationic surfactants show the largest effect in decreasing polymer JR uptake onto hair. The cationic myristyl benzyl dimethyl ammonium chloride probably functions via competitive inhibition. Anionic surfactants probably function by forming association complexes that neu tralize the cationic charge of the polymer. Nevertheless, small amounts of polymer JR were still detected on the hair even in the presence of excessive amounts of anionic surfactant [33].

Cationic Polymer–Surfactant Complexes

Polymers have been shown to form association complexes with surfactants in solution [36–40], and the interaction of polymer JR with anionic surfactants such as SDS has been studied by Goddard and Hannah [31]. These authors concluded that this interaction occurs in two stages with increasing concentration of anionic detergent relative to polymer. The first stage involves adsorption of surfactant to the polymer, forming a primary layer that neutralizes the cationic charge of the polymer. A decrease in solubility occurs at this stage, and the new polymer complex is highly surface active. As the ratio of anionic surfactant to cationic polymer increases, adsorption of a secondary layer results, accompanied by reversal of the net charge of the total polymer complex species. Increased solubility also occurs.

Hannah et al. [32] have shown that this type of polymer complex, formed from 0.1% polymer JR and 1% SDS does indeed sorb to the hair. Water can remove only some 30% of this JR–SDS complex, and SDS and salts are no more effective, leaving some 60% (~0.1 mg complex per gram hair) strongly bound to the hair. Analogous complexes with other cationic polymers have been used for binding or for increasing the substantivity of ingredients to the hair [3,4].

Polyethyleneimine

Polyethyleneimine (PEI) was used commercially in the 1960s and then removed from hair products. However, it may be making a comeback because of new and improved synthetic techniques. Furthermore, several interesting scientific studies have been conducted with this polymer that illustrate some useful principles relevant to the adsorption of cationic polymers to keratin fibers. There are two significant structural differences between PEI and polymer JR: PEI has a higher charge density than polymer JR, and PEI is not quaternized, but is a polyamine.

Polyethyleneimine is formed from the aziridine ethyleneimine; its chemistry has been reviewed by Woodard [41]. Although PEI is not

```
CH2–CH2        –(CH2–CH2–NH–)x
   \ /
   NH
```

Ethyleneimine *Polyethyleneimine*

quaternized, it is highly cationic, because a large number of its amine groups are protonated even near neutral pH. Woodard [41] indicates that 4% of the amine nitrogens are protonated at pH 10.5; 25% at pH 8; and at pH 4, 50%. Therefore, PEI would have a charge density of approximately 176 at pH 8, or nearly four times the frequency of cationic sites as polymer JR.

Three different polyethyleneimines have been described with regard to their interactions with human hair: PEI-6 (molecular weight, 600), PEI-600 (molecular weight, 60,000), and PEI-600E, which is PEI-600 reacted with an almost equivalent amount of ethylene oxide. This reaction with ethylene oxide forms quaternary nitrogen groups and increases the molecular weight to approximately 100,000 [28,41].

Chow [28] has provided evidence for penetration of the lower molecular weight PEI-6 into hair, and Woodard [41] has shown that sorption increases with concentration and with bleaching of hair similar to polymer JR. Sorption of PEI-6 was also slightly greater at neutral pH compared with acidic or alkaline pH. However, because the charge density of PEI decreases with increasing pH, this result is not unexpected. This PEI study was conducted in an unbuffered medium.

Although a direct comparison has not been made, PEI appears to be even more substantive to hair than polymer JR, probably because of its higher charge density. PEI-600 was sorbed onto hair and tested for desorption toward a 10% shampoo system. After 30 min, less than 20% of the PEI was removed, and only approximately 30% after 6 h [41].

The rate of PEI desorption has also been examined with radiolabeled PEI on the hair, desorbing with unlabeled PEI. Most of the polymer, more than 60%, could not be removed in 24 h by this procedure [28,41]. This result is consistent with a slow degree of release of PEI because of the multiple ionic binding sites to the hair.

PEI polymers, like polymer JR, also interact with anionic ingredients, and they have been used for increasing the substantivity of other molecules to hair [3,4]. PEI was formerly used in one commercial shampoo but is no longer being used, presumably because ethyleneimine monomer impurity has been found to be a carcinogen.

Polyquaternium-6 and Polyquaternium-7 (Merquats)

Merquat polymers formerly quaternium-40 and quaternium-41 (*Merquats*) are another type of cationic polymer finding use in hair care products [42]. There are two different Merquat polymers: one is a homopolymer of dimethyldiallylammonium chloride (DMDAAC), and the second is a copolymer of DMDAAC and acrylamide. The homopolymer is Merquat 100 (polyquaternium-6) with an average molecular weight of approximately 100,000, and the copolymer is Merquat 550 (polyquaternium-7) with an average molecular weight of approximately 500,000. Polyquaternium-7 has been used commercially in several different conditioning shampoos.

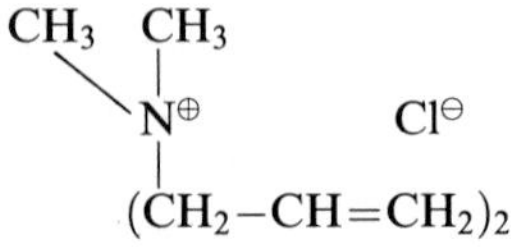

$$(CH_3)_2N^{\oplus}(CH_2{-}CH{=}CH_2)_2\ \ Cl^{\ominus}$$

Dimethyldiallylammonium chloride

$$CH_2{=}CH{-}CO{-}NH_2$$

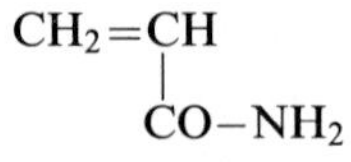

Acrylamide

The homopolymer has a charge density of approximately 126 and the copolymer 197. Thus, both these polymers have a relatively high charge density compared with polymer JR or many other cationic polymers now available for use in hair care.

Sykes and Hammes [42] have described the adsorption of both these cationic polymers onto hair from solutions of different amphoteric and anionic surfactants. Analogous to the adsorption of polymer JR, uptake values were greater onto bleached hair than onto unbleached hair, and greater from amphoteric systems such as cocobetaine or cocoamphyglycinate, than from anionic surfactants like sodium lauryl sulfate or triethanolammonium lauryl sulfate.

The following rationale accounts for these experimental findings. Sodium lauryl sulfate interacts with polymer primarily through an electrostatic interaction, and the net result is to neutralize the charge of the polymer and thereby reduce its affinity for keratin. Amphoterics do not neutralize the charge of the cationic polymer as effectively as do anionics. Therefore, cationic polymers demonstrate a greater affinity for keratins in an amphoteric surfactant system than in an anionic surfactant system.

Other Cationic Polymers

A large number of other cationic polymers have been made available during the past few years by chemical suppliers for use in hair care products. Some of these polymers are polyquaternium-4 (a grafted copolymer with a cellulosic backbone and quaternary ammonium groups attached through the allyl dimethyl ammonium chloride moiety; see the section on mousses); cationic guar gums such as guar hydroxypropyl trimonium chloride (e.g., Jaguar C-13-S or Guar C-261); Ucare polymer LR (a lower charge density cationic cellulose derivative of polymer JR); copolymer 845 (PVP/dimethyl aminoethyl methacrylate copolymer derivative of polyquaternium-11, but of lower charge density); copolymers of vinyl imidazole and vinyl pyrrolidone (of varying charge density); and even quaternized and amino silicone polymers and copolymers, also with varying charge density. For additional details on some of these cationic polymers, see the article by Idson and Lee [43].

Other Polymers

Polypeptides and Proteins

Polymeric collagen peptides should be somewhat substantive to hair, because they contain multiple ionic and polar sites for bonding in addition to offering large molecular surfaces with many sites for van der Waals bonding. Methionine, tyrosine [44], and tryptophan [45] are monomeric species of proteins, and they have been shown to sorb onto hair from aqueous solution. Collagen-derived polypeptides, or polymers of amino acids, have also been shown to have an affinity for hair [46–48], and one would predict that they should be more substantive to hair than their monomers.

Uptake of this type of species by hair has been shown to increase with either increasing hair damage or increasing polypeptide concentration. An average molecular weight ($\overline{M}_n$) of about 1,000 provides optimum pickup, which decreases with higher molecular weight [49]. Bleaching produces an increase in uptake at neutral pH, whereas thioglycolate-treated hair sorbs more polypeptide at alkaline pH, as does unaltered hair [45]. Bleaching should lower the isoionic point of hair more than thioglycolate, producing more swelling at neutral pH and additional anionic sites to bind the polypeptides. Penetration of polypeptide mixtures into hair has been shown by Cooperman and Johnson [48] and is described earlier in this chapter.

The desorptive action of surfactants and salts on polypeptides already sorbed to hair has not been examined as fully as for polymer JR. On the basis of theory, one would not expect collagen-derived polypeptides to be as substantive to hair as a high charge density cationic polymer.

Neutral and Anionic Polymers

The low isoelectric point of hair, near pH 3.6 [24], suggests that the net charge on the hair fiber surface is negative in the presence of most hair care

TABLE 7–3. Some neutral and anionic polymers used in hair products.

```
–CH2–CH–
      |
      N
     / \
  H2C   C=O
   |    |
  H2C – CH2
```

Polyvinylpyrrolidone

```
          O–CH3
          |
–CH2–CH–CH–CH–
          |    |
        O=C   C=O
          |    |
          O    O–CH2–CH3
          |
          H
```

Copolymer of methyl vinyl ether and ethyl ester of maleic anhydride

```
                  O
                  ‖
              O–C–CH3
              |
–CH2–CH–CH–CH2
      |
      N
     / \
  H2C   C=O
   |    |
  H2C – CH2
```

Copolymer of polyvinylpyrrolidone and vinyl acetate

```
    O
    ‖
R–C–(–O–CH2CH2–)x–OH
```

An ethoxylate ester polymer

products (at any pH above the isoelectric point). Table 7–3 describes structural formulas for some of the neutral and anionic polymers that have been used in hair products. These structures and the isoelectric point of hair suggest that the primary binding of these molecules to hair is by polar and van der Waals interactions. Because shampooing is in an aqueous system and water is a good hydrogen bond-breaking agent, the principal binding to resist shampooing for this kind of structure comes from van der Waals attractive forces. Anionic and neutral polymers are therefore not highly substantive to hair, and they have been used in applications where their ease of removal by shampoos is almost as important as their adhesional and film properties.

Hair Fixatives

Hair Sprays

Aerosol hair sprays were introduced into the marketplace by the Liquinet Corporation in Chicago in 1949 [50], and they have enjoyed considerable commercial success for more than two decades. However, hair spray sales peaked during 1969 and began to decline owing to public acceptance of more natural hair styles not requiring hair fixatives.

In the early to mid-1970s, hair spray sales declined even further because of environmental pressures to restrict the use of fluorocarbons in aerosol products. The large drop in hair spray sales occurred in 1975, after Roland and Molina theorized that fluorocarbons deplete the ozone layer in the stratosphere. For additional details on the rise and the decline in hair spray sales, see the article by Root [51]. As indicated in the introductory section to this chapter, there is a great deal of current research concerned with lowering VOCs in hair care products, and in particular in hair sprays, and the California CARB regulations outlined in the introduction are the driving force behind these efforts. Although no significant commercial impact has occurred in new hair care products as of this writing, within the next 5 years significant commercial impact should result in either new resins, new formulations, new devices, or some combination of these factors in the marketplace.

Three types of hair sprays are being produced today: pump hair sprays, hydrocarbon aerosols, and carbon dioxide aerosols. The first two of these products account for the major sales of hair spray.

Hydrocarbon aerosol hair sprays contain an alcohol-hydrocarbon solvent-propellant system, a synthetic polymeric resin, a base to neutralize the resin if it is a carboxylic acid-containing resin, plasticizer(s), fragrance and, in some cases, surfactant(s) to improve the spreading characteristics of the polymer.

Pump sprays are very similar to aerosol sprays and consist of solvent(s), a synthetic polymeric resin, a base to neutralize the resin (for carboxylic acid-containing resins), plasticizer(s), and in some cases a surfactant and a fragrance.

The resin is usually a synthetic polymer and is the primary ingredient that determines the holding properties of the hair fixative product. Nevertheless, considerable control over the properties of the resin may be achieved by altering the spray characteristics. The degree of neutralization and the type of neutralizer, plasticizer, or type of surfactant employed are also important to the properties of the hair fixative.

Set-holding under conditions of changing humidity (especially changes to a higher humidity) is critical to hair spray performance. At the same time, the hair spray fixative system must also be capable of being washed out of the hair with an aqueous detergent system. Thus a careful balance of properties is required to ensure good set stability at high humidity with good washout characteristics.

Synthetic polymeric resins selected for hair spray use are generally anionic or neutral rather than cationic resins to help ensure good washout characteristics. Before development of synthetic polymers with these properties, "hair lacquers" were generally alcoholic solutions of benzoin, rosin, or shellac [52]. These products provided excellent style retention, but they were difficult to wash out of the hair. In the early 1950s, polyvinyl pyrrolidone (PVP) was introduced as a hair-setting agent. This polymer exhibited good washout characteristics but was lacking in set retention. A few years after the introduction of PVP, even better hair-setting resins were developed.

Hair sprays, setting lotions, and mousses are related in the sense that each of these products applies a resinous material to the hair and helps to maintain style retention by enhancing interfiber interactions. If the hair is not combed through after the product has set (after the solvent evaporates), rigid contact sites of resin are formed between fibers, analogous to strip welding (see Figure 7–1). This type of interfiber bonding provides the mechanism for set retention of a hair spray.

When the product is applied to wet hair and the hair is set and combed, after the solvent evaporates (as for styling lotions and mousses), the deposited polymer still influences the hair assembly character by increasing the interfiber forces but not through rigid contacts (Figure 7–2). Many different types of synthetic polymers have been introduced into the marketplace for hair spray use. Chemical structures for three functional and popular types of resins in use today are described in Table 7–4. The polyvinyl pyrrolidone-type resins are usually copolymers of polyvinyl pyrrolidone and vinyl acetate (PVP-VA), because these copolymers are more functional than PVP itself, although PVP is widely used in setting lotions and in cheaper hair sprays. A second system widely used for several years is a copolymer of vinyl acetate and crotonic acid. This resin is generally superior to the PVP-type resins for hair spray use. However, one of the most func-

FIGURE 7–1. Two hair fibers bonded together with hair spray.

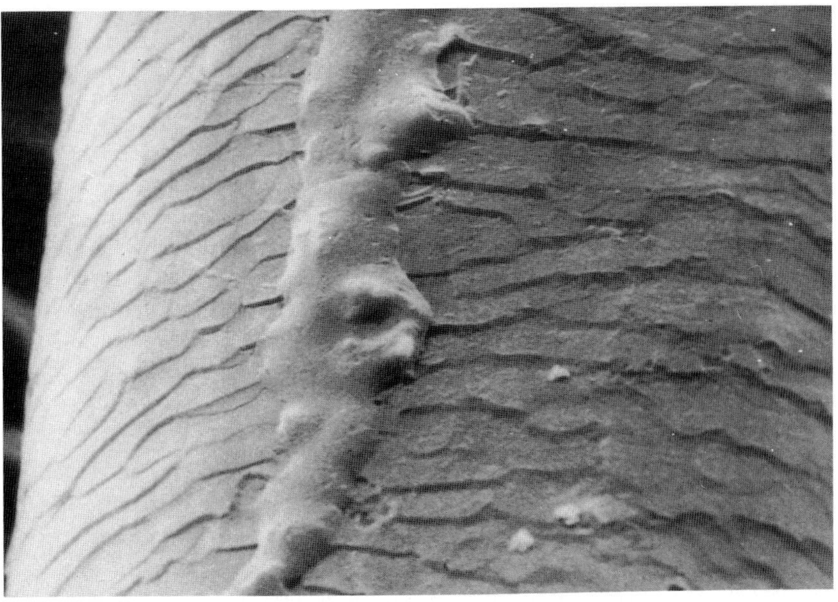

FIGURE 7–2. Setting lotion deposit on hair surface.

TABLE 7–4. Some resins used in hair sprays.

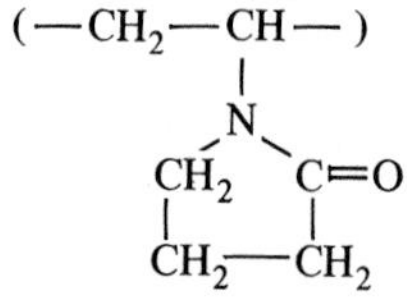

$(-CH_2-CH-)$
N
$CH_2 \quad C{=}O$
CH_2-CH_2

Polyvinylpyrrolidone (PVP)

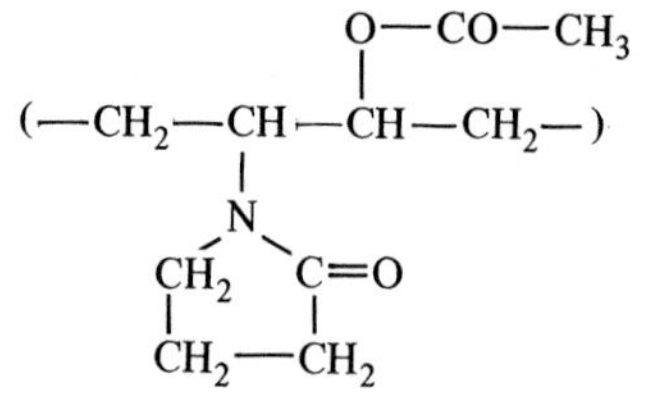

$O-CO-CH_3$
$(-CH_2-CH-CH-CH_2-)$
N
$CH_2 \quad C{=}O$
CH_2-CH_2

Copolymer of PVP-vinylacetate

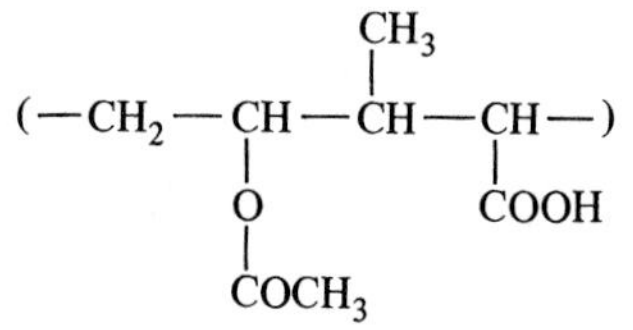

CH_3
$(-CH_2-CH-CH-CH-)$
$O \qquad COOH$
$COCH_3$

Copolymer of vinylacetate and crotonic acid

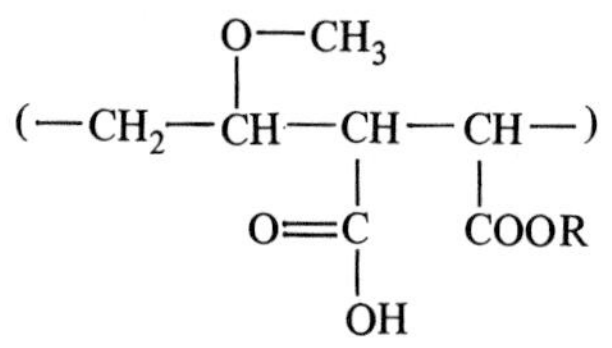

$O-CH_3$
$(-CH_2-CH-CH-CH-)$
$O{=}C \qquad COOR$
OH

Copolymer of methyl vinyl ether and a half-ester of maleic anhydride

tional and popular hair spray resins in retail sale today is the ethyl ester of the copolymer of polyvinyl methyl ether and maleic anhydride (see Table 7–4). The butyl ester of this copolymer has also been used and is generally superior to the ethyl ester for style retention; however, because of some limitations in fragrance selection for the butyl ester, the ethyl ester is the derivative of this copolymer most frequently used.

Other, more structurally complex resins that are highly functional and in use today are a copolymer of three monomers (vinyl acetate, crotonic acid, and vinyl neodecanoate), a polymer formed from octylacrylamide, *t*-butylaminoethyl methacrylate, and two or more monomers consisting of acrylic acid, methacrylic acid, or their simple esters (Table 7–5).

The solvents in pump sprays are limited to alcohol-water mixtures and are therefore not as complex as the solvent-propellant mixtures of aerosols. Generally, ethyl alcohol is the primary solvent and water the secondary solvent. In some cases, small quantities of propanols are also used. The solvent and of course the pump spray system largely determine the spray characteristics of a given system, which are obviously important to the

TABLE 7–5. Two structurally complex but effective hair spray polymers.

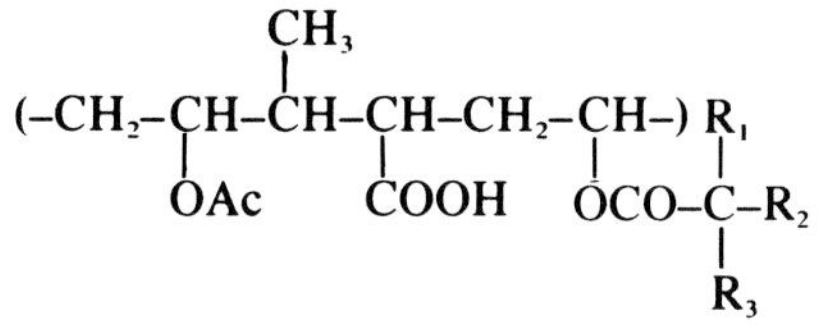

```
           CH3
           |
(–CH2–CH–CH–CH–CH2–CH–) R1
      |        |        |   |
      OAc     COOH   OCO–C–R2
                            |
                            R3
```

Simplified structure for copolymer of vinyl acetate, crotonic acid, and vinyl neodecanoate

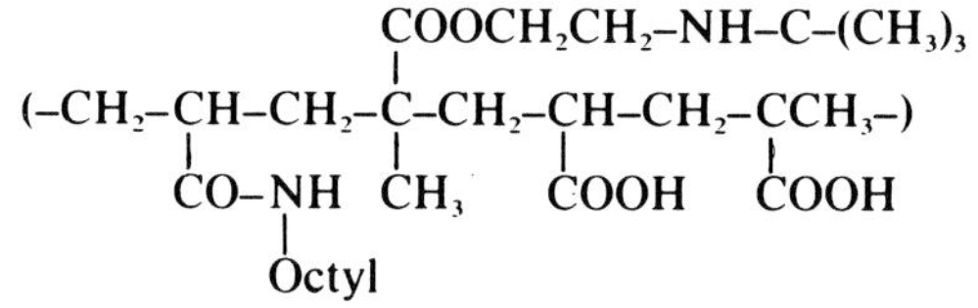

```
                      COOCH2CH2–NH–C–(CH3)3
                      |
(–CH2–CH–CH2–C–CH2–CH–CH2–CCH3–)
      |              |        |            |
     CO–NH   CH3     COOH    COOH
           |
          Octyl
```

Simplified structure for copolymer of octylacrylamide, *t*-butylaminoethyl methacrylate, acrylic acid (or ester), and methacrylic acid (or esters).

functional character of the product. The solvent-propellant systems of today's hydrocarbon aerosol hair sprays generally consist of alcohol combined with hydrocarbons such as isobutane and propane. In some aerosol sprays, *n*-butane and methylene chloride are also added. For additional details on aerosol propellants for hair sprays, see the article by Root [51].

The solvent-propellant in both aerosol and pump sprays contains the VOC and therefore presents the apparent environmental problem. The CARB regulations for 1998 of 55% VOC present the target that is stimulating research and development in this area, because to deliver a product that is equivalent to today's hair sprays with 55% or less VOC will require new creative technology.

The ability of the resin to spread over the hair surface is largely a function of the resin or its surface tension and the viscosity of the system; however, spreading characteristics of resins can be somewhat improved by adding surfactants [53]. Nonionic or cationic surfactants are generally preferable to anionics for this purpose.

Film properties and the solubility of resins containing carboxylic acid groups may be altered by the type of base used to neutralize the resin [53] and the degree of neutralization (Figure 7–3). Organic bases such as aminomethyl propanol, triisopropanol amine, or aminomethyl propanediol are generally preferable to inorganic bases as hair spray resin neutralizers.

The structures in Table 7–5 are simplified descriptions of these copolymers and are provided primarily to illustrate the monomer structures forming the basic structure of these polymers.

Plasticizers such as dimethicone, cetyl alcohol, dioctyl sebacate, or re-

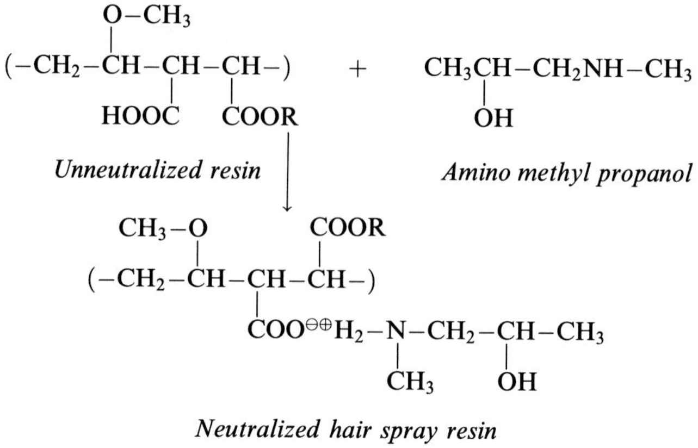

$$O{-}CH_3$$

$$({-}CH_2{-}CH{-}CH{-}CH{-}) \quad + \quad CH_3CH{-}CH_2NH{-}CH_3$$

$$HOOC \quad COOR \qquad\qquad OH$$

Unneutralized resin ↓ *Amino methyl propanol*

$$CH_3{-}O \qquad COOR$$

$$({-}CH_2{-}CH{-}CH{-}CH{-})$$

$$COO^{\ominus\oplus}H_2{-}N{-}CH_2{-}CH{-}CH_3$$

$$CH_3 \qquad OH$$

Neutralized hair spray resin

FIGURE 7–3. Neutralization of a hair spray resin.

lated ingredients are often added to provide more flexible, less brittle films to minimize flaking and thus to maximize luster and set-holding.

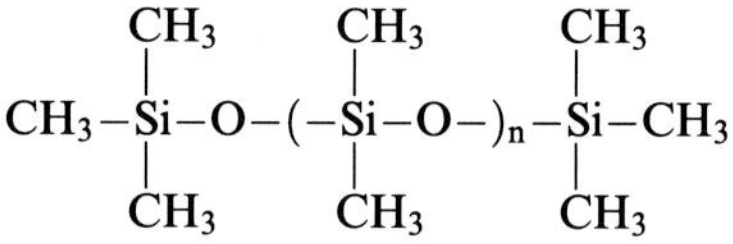

$$CH_3{-}Si(CH_3)_2{-}O{-}({-}Si(CH_3)_2{-}O{-})_n{-}Si(CH_3)_2{-}CH_3$$

Dimethicone

$$CH_3{-}(CH_2)_{15}{-}OH$$

Cetyl alcohol

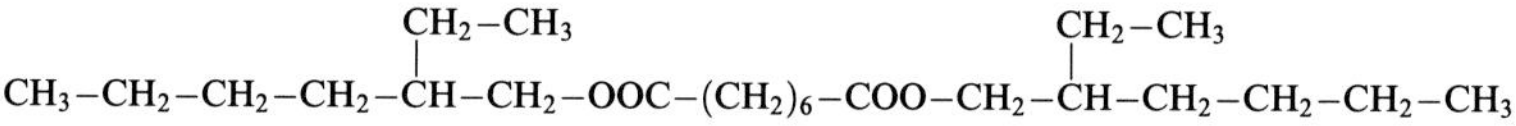

$$CH_3{-}CH_2{-}CH_2{-}CH_2{-}CH(CH_2{-}CH_3){-}CH_2{-}OOC{-}(CH_2)_6{-}COO{-}CH_2{-}CH(CH_2{-}CH_3){-}CH_2{-}CH_2{-}CH_2{-}CH_3$$

Dioctyl sebacate

Some Hair Fixative Formulations

This section, as well as the section on the testing of hair fixative products and the beginning of this chapter describing the status of VOC regulations (see Table 7–6) may be of some use to the beginning formulator.

Aerosol Hair Sprays

The formulas are hair spray concentrates and represent 85% of the total product. The concentrates are diluted with 15% A-46 Propellant (isobutane and propane) to obtain an acceptable pressure, and the spray character is adjusted with the appropriate valve system. This system contains 80% VOC and meets the CARB standards for 1993.

TABLE 7–6. Hair spray concentrates.

Ingredient	Holding power		
	Regular (%)	Extra (%)	Super (%)
Octylacrylamide/acrylates/ butylaminoethyl methacrylate copolymer (Amphomer)	1.5	2.0	3.0
Aminomethylpropanol (AMP)	0.23	0.3	0.4
Glycerine	0.4	0.4	0.4
Dimethicone copolyol (SF-193)	0.15	0.2	0.3
Octadecyl trimethyl ammonium chloride	0.06	0.06	0.06
Fragrance	0.15–0.5	0.15–0.5	0.15–0.5
Alcohol	q.s.[a]	q.s.	q.s.
Water (deionized)	22.5	22	21
Total formula %	85%	85%	85%

[a] q.s., add alcohol to a total of 85%.

TABLE 7–7. Pump hair spray formulation.

Ingredient	Percentage
Alcohol	q.s.[a]
Octylacrylamide/acrylates; butylaminoethyl methacrylate copolymer	4.5
Lauramide DEA	0.1
Dimethicone copolyol	0.4
Fragrance	0.2
Water	15.4

[a] q.s., add alcohol to 100%.

Hair Spray Making Procedure

Add the alcohol into the main mixing tank and begin stirring. Add the resin and stir until a clear solution is obtained. Adjust the pH with AMP to 8.5. Add glycerine, silicone, quat, and perfume with stirring. Add the water with stirring, and when a clean uniform solution is obtained transfer the concentrate to the aerosol filling line for filling and pressurizing.

This product (Table 7–7) is made by the same procedure as for the aerosol hair spray, except that no propellant is used and the listed ingredients represent 100% of the formula. A spritz product can be made from the same formula by proportioning all ingredients except the solvents by about 150%.

Mousses

Mousses are related to both hairsprays and setting lotions. Their formulations are generally more complex, sometimes consisting of two or more polymeric resins and additives in a water-hydrocarbon (solvent-propellant) system similar to shaving creams. As indicated in the introduction, the CARB standard for January 1, 1994, for mousse hair products is 16%.

The resins of mousses are generally cationic, such as polyquaternium-4 or polyquaternium-11, or cationic and neutral, or cationic and anionic combined, or amphoteric. Polyquaternium-4 is a copolymer of hydroxyethyl cellulose and diallyldimethylammonium chloride. This polymer is related to the quaternized hydroxyethyl cellulose polymer depicted in Table 7–2, except the cationic groups are formed from the diallyldimethylammonium chloride moiety attached to the cellulosic backbone.

Polyquaternium-11 (quaternium-23; Gafquat) is a copolymer of vinyl pyrrolidone and dimethylaminoethyl methacrylate quaternized with dimethyl sulfate. This polymer is related to the quaternized copolymer of PVP depicted in Table 7–2. One mousse product in retail sale contains both these cationic polymers. Others contain one cationic polymer and either a neutral or an anionic resin such as polyvinyl pyrrolidone or vinyl acetate copolymer, or a resin such as the butyl ester of polyvinyl methyl ether, maleic anydride copolymer.

Surfactants and oils are also present in most mousse formulations (Table 7–8). Surfactants function to lower the surface tension to help spread the polymer film over the hair surface and to support the foam character of the product. Mousses are added to wet hair. They are combed through without rinsing, to distribute the product through the hair. The hair is then styled and dried. Mousses can improve wet hair manageability by lowering combing forces. They also enhance body and style retention.

With these products, the hair is combed dry, after the solvent has evapo-

TABLE 7–8. A styling mousse formulation.

Ingredient	Percentage
GAF Quat 755	0.4
PVP/VA 70/30	3.0
PVP K-30	1.0
Cetyl trimethyl ammonium chloride (CTAC)	1.5
Cocodiethanolamide (Standamid KD)	1.0
Germaben II	0.5
Fragrance	0.25
Water	q.s.[a]
Propellant 12	6.0
Propellant 114	4.0

[a] q.s., add water to 100%.

rated. Therefore, they cannot provide rigid contacts as hair sprays do; however, they do enhance interfiber attractive forces to provide increased body and improved style retention.

Mousse Making Procedure

For the formulation described by Table 7–8, charge the water into the mixing vessel and with stirring dissolve the CTAC and the diethanolamide. After the system is homogeneous, slowly dissolve the GafQuat, the PVP/VA, and the PVP, then add preservative and fragrance. When the solution is homogeneous, transfer this concentrate to the aerosol filling line for filling and pressurizing.

Setting/Styling Lotions and Gels

These lotions and gels are related to both hair sprays and mousses. They are more similar to mousses, because they are aqueous based or alcohol-water solvent systems, and they are applied to wet hair (or sometimes to dry hair) before setting as opposed to spraying onto dry hair (see Figures 7–1 and 7–2). The CARB standard for hair styling gels for January 1, 1994, is 6% VOC.

The resins of styling or setting products (Tables 7–9 and 7–10) are often cationic, for example, polyquaternium-11 (copolymer of vinyl pyrrolidone and dimethylaminoethyl methacrylate quaternized with dimethyl sulfate, which is similar to the quaternized copolymer of PVP depicted in Table 7–2); anionic, for example, butyl ester of polyvinylmethyl ether and

TABLE 7–9. A setting lotion formulation.

Ingredient	Percentage
Part A	
CTAC	0.5
Distearyl dimmonium chloride	0.5
Ethylene glycol distearate	0.5
Water	40.5
Part B	
PVP-VA (70/30) (50%)	5.0
PVP K-30	1.0
Arosurf-20 (isosteareth-20)	1.0
Polymer JR-30M	0.5
Water	48.5
Part C	
Germaben II	0.5
Colors	q.s.
Fragrance	0.25

TABLE 7–10. Setting/styling gel.

Ingredient	Percentage
Alcohol	5.0
PVP/VA 64	2.2
Triethanolamine	0.6
Carbomer 940	0.6
Dimethicone copolyol	0.3
Fragrance	0.3
PEG-40 hydrogenated castor oil	0.3
Tetrasodium EDTA	0.1
Colors	0.02%
Water (deionized)	q.s.[a]

[a] q.s., add water to 100%.

maleic anhydride; or neutral, such as a copolymer of polyvinyl pyrrolidone, vinyl acetate, or even polyvinyl pyrrolidone itself with polyacrylate thickeners.

Because the mode of application of these products is similar to that of mousse, and involves combing the product through wet hair before setting and drying and then recombing the hair after it has dried, these products do not function by forming rigid bonds exactly as hair sprays do. Nevertheless, these products do enhance interfiber forces to provide increased hair body and improved style retention.

Making Procedure for Setting Lotions

Slowly dissolve the isosteareth-20 into the water; after it is completely dissolved, slowly add the PVP-VA. Add the PVP until it is completely dissolved, then add JR and hold Part B. Add the CTAC to the water, then heat to 50°C and add the melted distearyl dimmonium chloride and EGDS. Stirr and cool to 40°C and add Part B to Part A. Add preservative, colors, and fragrance.

Making Procedure for Setting Gel

Dissolve the PVP/VA resin in the alcohol and then add about 10% of the formula amount of water. Make a homogeneous solution of the Carbomer in about 80% of the formula amount of water. Deaerate this solution and then add the EDTA and the TEA. In a separate container add, with heating (to ~50°C) and stirring, the PEG-40 hydrogenated castor oil to about 10% of the formula amount of water and stirr until homogeneous. Add dimethicone copolyol with stirring until homogeneous. Cool and add the benzophenone and fragrance to this solution, and then slowly mix it into the PVP/VA solution and stir until the system is homogeneous. Add this

TABLE 7–11. Spray-on gel formulation.

Ingredient	Percentage
PVP/VA Copolymer [Luviskol VA 73E (50%)]	8.0
Isosteareth-20 (Arosurf 66 E20)	1.0
DMDM hydantoin	0.7
Octyl salicylate	0.3
Disodium EDTA	0.2
Fragrance	0.2
Water (deionized)	q.s.

solution to the Carbomer solution and mix until a clear gel is obtained. Be careful to not aerate the system too much.

The actual product is in the form of a gel. However, products have recently appeared in the marketplace that provide hair-styling properties like those of a gel but are not gel in form; nevertheless, they are called gels. For example, the "spray gel" or "spray-on gel" is such a product (Table 7–11).

Making Procedure for Spray-On Gel

Charge the water into the mixing vessel and add the isosteareth-20 with stirring until it dissolves completely. Slowly add the PVP/VA polymer with stirring; when the system is homogeneous add the remaining ingredients.

Evaluation of Hair Fixative Products

To develop and evaluate hair sprays, setting products, and mousses, a variety of methods have been developed. The methods described in this section have been developed primarily for hair spray formulation and evaluation. However, many of these methods are also used to develop and to evaluate hair-setting products and mousses. Lang and Sendelback [54] have reviewed a large number of test methods for the evaluation of hair spray polymers and products. One of their primary conclusions was that although several interesting and useful methods have been introduced into the scientific literature for fixative polymer evaluation, the well-known curl retention test is still the fastest and most accurate method for obtaining useful information about the holding properties of hair spray polymers.

In addition to style or curl retention, however, laboratory tests are helpful to characterize the product spray characteristics, and film properties of hair sprays. Among the more important spray characteristics are spray rate, spray pattern, and droplet size. Of course, safety considerations related to flammability are also important.

For mousses, measurement of foam properties including foam volume, foam quality, and foam stability is critical to the performance of this type of product. Although such tests have not been described for mousse evaluation per se, minor modification to shave cream foam tests should provide satisfactory procedures.

Film properties of these products are crucial to performance, and several methods to evaluate film properties of hair spray products have been developed [55–57]. Erlemann [55] has described a variety of methods, both subjective and objective, to evaluate hair spray films formed on different substrates including metal plates or glass, flexible foils or tissues, and hair. Ayer and Thompson [57] have described evaluation of hair spray properties by scanning electron microscopy.

Style retention is without question the most important property of hair sprays, and several approaches to evaluating hair spray holding power have been described in the literature [55,58,59]. One novel approach by Ganslaw and Koehler [60] involves measurement of the rate of untwisting of hair swatches treated with hair fixative solution. This parameter, which these authors call twist retention analysis, correlates with curl retention and allows for more rapid evaluation of data.

Another approach, by Frosch and Vogel [61], involves measuring the force required to break polymer-treated hair strands. A more recent and novel approach, on single hairs, is the method by Wickett and Sramek [62,63] that determines the adhesive strength of hair/hairspray junctions.

In Situ Polymerizations in Hair

With the exception of oxidation hair dyes, in situ polymerizations in hair have been only laboratory and concept curiosities. However, remarkable changes to the chemical [64] and physical properties [65] of the fibers have already been achieved using this technology, and in the future, through the combination of science and imagination, other in situ polymerization hair treatments may end up in the marketplace.

The remaining sections of this chapter describe oxidation dyes as in situ polymerization hair treatments and in situ vinyl polymerization reactions in human hair.

Oxidation Dye Reactions as In Situ Polymerization Reactions

Although not generally described as such, certain reactions of oxidation hair dyes are examples of in situ polymerizations in hair. They consist of the oxidation of electron-rich aromatic amine and phenol monomers that condense with each other and perhaps even attach to amino acid residues of hair. The net result, at least with products containing *p*-phenylenediamine (PPD) and resorcinol, is the formation of polyindophenol-type polymeric pigments [66–68] that render color to the hair. (See the discussion on oxidation hair dyes in Chapter 6 and its references for additional details.)

In Situ Polymerization of Vinyl Monomers in Hair

Several techniques have been employed for the polymerization of vinyl monomers on and in wool fiber, including:

- Reduction of the fibers, followed by reaction with vinyl monomer and oxidizing agent in an inert atmosphere [69,70]
- Radiation Grafting [71,72]
- The Wurlan Process [73] is the condensation of diamines and diacid chloride in the presence of wool fiber.

The procedure that has been most thoroughly studied for polymerization into human hair is related to the first procedure described for wool fiber, reduction of the fibers (an air atmosphere was employed in these studies) [74,75].

Mechanism of Action

Polymerization of a vinyl monomer into human hair is a complicated, multistep process that may be summarized by the following reaction scheme:

1. Diffusion of reducing agent into the fibers
2. Nucleophilic cleavage of the sulfur-sulfur bond by reducing agent
3. Water rinse
4. Diffusion of oxidizing agent into the fibers
5. Reaction of reduced hair and oxidizing agent
6. Diffusion of vinyl monomer into the fibers
7. Chain-initiating reactions
8. Chain-propagating reactions
9. Termination of free radical chains.

Steps 1 Through 3

Among the reducing agents that have been used in this type of process are thioglycolic acid (TGA) [74], bisulfite [65], and tetrakis (hydroxymethyl) phosphonium chloride (THPC) [75]. The critical point is the rate or extent of reduction, because each of these reducing agents provides increasing polymer add-on with increasing time of reduction (Figure 7–4). Comparison of the TGA system with the bisulfite system shows a faster rate of polymerization with TGA than with bisulfite, and this rate difference is consistent with a faster rate of reduction of hair by TGA. Because

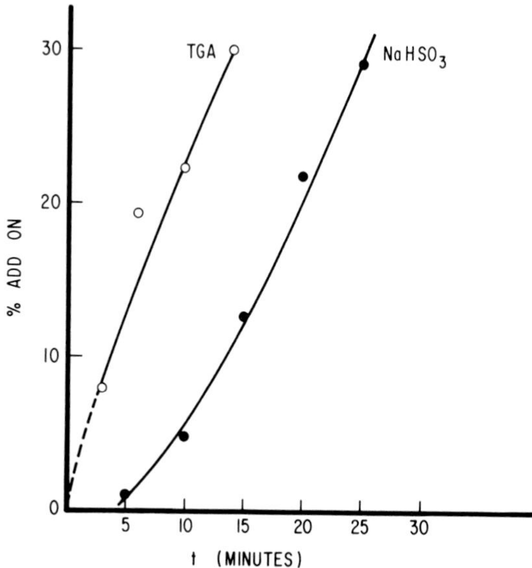

FIGURE 7–4. Influence of the reduction step on polymer add-on.

the reaction of TGA with human hair is diffusion controlled [76], step 2 is extremely important to the overall kinetic scheme, and since cleavage of the sulfur-sulfur bond by TGA is faster than diffusion, it seems of lesser importance to the overall kinetic scheme. However, the extent of disulfide fission is a controlling factor in the remaining steps: the diffusion of initiator (oxidizing agent) and monomer into the fibers.

The effect of pH during reduction is very important to both the TGA [69] and the THPC [75] systems. In the case of the TGA system, reduction rate increases with pH, and polymer add-on increases similarly. For the THPC system, pH is also critical, but not for entirely the same reason. According to Jenkins and Wolfram [77], THPC dissociates in aqueous solution to Tris (hydroxymethyl) phosphine (THP), and THP is the actual ingredient that reduces the hair.

$$\underset{(THPC)}{P{-}(CH_2{-}OH)_4^{\oplus}\,Cl^{\ominus}} \rightleftharpoons \underset{(THP)}{P{-}(CH_2{-}OH)_3} + CH_2O + HCl$$

Wolfram [75] has shown that polymer add-on increases with the pH of the reducing solution to pH 7, where it appears to level, and THPC dissociation and sulfur–sulfur bond cleavage also increase with pH. The fact that polymer add-on is not higher at pH 9.2 than at pH 7.0 suggests interference of alkalinity in a subsequent reaction step. The most probable complications are in the chain initiation or propagation steps, because Wolfram

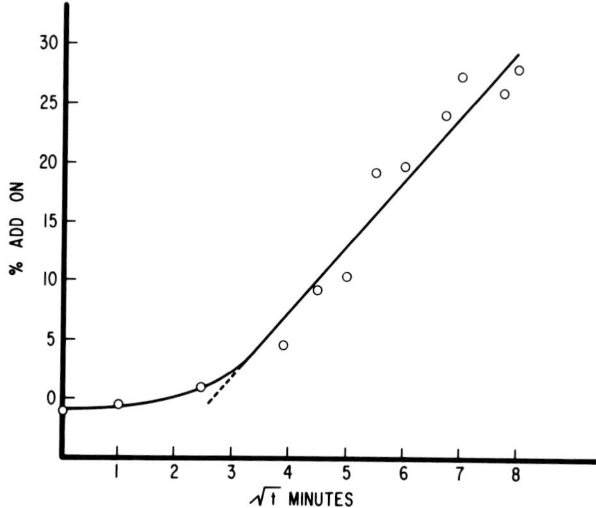

FIGURE 7–5. Influence of reaction time (steps 4–9) on polymer add-on.

indicates that the cysteine-persulfate redox system is at an optimum in the pH region of 1.5 to 3.5.

Steps 4 Through 9

Both TGA-hydroperoxide and THPC-persulfate systems show a linear relationship between polymer add-on and the square root of time (Figure 7–5) [78]. In both these systems, the amount of polymer add-on can be influenced by concentration changes in both oxidizing agent and vinyl monomer. Therefore, the diffusion of vinyl monomer and oxidizing agent into the fibers can also be rate limiting. As a result, any variable that can accelerate diffusion of oxidizing agent and/or vinyl monomer into the fibers is capable of increasing the rate of polymer add-on.

Step 5, the reaction between reduced hair and oxidizing agent, generates the free radical species that initiates polymerization. For the system of Wolfram [75], this step involves reaction of THP or mercaptan of reduced hair with persulfate, whereas for the TGA system it involves reaction of mercaptan with cumene hydroperoxide.

The remaining reactions, steps 8 and 9, are classical free radical propagation and termination reactions (Figure 7–6). If the chain-initiating radicals are from cysteine residues or other residues of the hair, then the resultant polymer is grafted (i.e., covalently bonded) to the hair protein. However, if the chain-initiating radicals are from oxidizing agent or reducing agent, then the polymer, if inside the hair, may become entrapped as it grows.

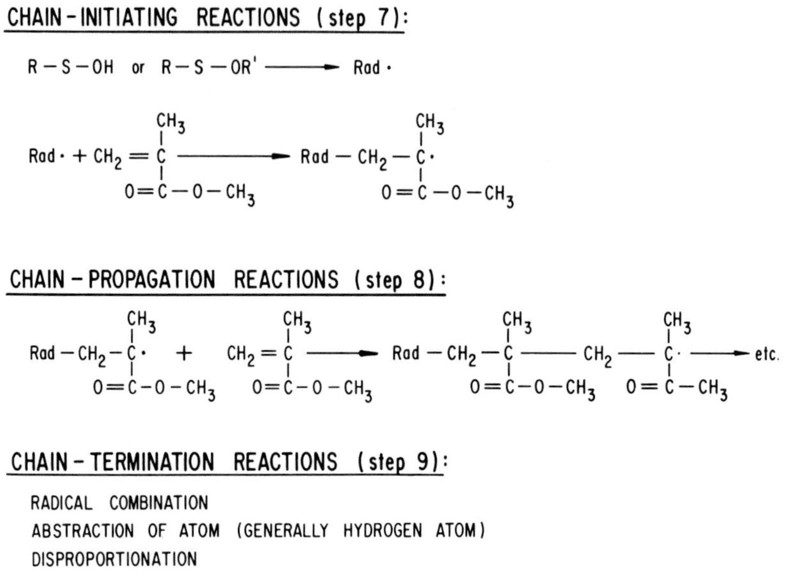

CHAIN-INITIATING REACTIONS (step 7):

$$R-S-OH \text{ or } R-S-OR' \longrightarrow Rad\cdot$$

$$Rad\cdot + CH_2=C(CH_3)(O=C-O-CH_3) \longrightarrow Rad-CH_2-C\cdot(CH_3)(O=C-O-CH_3)$$

CHAIN-PROPAGATION REACTIONS (step 8):

$$Rad-CH_2-C\cdot(CH_3)(O=C-O-CH_3) + CH_2=C(CH_3)(O=C-O-CH_3) \longrightarrow Rad-CH_2-C(CH_3)(O=C-O-CH_3)-CH_2-C\cdot(CH_3)(O=C-CH_3) \longrightarrow \text{etc.}$$

CHAIN-TERMINATION REACTIONS (step 9):

RADICAL COMBINATION
ABSTRACTION OF ATOM (GENERALLY HYDROGEN ATOM)
DISPROPORTIONATION

FIGURE 7–6. Chain initiation, propagation, and termination steps.

Solvent System

For the TGA cumene hydroperoxide system, an ethanol water solvent was employed for monomer and initiator. The data clearly demonstrate maximum add-on with larger proportions of water in the system, providing greater hair swelling, consistent with diffusion rate control [74].

Polymerization into Chemically Altered Hair

Bleaching increases the permeability of hair, but it also decreases the disulfide content, and the disulfide bonds are potential sites to form mercaptan groups, a part of the redox system. Therefore, it was of interest to determine which of these two parameters, permeability or mercaptan content, might contribute more to the rate of polymer add-on. Hair was bleached to varying extents with alkaline hydrogen peroxide and then treated with a bisulfite-cumene hydroperoxide system, using methyl methacrylate monomer. The data clearly show increasing polymer add-on with decreasing disulfide cross-links (Table 7–12), once again emphasizing the importance of diffusion rate control to this process [74].

A similar effect was demonstrated for reduced-oxidized hair, that is, hair that was permanent waved to varying extents prior to treatment with the polymerization system [74].

TABLE 7–12. Polymerization into bleached hair with methyl methacrylate.

Cystine in hair (%)	Add-on (%)
18.1	12
15.9	18
15.4	21
14.5	25
10.9	38

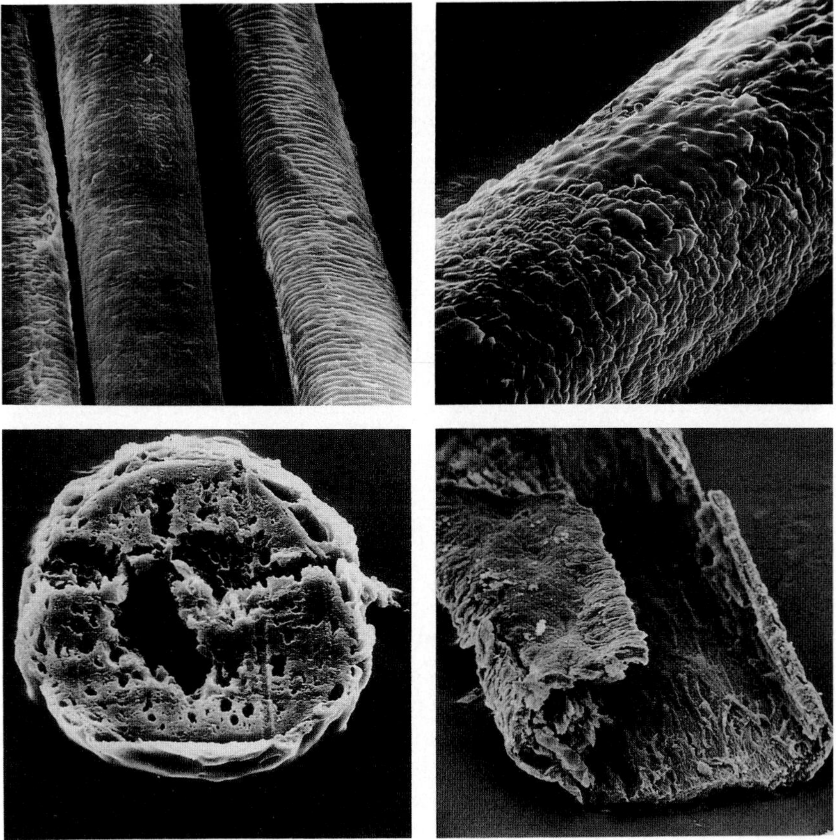

FIGURE 7–7. Scanning electron micrographs of polymethyl methacrylate-PMM-containing hair fibers. Upper left: Surface of hair fibers containing 16% PMM add-on. Upper right: Surface of hair fibers containing 119% PMM add-on. Lower left: Cross section after hydrolysis of hair fibers containing 119% add-on. Lower right: End view after hydrolysis of hair containing 16% add-on. (Reprinted with permission of the *Journal of the Society of Cosmetic Chemists.*)

Evidence for Polymer in the Hair

Hair after treatment with TGA cumene hydroperoxide using methyl methacrylate monomer was hydrolyzed with 5 N hydrochloric acid to dissolve away the keratin from the polymer, and part of the resultant fiber-like residue was dissolved in organic solvents. The solute (in the organic solvents) was shown to be polymethyl methacrylate by refractive index and infrared spectroscopy [74].

Viscosity average molecular weights of this polymer were also determined and found to be relatively constant, near 90,000 daltons, from bleached, waved, and chemically unaltered hair. This suggests an average degree of polymerization of approximately 900.

Scanning electron micrographs of the surface of hair fibers treated with the TGA cumene hydroperoxide system using methyl methacrylate monomer show a thick coating of polymer on the hair fiber surface. However, the fibers still retain repeating irregularities perpendicular to the fiber axis, which correspond to scale edges covered with a thick coating of polymer (Figure 7–7) [74].

As indicated earlier, fiber-like fragments were isolated from polymethyl methacrylate after hydrolyzing away the keratin with 5 N hydrochloric acid. These "synthetic fibers" were examined in cross section and at the ends. From hair with large amounts of polymer add-on (near 100% weight gain), these "fibers" were almost complete cylinders of porous polymethyl methacrylate threads, whereas hair with lower polymer add-ons (10%–20% weight gain) yielded thin-walled hollow cylinders (see Figure 7–7). These observations are also consistent with diffusion rate control for this in situ vinyl polymerization process.

Safety Considerations for Polymers

Safety considerations for products containing polymers are often related to the components of the products rather than the polymers, which are usually relatively safe ingredients. Protein polymers should be tested for sensitization; however, such problems are not frequent for the types of protein hydrolysates used in hair care. "Pure" synthetic polymers (with no monomer contaminant) are generally mild ingredients of relatively low toxicity, one example being polyvinyl pyrrolidone, originally used as a blood plasma extender in medicine.

Safety concerns for synthetic polymers sometimes result from contamination with monomeric impurity. Monomers are sometimes highly toxic, such as acrylamide [79], or even carcinogenic, such as ethyleneimine [80] or vinyl chloride [81], or highly irritating to skin, such as acrylamide [79] or acrylic acid [82]. Therefore, control of unreacted monomer level can sometimes be critical to the safety performance of synthetic polymers.

References

1. Gerstein, T. U.S. patent 3,990,991 (1976).
2. Lang, E.W. U.S. patent 3,400,198 (1968).
3. Parran, J., Jr. U.S. patent 3,489,686 (1970).
4. Parran, J., Jr. U.S. patent 3,580,853 (1971).
5. Hough, P.S.; Huey, H.; Tolgyesi, W.; et al. J. Soc. Cosmet. Chem. 27:571 (1976).
6. Kitazawa, T.; et al. Cosmet. Perfumery 89(1):33 (1974).
7. Gillette Co. Br. patent 944,439 (1963).
8. Rufe, R.G. Cosmet. Perfumery 90:93 (1975).
9. Knechtel, A. Am. Perfumer 78:95 (1963).
10. Kamath, Y.; et al. J. Soc. Cosmet. Chem. 28:273 (1977).
11. Miller, B.; Young, R. Text. Res. J. 45:359 (1975).
12. Fawkes, F.M. In Chemistry and Physics of Interfaces, Gushee, D.E., ed., pp. 1–12. American Chemical Society, Washington, DC (1965).
13. Wu, S. J. Polymer Sci. (Part C) 34:19 (1971).
14. Steinhardt, J.; et al. J. Res. Natl. Bur. Stand. 28:201 (1942).
15. Steinhardt, J.; Zaiser, E. J. Res. Natl. Bur. Stand. 35:789 (1949).
16. Pauling, L. The Nature of the Chemical Bond, p. 3. Cornell University Press, Ithaca, NY (1948).
17. Maron, S.H.; Prutton, C.F. Principles of Physical Chemistry, p. 742. Macmillan, New York (1958).
18. Gilreath, E.S. Fundamental Concepts of Inorganic Chemistry, p. 219. McGraw-Hill, New York (1958).
19. Morrison, R.T.; Boyd, R.N. Organic Chemistry, p. 19. Allyn and Bacon, Boston (1960).
20. Mark, H. In The Chemistry of Large Molecules, Frontiers in Science I, Burk, Grummitt, eds., p. 66. Interscience, New York (1943).
21. Mark, H. Ind. Eng. Chem. 34:1343 (1942).
22. Pauling, L. The Nature of the Chemical Bond, p. 160. Cornell University Press, Ithaca, NY (1948).
23. Faucher, J.A.; Goddard, E. J. Colloid Interface Sci. 55:313 (1976).
24. Wilkerson, V. J. Biol. Chem. 112:329 (1935).
25. Faucher, J.A.; et al. Text. Res. J. 47:616 (1977).
26. Paul, D.R.; McSpadden, S.K. J. Membrane Sci. 1:33 (1976).
27. Cooperman, E.S.; Johnson, V.L. Cosmet. Perfumery 88:19 (1973).
28. Chow, C. Text. Res. J. 41, 444 (1971).
29. Crawford, R.J.; Robbins, C.R. J. Soc. Cosmet. Chem. 31:273 (1980).
30. Scott, G.V.; et al. J. Soc. Cosmet. Chem. 20:135 (1969).
31. Goddard, E.; Hannah, R.B. J. Colloid Interface Sci. 55:73 (1976).
32. Hannah, R.B.; et al. Text. Res. J. 48:57 (1978).
33. Goddard, E.D.; et al. J. Soc. Cosmet. Chem. 26:539 (1975).
34. CTFA Cosmet. Ingredient Dictionary, 4th Ed., Nikitakis, J.; McEwan Jr., G.N.; eds., Cosmet., Toiletry and Fragrance Association, Washington, DC (1991).
35. Robbins, C.R.; Scott, G.V. Text. Res. J. 40:951 (1970).
36. Schwuger, J.J. J. Colloid Interface Sci. 43:491 (1973).
37. Putnam, F.W.; Neurath, H.J. J. Am. Chem. Soc. 66:692 (1944).

38. Putnam, F.W.; Neurath, H.J. J. Am. Chem. Soc. 66:1992 (1944).
39. Isemura, T.; Imanishi, J. J. Polymer Sci. 33:337 (1958).
40. Murata, M.; Arai, H. J. Colloid Interface Sci. 44:474 (1973).
41. Woodard, J. J. Soc. Cosmet. Chem. 23:593 (1972).
42. Sykes, A.R.; Hammes, P.A. Drug Cosmet. Ind. 62 (February 1980).
43. Idson, B.; Lee, W. Cosmet. Toiletries 98:41 (1983).
44. Herd, J.; Marriot, R. J. Soc. Cosmet. Chem. 10:272 (1959).
45. Newman, W. Text. Res. J. 42:214 (1972).
46. Karjala, S.A.; et al. J. Soc. Cosmet. Chem. 18:599 (1967).
47. Karjala, S.A.; et al. J. Soc. Cosmet. Chem. 17:513 (1966).
48. Cooperman, E.; Johnson, V. Cosmet. Perfumery 88:(7):19 (1973).
49. Stern, E.S. and Johnson, V. 9th Int. Fed. Soc. Cosmet. Chem. USA., p. 753 (1976).
50. Johnsen, M.A. Cosmet. Toiletries 97:27 (1982).
51. Root, M. Cosmet. Toiletries 94:37 (1979).
52. Berger, F.J. In Cosmetics Science and Technology, Sagarin, E., ed., p. 531. Interscience, New York (1957).
53. Bohac, S. J. Soc. Cosmet. Chem. 23:125 (1972).
54. Lang, G.; Sendelback, G. 8th International Hair Science Symposium of the German Wool Research Institute, Kiel Germany, September (1992).
55. Erlemann, G.A. J. Soc. Cosmet. Chem. 22:287 (1971).
56. Eckardt, W. J. Soc. Cosmet. Chem. 21:281 (1970).
57. Ayer, R.; Thompson, J. J. Soc. Cosmet. Chem. 23:617 (1972).
58. Reed, A.B., Jr.; Bronfein, I. Drug Cosmet. Ind. 94:178 (1964).
59. Micchelli, A.; Koehler, F.T. J. Soc. Cosmet. Chem. 19:863 (1968).
60. Ganslaw, S.; Koehler, F.T. J. Soc. Cosmet. Chem. 29:65 (1978).
61. Frosch, F.; and Vogel, F. 6th International Hair Science Symposium of the German Wool Research Institute, Luneburg Germany (1988) German Wool Res. Inst. Publ. of Abstracts.
62. Wickett, R.; Sramek, J. 7th International Hair Science Symposium of the German Wool Research Institute, Bad Neuenahr, Germany (1990) German Wool Res. Inst. Publ. of Abstracts.
63. Wickett, R.; Sramek, J.; Trobaugh, C. J. Soc. Cosmet. Chem. 43:169 (1992).
64. Robbins, C.R.; Anzuino, G. J. Soc. Cosmet. Chem. 22:579 (1971).
65. Robbins, C.; et al. U.S. patent 3,634,022 (1972).
66. Brody, F.; Burns, M. J. Soc. Cosmet. Chem. 19:361 (1968).
67. Corbett, J. J. Soc. Cosmet. Chem. 20:253 (1969).
68. Corbett, J. J. Soc. Cosmet. Chem. 24:103 (1973).
69. Madaras, G.W.; Speakman, J.B. J. Soc. Dyers Col. 70:112 (1954).
70. Negishi, M.; et al. J. Appl. Polymer Sci. 11:115 (1967).
71. Ingram, P.; et al. J. Polymer Sci. 6:1895 (1968).
72. Campbell, J.; et al. Polymer Letters 6:409 (1968).
73. Fong, W.; et al. Proc. 3rd Int. Wool Text. Res. Conf. III:417 (1965).
74. Robbins, C.R.; Crawford, R.; Anzuino, G. J. Soc. Cosmet. Chem. 25:407 (1974).
75. Wolfram, L.J. J. Soc. Cosmet. Chem. 20:539 (1969).
76. Hermann, K.W. Trans. Faraday Soc. 59:1663 (1963).
77. Jenkins, A.D.; Wolfram, L.J. J. Soc. Dyers Col. 79:55 (1963).

78. Crank, J. Mathematics of Diffusion, p. 71, Oxford University Press, Oxford, UK (1967).
79. Merck Index, 9th Ed., Windholz, M.; ed., p. 127. Merck & Co., Rahway, NJ (1976).
80. Merck Index, 9th Ed., Windholz, M.; ed., p. 500. Merck & Co., Rahway, NJ (1976).
81. Ann. N.Y. Acad. Sci. 246:1 (1975).
82. Merck Index, 9th Ed., Windholz, M.; et al. ed., p. 127. Merck & Co., Rahway, NJ (1976).

8

The Physical Properties and Cosmetic Behavior of Hair

One of the roadblocks to both communication between the cosmetic scientist and consumers and scientific investigations in hair care has been the lack of meaningful definitions for the important cosmetic properties of hair conditioning, hair body, and manageability. A definition for hair conditioning has emerged that allows this important cosmetic property to be approached by considering ease of combing as the first boundary condition. In addition, progress has been made for the definitions of hair body and manageability over the past two decades. Admittedly, there is still room for improvement in the understanding of these hair assembly characteristics, but even more progress awaits those important properties associated with hair feel, such as clean hair feel, conditioned feel, and dryness–oiliness.

During the past few years, some progress has been made in the understanding of those properties that we call single-fiber properties. However, as indicated, more progress has been made in the area of definitions and test methods for those hair assembly assessments that are important to consumers, that is, cosmetic assembly charactistics.

For the main discussion in this chapter, the physical properties of human hair have been divided into three categories:

1. elastic deformations
2. other important physical properties
3. some important consumer assessments.

Another important classification that is referred to routinely is single-fiber and fiber assembly properties. The more important single-fiber properties are elastic deformations, friction, cross-sectional area (diameter), static charge, luster or hair shine, and cohesive forces.

Elastic deformations include stretching (longitudinal strain), bending (including stiffness or the resistance to bending), and torsion (twisting) and its resistance, rigidity. Cross-sectional shape, friction, density, and static charge are described as other important physical properties, and hair shine, combing ease, body, style retention, manageability, and hair conditioning are the primary important consumer assessments described in this chapter.

Several years ago, Robbins and Scott [1] hypothesized that most consumer assessments of hair (properties of fiber assemblies such as combing ease, style retention, flyaway, body, and manageability) may be approximated by algebraic expressions involving the single-fiber properties of friction, stiffness, static charge, fiber curvature, weight, diameter, luster, and color. A somewhat similar analysis of hair body has been described by Hough et al. [2]. Robbins and Reich [3] have determined empirical relationships between combing ease and the fiber properties of friction, stiffness, fiber curvature, and diameter, and this work demonstrates that hair assembly properties can indeed be defined by a few fundamental single fiber properties. Robbins has taken the conclusions from this study and proposed a general hypothesis for hair behavior. This general approach of relating fiber assembly behavior to single-fiber properties is the basis for the discussion on consumer assessments in this chapter.

Elastic and Tensile Deformations

For every "strain" (deformation) of an elastic substance, there is a corresponding "stress" (the tendency to recover its normal condition). The units of stress are force per unit area (F/A).

The most common types of strain are stretching or elongation (the ratio of an increase in length to the original length), linear compression (the ratio of a decrease in length to the original length), shear (the ratio of the displacement of one plane relative to an adjacent plane), bending, and torsion [4]. These latter two strains are combinations of the former three. Only stretching, bending, and torsional strains are considered here.

Each type of stress and strain has a modulus (the ratio of stress to strain) that also has units of F/A. The elastic modulus for stretching is commonly called Young's modulus. The bending modulus has also been called Young's modulus of bending, and the torsional modulus is called the modulus of rigidity.

Tensile Properties

Human hair has been referred to as a substrate with only one dimension, "length," which suggests why its tensile properties have been studied more than its other elastic properties.

The usual procedure for evaluating the stretching properties of human hair is via load elongation (stress–strain) methods; that is, a fiber of known length (we usually use 5-cm fibers) is stretched at a fixed rate (a convenient rate is 0.25 cm/min) in water, in buffer, or at a fixed relative humidity [approximately 60% relative humidity (RH)], near room temperature on an automated instrument such as an Instron Tensile Tester (Instron Corporation).

Tensile properties are obviously whole-fiber properties, as opposed to surface properties, and have been suggested as being cortical properties not related to the cuticle. This is because of experimental evidence and in part the importance of the alpha to beta transformation that occurs on stretching [5]. Wolfram and Lindemann [6] suggested that the cuticle does contribute to the tensile properties, especially in thin hair. However, Scott [7] has provided support for the "no cuticle involvement" hypothesis by evaluating the tensile properties of hair fibers that were abraded under controlled conditions. In no instance could he demonstrate a significant change in tensile properties where only cuticle had been abraded.

Robbins and Crawford [8] provided additional evidence that the cortex and not the cuticle is responsible for the tensile properties of human hair, and that severe damage can occur in the cuticle which cannot be detected by tensile property evaluation. This study involved selective oxidation of the cuticle with diperisophthalic acid, which produced extensive cuticular damage that was easily detectable microscopically but could not be detected through either wet or dry tensile property evaluation. These results are consistent with the fact that wet extension of hair fibers to 30% can damage the cuticle [9], but tensile recovery occurs on relaxation in water, producing virtually identical elongation–recovery curves in a before-and-after type of evaluation that is commonly used throughout the industry.

When a keratin fiber is stretched, the load elongation curve shows three distinct regions (Figure 8–1). The lower curve in Figure 8–1 represents stretching a hair fiber in water and the curve at the top of the chart represents stretching at 65% RH. In the Hookean region of the load elongation curves, the stress (load) is approximately linearly related to the strain (elongation), and the ratio of stress to strain in this region is the elastic modulus (E_s), commonly called Young's modulus. The elastic modulus is usually expressed in units of dynes/cm^2 and may be calculated from this simple expression:

$$E_s = HgL/A\Delta L$$

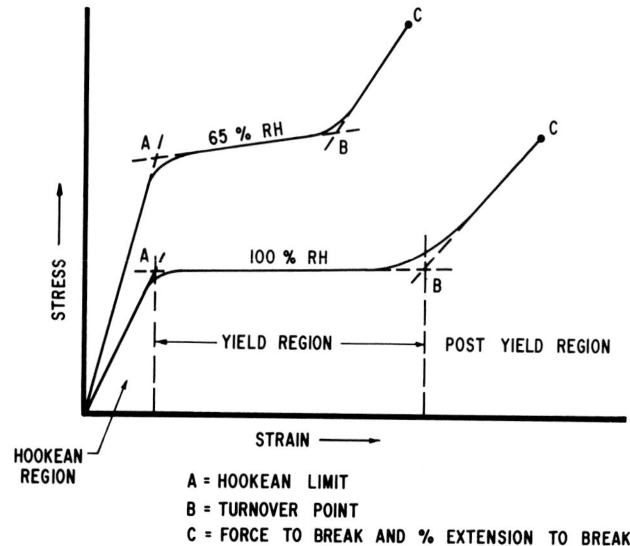

FIGURE 8–1. Schematic diagram for load elongation curves for human hair fibers.

where H = Hookean slope in grams per millimeter elastic extension, g = gravitation constant (980.6 cm/sec^2), L = fiber length in centimeters, ΔL = fiber extension in centimeters, and A = fiber cross-sectional area in centimeters squared.

The elastic modulus for stretching human hair, determined in our laboratories at 60% RH and room temperature, is 3.89×10^{10} dynes/cm.^{2} Methods other than load elongation used to determine the elastic modulus of hair fibers [10,11] are described later in this chapter.

Other important parameters of load elongation curves are the Hookean limit (Figure 8–1, point A), the turnover point (Figure 8–1, point B), the percentage extension to break, the stress to break (the tensile strength), the postyield modulus (stress–strain) in the postyield region, and the work of elongation (the total area under the load elongation curve). An interesting but limited study by Hamburger et al. [12] suggested that the pullout energy of cosmetically unaltered hair at 65% RH is approximately equal to the Hookean limit and considerably less than the stress to break. This suggested that most undamaged hair fibers under stress should pull out before breaking. Scott found that approximately one-half of the fibers collected from combing a few heads in our beauty salon contained bulbs and therefore were pulled out. However, some fibers on heads actually exhibit different types of fracture patterns, and in some cases evidence for cuticle fracturing also can be found on hairs growing on the scalp.

Stretching hair fibers under ambient conditions can cause damage well before the fiber breaks apart. For example, during stretching or extension

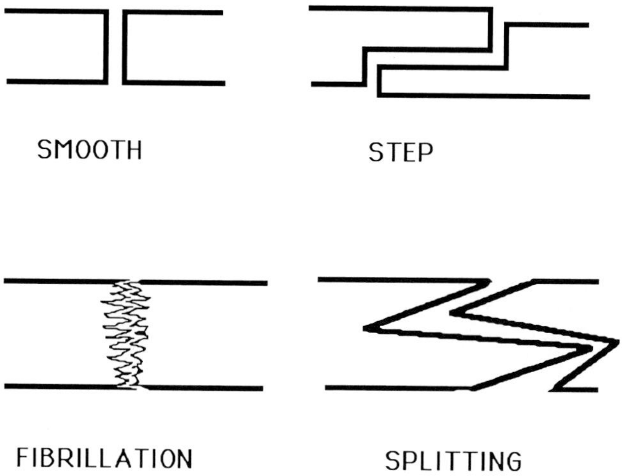

FIGURE 8–2. Schematic depicting some of the more important fracture patterns for human hair.

at 45% RH, signs of cuticle separation and damage occur in the nonkeratinous regions of the hair cuticle (endocuticle), and damage occurs sooner in the cuticle of tip ends (~10% extension) of hair fibers than in the root ends (~20% extension). Further, this cuticle damage or fracturing [9] occurs much sooner than fiber breakage (~40%–50% extension at moderate humidities and >50% extension in the wet state).

Three important papers on the fracturing of human hair by Henderson et al. [13] and Kamath and Weigmann [14,15] show that breaking or fracturing of hair occurs differently in the cuticle compared to the cortex and that fracturing of hair fibers occurs in different patterns; Figure 8–2 shows the four most common fracture patterns for human hair. The fracture pattern observed generally depends on the extent of hair damage, RH (whether the hair is wet or dry), and whether the fiber is twisted or contains flaws.

If the hair and its cuticle are in good condition, for example, near the root end, and the fiber is wet, a smooth break tends to occur (Figure 8–3). As the fiber becomes drier, below 90% RH, step fractures are the pattern most commonly observed (Figure 5–17). Fibrillation and splitting describe distinct cortical fracturing patterns and tend to occur more often in twisted or kinky fibers [15] and when the RH is lower, rather than when the fiber is wet, because the cortex is less extensible than the cuticle when the fiber is dry (below 90% RH) [13].

Split ends can occur from breaking if the cuticle is in poor condition; however, the more common fracture patterns such as the step fracture and fibrillation are also capable of producing split ends through the subse-

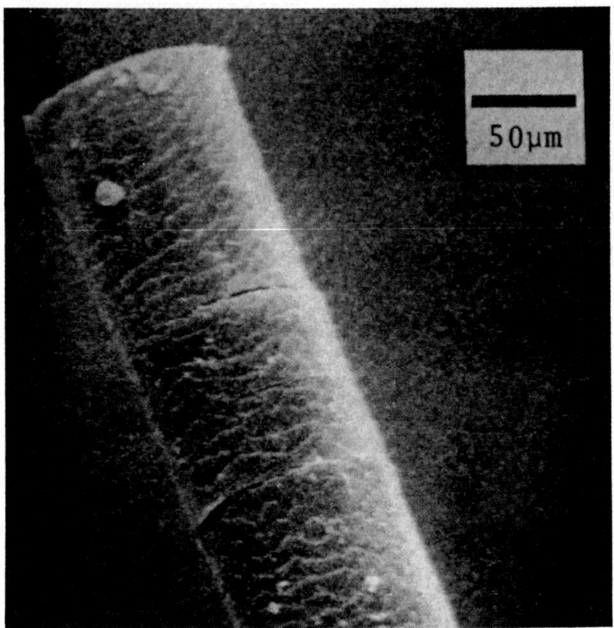

FIGURE 8–3. A hair fiber broken by stretching in the wet state. Reprinted with permission of the *Journal of the Society of Cosmetic Chemists*).

quent abrasive actions of combing and brushing; see the discussion on damage to hair from shampoos, grooming and weathering in Chapter 5.

Stretching is very different in the wet state than the dry state. As indicated, stretching to break in the wet state (~50% extension) produces what appears to be a clean break, that macroscopically resembles a razor cut (see Figure 8–3). However, on closer examination of the fiber, we see that fracturing has occurred in the cuticle (Figure 8–4) even before fiber breakage, and the cortex is also not cleanly broken. Extension of hair fibers to only 10% to 20%, at 45% RH, can induce failure in the endocuticule, which is the weakest region of the cuticle; this causes separation of the surface scales from the underlying layers, producing an uplifting of scales (Figure 8–5). Extensions to about 30% or slightly higher, in the wet state, can produce multiple circumferential fracturing of the cuticle with separation of cuticle sections from the cortex [13] (see Figure 8–4).

Because stretching to break in the dry state (below 90% RH) is more prone to induce uneven cortical fracturing, such as a step fracture or fibrillated ends or even split ends, stretching or bending hair fibers during combing or grooming operations can produce stress cracking in the nonkeratinous regions, the endocuticle, and the intercellular regions, and scale lifting can then occur in the damaged region of the hair fiber.

When a keratin fiber is stretched to as much as 30% more than its origi-

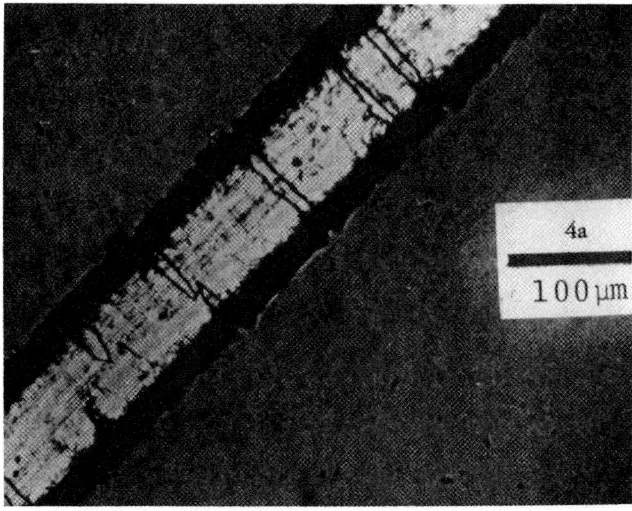

FIGURE 8–4. Damage to cuticle by stretching hair fibers to break in the wet state [13]. (Reprinted with permission of the *Journal of the Society of Cosmetic Chemists*).

nal length in water and then allowed to return to its original length, in spite of cuticular damage, a curve similar to that in Figure 8–6 is obtained. The work of elongation is always greater than the work of recovery; thus, a hysteresis exists, and the ratio of these two work values is called the resilience ratio (or hysteresis ratio) [16], another important load elongation parameter.

As indicated earlier, we normally stretch fibers 5 cm in length to 20% more than their original length at an extension and recovery rate of 0.25 cm/min. The rate will influence the results [17]. Sikorski and Woods [18] and Simpson [10] suggest an increase of approximately 5% in the elastic modulus for a 10-fold increase in extension rate. Stretching hair fibers to 120% (20% more than original) and holding for 4 h produces a temporary increase in length [19].

After stretching the fibers and relaxing them, as just described, we allow the fibers to relax overnight in water, then treat and restretch them, thus making before-and-after treatment comparisons on the same fibers. Such a procedure, without treatment, provides reproducible load elongation curves, confirming the validity of the before-and-after comparison. This type of test procedure was first suggested by Speakman [20] and Sookne and Harris [21]. Speakman referred to percentage changes in the work of extension, and Harris coined the term "30% index" as the ratio of extension values. An implicit assumption in this procedure is that calibration, that is, stretching before treatment, does not alter the reactivity of the fibers. Wolfram and Lennhoff [22] have provided evidence that supports

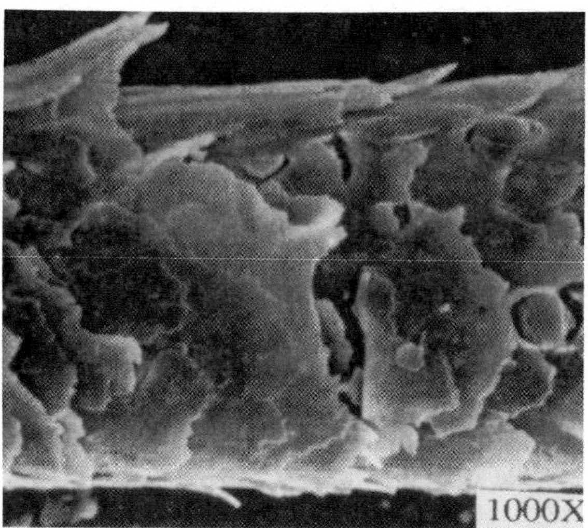

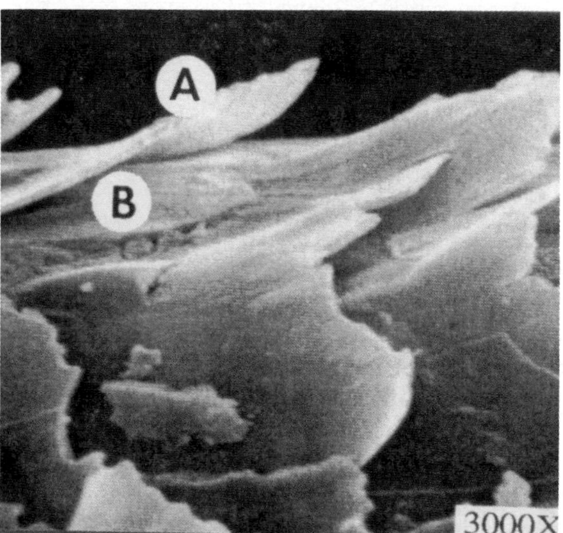

FIGURE 8–5. Damage to scales of human hair by stretching fibers at 45% relative humidity. [Scanning electron microscopy (SEM) photograph kindly provided by Dr. H.-D. Weigmann].

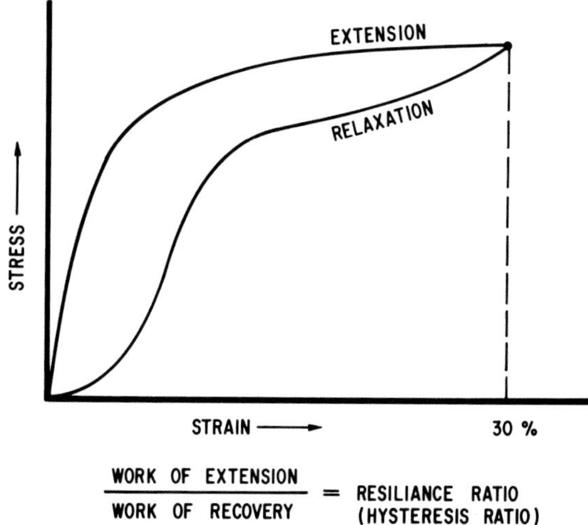

FIGURE 8–6. Schematic diagram for load elongation and recovery curves for human hair fibers.

the validity of this assumption. Work and force values to 15% and 25% elongation have also been used [23,24].

Some Factors That Influence Load Elongation Properties

Relative Humidity or Moisture Content

The moisture content of human hair varies with relative humidity (RH), increasing with increasing RH (see Tables 8–15 and 8–16, later in this chapter). Relatively extensive load elongation studies of wool fiber at different relative humidities have been carried out by Speakman [25] and others [24], showing a regular increase in extensibility (percent extension to break) with increasing RH. Although such extensive studies could not be found in the literature for human hair, a similar relationship undoubtedly exists for these reasons:

1. Tensile properties of hair fibers at 55% to 65% RH and at 100% RH (in water) show greater extensibility at the higher humidity (see Figure 8–1), and a lower elastic modulus (Table 8–1), and lower work and force values in general [26].
2. The dynamic elastic modulus of human hair has been reported to respond similarly to changes in RH [27].
3. The quantitative binding of water to wool is virtually identical to the binding of water to hair as a function of RH (see Table 8–16) [28,29], and therefore we conclude a similar stress–strain RH relationship (Table 8–1).

TABLE 8–1. Elastic modulus versus relative humidity of hair and wool fiber.

	65% RH		100% RH[a]		Ratio
	Human hair	Merino wool	Human hair	Merino wool	(E_S 65% RH/ E_S 100% RH)
Elastic modulus (mg/cm^2) [22]	55	31	21	12	2.62 (hair) 2.58 (wool)

% RH	Wool E_S At Given RH[c]/ E_S At 100% RH
0[b]	2.76
32	2.44
44	2.27
65	2.10
78	1.85
91	1.41
100	1.00

[a] In pH 7 buffer.
[b] In dry glycerine, which approximates 0% RH.
[c] From Feughelman and Robinson [24].

Fiber Diameter

Both wet and dry tensile properties of keratin fibers are directly proportional to fiber diameter. Figure 8–7 summarizes this relationship with a plot of the Hookean slope versus fiber linear density at 62% RH. Because linear density is proportional to cross-sectional area and diameter, the tensile properties are also proportional to fiber diameter. Robbins and Scott [30] have reported a procedure for determining both wet and dry tensile properties on the same fibers. This procedure depends on the fundamental relationship that wet and dry tensile properties are proportional to fiber diameter, and, therefore, proportional to each other.

Temperature

Rebenfeld et al. [31] studied the effect of temperature on the load elongation properties of human hair in neutral buffer solution, comparing human hair with wool fiber (Table 8–2). Increasing temperature has an effect similar to increasing humidity on the shape of the load extension curve [31,32] (see also Table 8–5). The elastic moduli for both human hair and wool fiber decrease with increasing temperature, but are lower for wool at any temperature, probably because of its lower cross-link density. The post-yield modulus and fiber strength decrease with increasing temperature, although extensibility increases. The turnover point (extension to B in Fig-

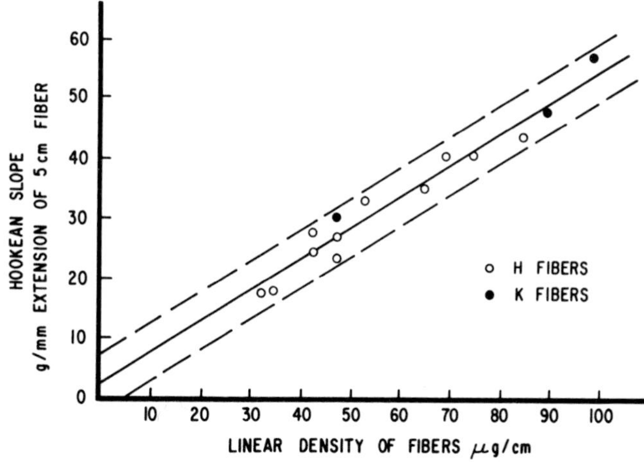

FIGURE 8–7. Hair fiber elastic extension versus linear density.

TABLE 8–2. The influence of temperature on the stress–strain properties of human hair at pH 7.[a]

Temperature (°C)	Elastic modulus (dynes × 10^{-10}/cm²)	Stress at Break[b] (dynes × 10^{-10}/cm²)	Percent of extention at break
21	2.08	0.168	48
35	1.77	0.129	—
50	1.67	0.125	50
70	1.64	0.140	—
90	1.36	0.099	72

[a] Calculated from data by Rebenfeld et al. [31].

[b] The stress at break provides a lower value than the elastic modulus. The equation given for E_S may be used to calculate both values, and by convention, the H term is in force/millimeter of elongation for the elastic modulus, while it is in force units for the stress to break.

ure 8–1) undergoes a transition at 85.5°C for unaltered hair and at 66°C for partially reduced hair. Rebenfeld explains these results in terms of a disulfide–sulfhydryl interchange mechanism whereby stressed disulfide bonds are relieved and transformed into stress-free positions at higher temperatures. Crawford [33] has examined the effect of temperature in a dry atmosphere on the wet tensile properties of hair fibers, and found significant changes in the force to 20% extension after heating to 100°C 25 times for 15-min intervals. Permanent-waved hair appeared to be more susceptible to heat, undergoing a decrease of approximately 7%, as compared to 4% for unaltered hair. A thermochemical technique for analysis of human hair has also been reported [34].

Bleaching

Chapter 4 describes the chemistry of bleaching and shows that a major side reaction is the oxidation of cystine cross-links to cysteic acid residues. This disruption of cross-links has a major influence on the wet tensile properties of hair.

Alexander et al. [35] oxidized wool fibers to different extents with peracetic acid and determined the work required to stretch these fibers, both wet and dry, and the cystine contents of the fibers. They concluded that the disulfide bonds contribute largely to the wet strength of the fibers, which decreases almost linearly with the cystine content, whereas the dry strength is virtually unaffected by disulfide bonds. In fact, a weakening in the dry state was found only after more than 60% of the cystine cross-links were broken [36]. Harris and Brown [37] reduced wool fibers to different extents, methylated them, and determined changes in the 30% index. Their conclusions were much the same as those of Alexander et al. [35]. Garson et al. [38] measured the dynamic elastic properties of hair fibers and found similar effects. These scientists found that the effects of oxidation are greater when measurements are made in water than at relative humidities ranging between 0% and 80%.

Oxidative bleaching of human hair on live heads provides similar results on tensile properties. Robbins examined both the wet and dry tensile properties of frosted and nonfrosted hair fibers from the same person (Table 8–3). Except for the resilience ratio, the loss in dry tensile properties was less than 10%, but the loss in wet tensile properties approached 60% at 48% disulfide cleavage in agreement with the oxidation of wool fiber by Alexander et al. [36]. These latter authors [35] suggest that both the dry and wet strength of wool fibers are greatly influenced by peptide bond cleav-

TABLE 8–3. Tensile properties[a] of frosted hair.[b]

Tensile parameter	Nonfrosted hair	Frosted hair	Loss in tensile property (%)	Loss in cystine (%)
Wet tensile properties				
Work to extend 20%	13.34	5.50	59	48
Hoookean slope	13.25	5.79	56	48
Hookean limit	13.30	5.19	61	48
Force to extend 20%	16.67	7.01	58	48
Resilience ratio	0.587	0.585	0	48
Dry (55% RH) tensile properties				
Work to extend 20%	32.06	30.84	4	48
Force to extend 20%	36.71	33.80	8	48
Resilience ratio	0.173	0.115	44	48

[a] Note: Work to 20% extension, grams × centimeters; Hookean slope, g load/mm elastic extension; Hookean limit, is grams; load to 20% extension, grams.
[b] Data are normalized to a 70-μm-diameter basis at a length of 5 cm.

age, but cleavage of the disulfide bond primarily affects the wet strength. The frosting treatment is an alkaline peroxide system capable of peptide bond cleavage, which may account for small losses in dry strength and for the small difference between the percentage decrease in cystine and the percentage loss in wet tensile properties.

Interestingly, the percentage change in the dry resilience ratio, an estimate of the ability of the fibers to recover from extension into the yield region, approximates the loss of cystine cross-links. Additional work is necessary to determine the significance of this observation.

The foregoing discussion suggests that the percentage loss in cystine, as estimated by cystine or cysteic acid analyses, is a good estimate of the loss in tensile properties of hair bleached by current "in use" treatments. Because frosting of hair is an extreme bleaching treatment, an obvious question is to what extent milder bleaching treatments affect the tensile properties of hair. Several papers [23,39–41] describe bleach damage to hair either by cystine or cysteic acid analyses [40] or by tensile properties [23,41,42]. In summary, these papers suggest that "in use" bleaching of hair commonly produces decreases in tensile properties of as much as 25%, with greater losses occurring when the fibers are frosted or stripped.

Permanent Waving

Chapter 3 describes the chemistry of the reactions of permanent waves with human hair. Permanent waving involves reduction of disulfide cross-links and molecular shifting by stressing the hair on rollers followed by reoxidation. These reactions produce extensive changes to the tensile properties of the fibers, not only during reduction but even after reoxidation.

Data by Crawford [33] (Table 8–4), using hair waved in the laboratory with a commercial home permanent at a 4:1 solution-to-hair ratio, show a decrease in the tensile properties of approximately 5% to 20%. Beyak et al. [23] confirms this finding for the wet tensile properties of permed hair. This amount of tensile damage appears to be typical for a "normal" permanent-wave treatment, in which approximately 20% of the disulfide bonds are ruptured during the reduction step [42].

Higher concentrations of mercaptan, higher pH [43], and higher solu-

TABLE 8–4. Effect of waving by commercial home permanent on the tensile properties of human hair.[a]

Stress to break hair		Stress to extend hair 20%	
Dry (65% RH)	Wet	Dry (65% RH)	Wet
−7%	−15%	−11%	−18%

[a] Data of Crawford [33].

tion-to-hair ratios all produce more extensive reduction [42,44] and ultimately more tensile damage. The decrease in dry tensile properties is less than in the wet state. This is in agreement with the work of Harris and Brown [37], who showed that as much as 60% elimination of disulfide bonds in keratin fibers by reduction and methylation produces only small effects on the dry tensile properties (65% RH). However, the wet tensile properties decrease almost linearly with the disulfide content.

Garson et al. [38], measuring the dynamic elastic properties of hair, showed that permanent waving, similar to bleaching, provides greater changes to the elastic properties of hair in water than at relative humidities between 0% and 80%. Certain hydrogen bonds are presumably inaccessible to water in "virgin" hair; however, the elimination of specific disulfide bonds by oxidation or permanent waving renders these hydrogen bonds accessible to water at high regains, and therefore lowers the wet tensile properties more than the dry tensile properties.

Stretching hair fibers in aqueous solutions of reducing agents, compared to water, results in lower required stresses to achieve a given strain (Figure 8–8) because of the rupture of disulfide bonds, hydrogen bonds, and molecular reorientation. Treatment of reduced hair with mild oxidizing agents (neutralization) increases the tensile properties, once again approaching that of the original untreated fibers.

Load elongation [45] and stress relaxation [46] measurements may be used to follow the course of the reduction of keratin fibers by mercaptans. Extension in the postyield region is resisted primarily by the disulfide bonds [45], and therefore, this region of the stress–strain curve holds spe-

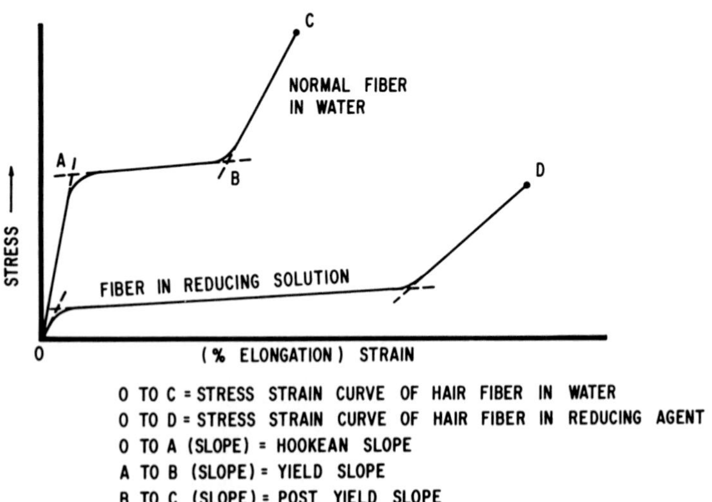

FIGURE 8–8. Schematic diagram representing a stress–strain curve for a hair fiber in reducing solution compared to water.

cial significance to the reduction reaction. The wet tensile properties thus serve as a valuable tool for studying the reduction of hair [43–47] its re-oxidation [45,48], and the effects of the total cold-wave process [23,41]. It should be noted, however, that the wet tensile properties will only reveal cortex damage, but not damage to the cuticle.

Dyes and Surfactants

Chapter 6 describes in detail the chemistry of oxidation dyes and ionic dyes. Oxidation dyes are by far the most prevalent, consisting of aromatic amines and phenols [49–51] that condense with each other and possibly with electron-rich side-chain groups in hair. Penetration of these species does occur, but it is limited, and reaction is primarily by addition rather than by cleavage. Therefore, tensile damage by these species must also be limited. Crawford [52] has found negligible losses in tensile properties in hair dyed from a lighter to a darker shade with a commercial oxidation dye. However, oxidative dyeing is done in the presence of hydrogen peroxide at a pH of up to 10 [50], and the concentration of peroxide and alkali required depends on the starting hair color and the desired shade. Therefore, the primary tensile damage to hair by oxidation dyes should depend on the extent of accompanying disulfide oxidation (bleaching). As a rule of thumb, dyeing hair from a darker to a lighter shade will produce limited tensile damage to hair, provided that additional bleaching is not required for the lighter shade. If additional bleaching or lightening is required, the damage will also be greater.

The tensile damage to hair by anionic and cationic surfactants or dyes in single treatments (i.e., short time intervals (hours) and moderate pHs) is-negligible. Zahn et al. [53] have examined the effect of anionic and cationic surfactants on the 25% index of wool and hair. Soaking wool fibers in sodium dodecyl sulfate for 7 to 41 days produces a decrease of only 6% in the 25% index, and shorter time intervals produce even less damage. Zahn [53] indicates that for human hair the decrease is approximately one-half that of wool fiber, and these effects are reversible with water. He also indicated slightly larger effects from cationic surfactants. Scott [54] examined the stress–strain properties of hair after treatment with a cationic surfactant and found negligible changes. These results test the effects of surfactants on "chemically unaltered" hair, in single treatments, with no mechanical stresses applied. Interactions with other treatments were not considered. We (Robbins and Reich) have recently conducted experiments showing that surfactants and mechanical stresses in combination [short-term treatments (min) and periodic (similar to "in use") conditions] can damage the cuticle and ultimately the cortex. This damage should be demonstratable through significant changes in the tensile properties of hair, but since cuticle damage is involved other techniques, for example abrasion resistance is preferable.

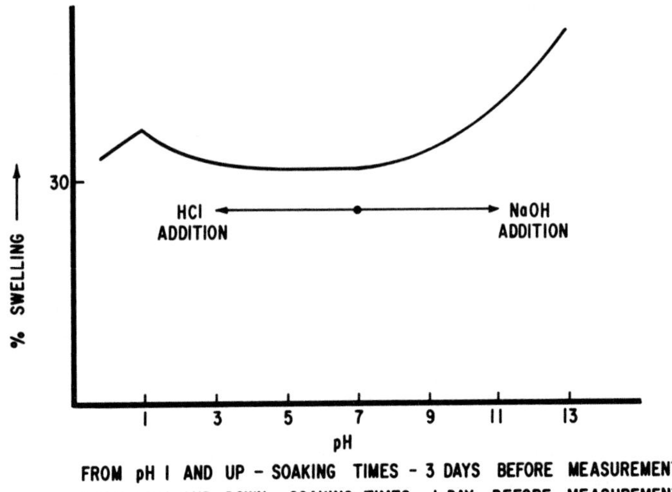

FIGURE 8–9. Swelling of human hair relative to pH.

pH

The influence of pH on the tensile properties of human hair should parallel the effects of swelling and pH of wool fiber [55]. Valko and Barnett [56] have shown that hair displays a minimum in swelling from pH 2 to 9 with a slight increase in swelling below pH 2 and a considerable increase in swelling above pH 9, using extended soaking times (Figure 8–9).

Breuer and Prichard [57] have shown that when human hair is exposed overnight to solutions of pH less than 2, it undergoes irreversible structural changes, evidenced by a weakening of the fibers of up to 30% in the 20% index. This effect is difficult to understand when one considers the decrease in swelling below pH 1.

Light Radiation

Sunlight and ultraviolet light have been shown to decrease the 15% index (in distilled water), and this effect relates to the total radiation on the hair [58]. These findings were interpreted as photochemical degradation of disulfide bonds and have been confirmed by Robbins and Kelly [59], who analyzed amino acids of both proximal and distal ends of human hair and found a significantly larger amount of cysteic acid and a significantly smaller amount of cystine in distal ends. Harris and Smith [60] have provided evidence for ultraviolet disruption of cystine in wool.

Physiological Abnormalities

Korastoff [61] has shown that human hair from patients with hypothyroidism and acromegaly shows a characteristic alteration in the yield region of the stress–strain curve (at low humidity) compared to hair from control groups. Swanbeck [62] has shown that patients suffering from congenital ectodermal dysplasia have hair of low tensile strength.

Monilethrix is a genetic anomaly in which hair fibers contain periodic constrictions along the fiber axis (see Figure 1–38). Monilethrix hairs tend to fracture at these constrictions [63] and therefore must exhibit abnormal stretching behavior. Trichorrhexis nodosa is another abnormal condition in which hair fibers contain nodes at irregular intervals along the fiber axis (see Figure 1–41). These nodes actually contain tiny fractures and tend to form broomlike breaks (see Figure 1–42) [63] under stress. Therefore, nodosa hair fibers should also exhibit abnormal stretching behavior.

Reductive Polymerization and Metal Salts

Anzuino and Robbins [64] have carried out in situ polymerizations of vinyl monomers in human hair and studied the reactions of the polymer-containing hair with metal salts via wet load extension evaluation. This reductive polymerization reaction decreased the wet tensile properties by approximately 15%. These scientists found that mercuric acetate treatment of hair containing polydimethylaminoethyl methacrylate, polyacrylonitrile, and polyethylene glycol monomethacrylate were the most effective systems for increasing wet load extension properties. The exact mechanism of action is not known at this time.

Other Approaches for Evaluating Stretching Properties of Hair

Several other approaches have been used for studying the stretching properties of human hair. Among these approaches are vibrational methods [10,11], stress relaxation [46], stretch rotation [65], and set and supercontraction [19].

Vibrational Methods

In this scheme, a fiber is attached to a beam with a known natural resonant frequency. Tension (within the Hookean region) is applied to the fiber. The beam is then deflected, and, from the change in oscillation frequency of the beam with the fiber attached compared with the natural resonant frequency of the freely vibrating beam, one can calculate the elastic modulus of the fiber. Huck and Baddiel [11] and Garson et al. [38] have used this type of system to evaluate the elastic properties of human hair fibers. Huck and Baddiel attached both ends of hair fibers to an oscillating beam with a third point of attachment in the middle of the fiber for applying tension. From the following expression, they calculated the elastic modulus.

$$E_s = \frac{8\pi^2 IL(B^2 - B^2{}_0)}{AZL^2}$$

where $B =$ the oscillation frequency with the fiber in position, $B_0 =$ the natural resonant frequency of the beam, $L =$ the fiber length, $I =$ the moment of inertia of the beam, $A =$ the fiber cross-sectional area, and $Z =$ the distance between the fiber ends.

The elastic modulus by this "dynamic" method is slightly higher than that by load extension, a "quasistatic" method. Tests involving elastic deformations where either stress or strain is held constant are called static tests. In quasistatic tests, stress or strain is changed slowly with time, and in dynamic tests, stress and/or strain are varied rapidly with time.

Stress Relaxation

Stress relaxation is a technique in which the fiber is stretched to a given length, treated, and maintained at the stretched length while the decaying stresses are followed with time. Kubu and Montgomery [46] have used this technique to follow the kinetics of the reduction of wool fiber. Robinson and Rigby [66] examined both wool and human hair fiber by stress relaxation, and found differences along the axis of the fibers that they attribute to a decreasing free mercaptan level further from the root. This effect provides for less disulfide–mercaptan interchange and a slower rate of stress relaxation as the distance from the scalp increases.

Stretch Rotation

Hirsch [65] studied the elongation of hair under a steadily increasing load, together with a rotational movement, and attempted to explain this combination of stretching and torsion in terms of molecular structure. (See also the section on the torsional properties of hair in this chapter.)

Set and Supercontraction

Set has been defined by Brown et al. [67] as a treatment that enables a keratin fiber to maintain a length greater than its original length. Chapter 3 described setting as it relates to the cold-waving process. Supercontraction, although not a stretching phenomenon, is the condition in which a keratin fiber is fixed at a length less than its original length; it is related mechanistically to setting and is also described in Chapter 3.

Bending and Fiber Stiffness

When a fiber is bent (Figure 8–10), the outer layers of the arc that is formed (A) are stretched and the inner layers (C) are compressed. A region in the center, the neutral plane (B), is unchanged in length. Stiff-

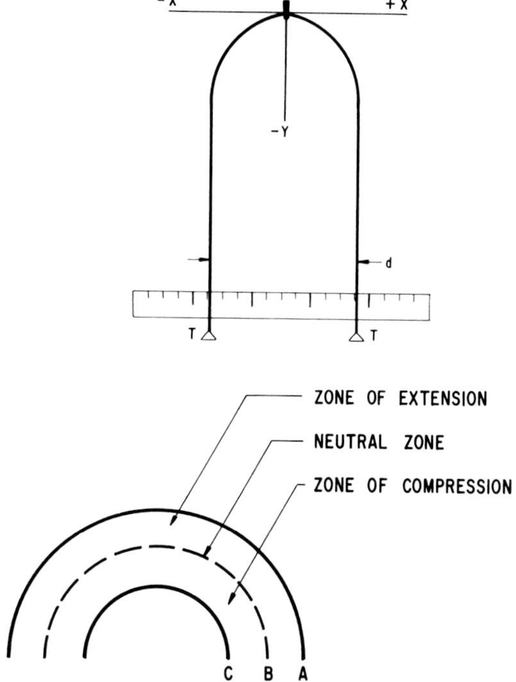

FIGURE 8–10. Schematic diagram of the hanging fiber bending test [69]. Top: Hair fiber photograph in the bending test. Bottom: Schematic of a bent hair. (Reprinted with permission of the *Journal of the Society of Cosmetic Chemists*).

ness, which is simply the resistance to bending, is a fundamental fiber property [64] that requires more attention than it has received.

Bending Methods

Several methods have been described for measuring the bending modulus of fibers. The balanced fiber method of Scott and Robbins [68,69] appears to be the easiest to handle experimentally (except for very curly hair) and provides less scatter than the other methods [69]. The vibrating reed method (oscillating fiber cantilever) has also been used with human hair [10]. The cantilever beam method [70], the loop deformation method [71], and the center load beam method [71] have been described for textile fibers.

The method of Scott and Robbins involves attaching small equal weights to each end of the fiber. Each end of the fiber is individually threaded through a short length of plastic tubing, and a tapered metal pin is inserted in the other end of the tubing (the combined weight of pin and tubing is known). The fiber is hung over a fine wire hook and the distance (*d*) between the two vertical legs of the hanging fiber is measured (see Figure 8–10).

The distance, *d*, is an index of stiffness of the fiber. The stiffness coefficient, G (ratio of applied force to bending deflection), may be calculated from *d* using this expression:

$$G = Td^2/8$$

where T is the force applied to each fiber leg in dynes ($g \times 980.6$ cm/s^2). The elastic modulus for bending, E_B, may also be calculated from *d*:

$$E_B = \frac{\pi T d^2}{2A^2}$$

where A is the fiber cross-sectional area, determined from diameter or linear density measurements.

Scott has developed an equation that describes the hanging fiber shape by conventional x, y coordinates and the *d* measurement; we have verified this equation by showing that calculated fiber shapes are exactly superimposible on those of enlarged photographs of actual balanced hanging fibers (see Figure 8–10).

The bending modulus E_B, by this method, at 62% RH and 75°F is approximately equal to Young's modulus for stretching (E_s) determined under similar experimental conditions by the load extension method, $E_B = 3.79 \times 10^{10}$ dynes/cm^2. These values have not been corrected for fiber ellipticity. Such a correction may be considered academic, but it should make E_B higher than E_s for human hair, because elliptical fibers orient to bend over their minor axis in this method.

Simpson [10] obtained higher values of E_B for human hair, 5.35×10^{10} dynes/cm^2. However, Simpson used the vibrating cantilever method, which produces E_B values that vary with vibrational frequency. He also used a lower RH (50%).

The stiffness index of hair fibers by the method of Scott and Robbins provides falling curves when plotted against weight attached to the fibers. Routine measurements are made at 0.2 g total weight (0.1 g per fiber leg), and with a wire hook of 0.77 mm diameter (0.19–1.28 mm). For thin fibers, less than 50 μm in diameter, less weights (0.05 g per fiber leg) is recommended. (For additional details, see references 68 and 69.)

Stiffness and Linear Density

The stiffness coefficient is directly proportional to fiber linear density. The data plotted in Figure 8–11 provide a correlation coefficient of 0.97 and an index of determination of 0.94 [69]. This demonstrates that 94% of the variation in stiffness (in this experiment) is accounted for by variation in linear density, and as theory predicts, stiffness increases with fiber diameter. Theory predicts a fourth-power dependence between fiber stiffness and diameter for a perfectly elastic system.

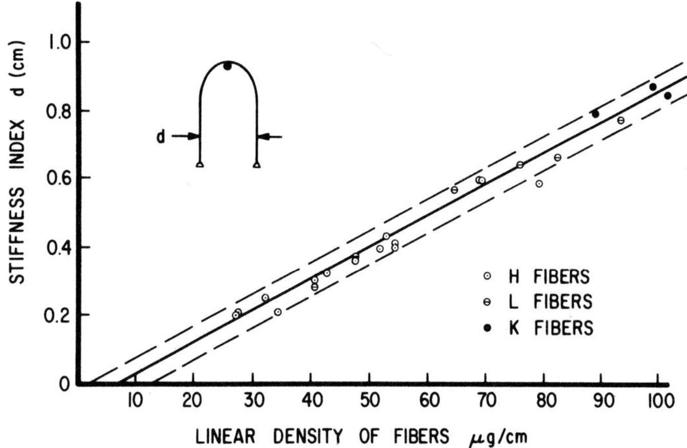

FIGURE 8–11. Hair fiber stiffness index and linear density [69]. (Reprinted with permission of the *Journal of the Society of Cosmetic Chemists.*)

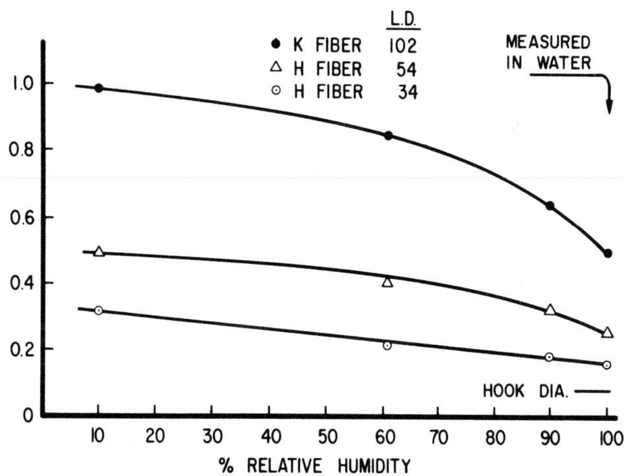

FIGURE 8–12. Hair fiber stiffness; effect of relative humidity [69]. (Reprinted with permission of the *Journal of the Society of Cosmetic Chemists.*)

Stiffness and Relative Humidity

As one might anticipate, hair fiber stiffness also varies with RH; it decreases with increasing RH (Figure 8–12). We might conclude that hair fiber stiffness generally parallels fiber-stretching properties with respect to treatments. This conclusion is probably correct; however, further empirical tests should be made before this conclusion becomes accepted.

Torsion and Fiber Rigidity

Hair fibers are routinely twisted during combing, brushing, and setting. The resistance to twisting is the torsional rigidity. By definition, rigidity is the torque required to produce a twist of one turn per centimeter [70]. Rigidity in twisting is analogous to stiffness in bending and is a fundamental property of great importance to cosmetic science.

Torsional Methods

Several methods have been described for determining the torsional rigidity and the torsional modulus (modulus of rigidity) of hair or textile fibers [29,70,71,73–78]. Basically, these methods are related to the torsion pendulum method [29,73,78]. The torsion pendulum method involves suspending a small pendulum from a fiber that can be set into free rotational oscillation. By determining the period (P) of oscillation (time of vibration, which generally averages from 10 to 20 complete oscillations), the fiber length (L), the fiber diameter (D), and the moment of inertia of the pendulum (I), the torsional modulus (E_T) may be calculated.

$$E_T = \frac{128\pi IL}{P^2 D^4}$$

The rigidity R (resistance to twisting) may be calculated from this expression:

$$R = 8\pi^3 IL/P^2$$

And the rigidity is related to its modulus by:

$$E_T = R/JA^2$$

where A is the fiber cross-sectional area and J is a shape factor, usually assumed to be 1 for human hair and wool fiber.

Another useful parameter is the logarithmic decrement (δ). This parameter describes the decay in amplitude of the untwisting pendulum with successive oscillations:

$$\delta = 1/n \ln a_1/a_n$$

where n = the number of oscillations and a_1 and a_n are the amplitude of the first and the nth oscillations. This parameter is an indication of the torsional elasticity of the system. When $\delta = 0$, the fiber is perfectly elastic, and as δ increases, the fiber becomes less elastic. δ is related to the torsional loss modulus ($E_{T'}$) in the following manner:

$$E_{T'} = E_T \delta/\pi$$

Previously, the major drawback to the simple torsion pendulum method was that it could not be used when the fiber was immersed in liquids because of the damping effect of the liquid. However, Wolfram and Albrecht [78] devised an ingenius scheme to overcome this obstacle by inserting the fiber into a small glass capillary tube, thus permitting the torsional properties of hair fibers (and other fibers) to be measured in both air and liquids. Other methods are available [74,75,79] for measurement of the rigidity of fibers immersed in liquids. For additional details, see the references indicated and the texts by Meredith and Hearle [79] and Morton and Hearle [71].

Rigidity and Moisture

The torsional modulus for human hair by the pendulum method is lower than either the stretching or the bending modulus at 60% to 65% RH. However, water has a greater effect on torsional properties than on either stretching or bending for both human hair and wool fiber (see data describing these three different elastic moduli in Tables 8–5 and 8–6).

For wool fiber, the effect of water on the torsional properties is almost three to four times as great as on the bending or stretching properties, whereas for human hair the effect is nearly a factor of 2 between 65% and 100% RH (Table 8–6).

Wolfram and Albrecht [78] found that the logarithmic decrement varies with fiber diameter. This parameter decreases with increasing fiber diameter in water, but it does not vary with diameter at 65% RH. This finding suggests that the hair is less elastic or more plastic in water. These scientists

TABLE 8–5. A comparison of the stretching, bending, and torsional moduli for human hair at 60% to 65% RH.[a]

$E_S \times 10^{-10}$	$E_B \times 10^{-10}$	$E_T \times 10^{-10}$
3.89	3.79	0.89

[a]All values are in dynes/cm^2. E_B and E_S were determined at 62% RH and room temperature. (From Bogaty [77] and Scott [80].)

TABLE 8–6. Influence of moisture on stretching, bending, and torsional moduli.

	E_S		E_B		E_T	
	0% RH/ 100% RH	65% RH/ 100% RH	0% RH/ 100% RH	65% RH/ 100% RH	0% RH/ 100% RH	65% RH/ 100% RH
Human hair	—	2.62 [26]	3.8[a]	2.4[a]	—	4.1 [77]
Wool fiber	2.76 [24]	2.58 [26]	—	—	16.1 [29]	9.2 [29]

[a]Calculated from unpublished data by Scott [80].

TABLE 8–7. Torsional moduli of waved and unwaved hair.[a]

%RH	Unwaved hair ($E_T \times 10^{-10}$ dynes/cm^2)	Waved hair ($E_T \times 10^{-10}$ dynes/cm^2)
41	1.19	1.25
58	1.06	1.13
65	0.89	0.99
81	0.73	0.76
93	0.42	0.40
100	0.22	0.14

[a] From data of Bogaty [77].

implicate the cuticle as primarily responsible for this increase in torsional plastic behavior for human hair in water.

Torsional Behavior of Damaged Hair

Bogaty [77] has examined the torsional properties of permanent-waved and unwaved hair. His results (Table 8–7) suggest that waved hair is more rigid at low RH and less rigid above 90% RH than unwaved hair. Wolfram and Albrecht [78] examined the torsional behavior of permanent-waved, bleached, and dyed hair. These scientists confirmed the finding of Bogaty that permanent waved hair (reduced hair) is less rigid than chemically unaltered hair in the dry state. These same scientists also found that the rigidity ratio ($R_{water}/R_{65\%\,RH}$) is lower for bleached hair than for dyed hair, consistent with the greater amount of disulfide bond cleavage in bleaching as compared to permanent dyeing. If one takes ratios of dry to wet torsional moduli in Table 8–7, it is apparent that there is a greater effect of moisture on the torsional properties of waved hair than on those of unaltered hair.

The torsional behavior of hair is more dependent than is the tensile behavior on the cuticle or the external layers of the fiber. Torsional behavior is also more sensitive to water than tensile properties, and waving and bleaching do change the torsional properties of hair, as demonstrated by Wolfram and Albrecht. Therefore, it is conceivable that torsional methods may in time prove to be more sensitive to whole-fiber hair damage than the currently used tensile methods.

Density of Hair

We have determined the density of human hair in solutions of benzene-carbon tetrachloride by the method of Abbott and Goodings [81]. The density of chemically unaltered hair at 60% RH varied from 1.320 to

TABLE 8–8. Variation in the density of wool fiber with RH.[a]

%RH	Density
0	1.304
15	1.3135
25	1.3150
68	1.3125
85	1.304
94	1.2915
100	1.268

[a] Data from King [82].

1.327, depending on lot (dark-brown European hair from DeMeo Bros., New York, and three samples taken from heads of volunteers). The density of our wool control was 1.320, identical with one lot of hair. Permanent waving did not change the density of hair. Bleaching (approximately 25% disulfide rupture) increased it, but only by 0.45%.

King [82] determined the density of wool fiber as a function of RH, and some of his results are summarized in Table 8–8. The data show that the density changes for wool fiber from 15% to 85% RH are negligible (normal room humidities), and one would expect the density versus RH relationship for wool and human hair to be similar, because their densities at 60% RH and their moisture binding–RH relationships are virtually identical (see Table 8–16, later in chapter). The increase in fiber density with moisture regain from 0% to 15% RH is contrary to expectations and is not fully understood [83].

The objective of these density experiments was not to determine the absolute density of human hair, which is elusive [83], but to determine the relative density of human hair and wool fiber and the influence of damaging cosmetic treatments on this important property. The results of these experiments confirm the conclusions of several others [70,84]: the densities of human hair and wool fiber are similar, and there is no appreciable change in the density of human hair from permanent-waving or bleaching treatments.

Dimensions and Swelling Behavior

Two of the most commonly measured hair fiber dimensions are its length and its diameter. Assuming that a hair fiber approximates a cylinder, its volume, cross-sectional area, radius, and surface area may be obtained from formulae that describe the volume of a cylinder (V), the area of a circle (A), and the surface area of a right cylinder (Su), in terms of its diameter (D) and length (L):

$$V = 0.7854\ D^2 L$$

$$A = 0.7854\ D^2$$

$$Su = D\pi L$$

Although human hair fibers vary in cross-sectional shape from nearly circular to elliptical, experimental scatter can be significantly reduced by normalizing most elastic and other properties to fiber thickness. Thickness is usually characterized as fiber diameter or cross-sectional area. Corrections to diameter for ellipticity are generally not employed. Hair fiber dimensions are also necessary to calculate fundamental elastic properties, and dimensional changes are often employed to follow the course of chemical reactions with hair.

For measurement of short lengths (in the millimeter range or less) a microscope may be used, but for longer lengths (several centimeters or longer) a cathetometer is useful. Although fiber diameter may be measured directly with a microscope, or more crudely with micrometer calipers, other excellent methods are available for determining cross-sectional dimensions of human hair.

Methods to Determine Hair Fiber Dimensions

Both single-fiber and multiple fiber methods are available for determining hair fiber cross-sectional dimensions or changes. Single-fiber methods include linear density, microscope (light or electron), vibrascope, micrometer caliper, and laser beam diffraction. For multiple fiber determinations, a centrifuge may be used.

Linear Density Method

The linear density method is the method of choice, in my opinion, for determining hair fiber thickness (diameter). A fiber is cut to a given length (10 cm is convenient), conditioned at 55% to 65% RH, and weighed on a microbalance sensitive to 2 μg or better. The fiber weight is in grams per centimeter which is divided by the fiber density, 1.32 g/cm^3, to obtain the cross-sectional area in centimeters squared (A).

$$A = \frac{\text{g/cm}}{1.32}$$

The area so calculated is independent of cross-sectional shape. The fiber diameter (D) may then be calculated, assuming circularity.

$$D = \sqrt{A/0.7854}$$

The volume (V, cm^3) for a given weight (M) of hair may be calculated from the fiber density.

$$V = M/1.32$$

Finally, the length (L) of a fiber of volume V and radius r may be rechecked (because precut at specified length and measured):

$$L = V/r^2\pi$$

The length may then be used to estimate the surface area (Su).

$$Su = 2\pi rL$$

This scheme assumes that the density of all hair fibers is the same. It requires a minimum of manipulations and is an excellent "averaging technique" for dry-state dimensions of hair fibers. Cross-sectional area and volume estimates for circular and elliptical fibers should be relatively accurate, as well as diameter and radius for round fibers. This method does not provide an indication of ellipticity, but provides an average diameter with respect to length as well as to cross section (average diameter, not maximum or minimum diameter). The deviation of fiber diameter with increasing ellipticity is described in Table 8–9 (an indication of the expected deviation from circularity by race is given in Table 8–13, later in this chapter).

Microscopic Methods

Several excellent articles [85–87] describe experimental details for measuring the diameter of human hair fibers with a light microscope. Once the diameter is obtained, calculation of radius, cross-sectional area, volume, and surface area may be made as described in the previous section.

The light microscope is an excellent instrument for determining dimensional changes in hair fibers while they react with either liquid or gaseous

TABLE 8–9. Deviation of the diameter of a circle from the major and minor axes of an ellipse.

		Ratio D_1/D_2 [a]							
		1.108		1.234		1.500		1.700	
Cross-sectional area ($cm^2 \times 10^6$)	Diameter if circular (μm)	D_1	D_2	D_1	D_2	D_1	D_2	D_1	D_2
12.57	40.0	42.1	38.0	44.4	36	49	32.7	53.3	30
113.1	120.0	126.3	114	133.3	108	147	98	156.5	92

[a] Note: D_1 = major axis of ellipse in micrometers; D_2 = minor axis of ellipse in micrometers. These calculations assume that the cross-sectional shape is that of a regular ellipse.

systems (including moisture in air). The light microscope can also be used to measure deviation from circularity. For accurate measurements, however, extensive manipulation and multiple measurements (10–20) along each fiber axis must be made.

A modern scanning electron microscope, with vernier, is also an excellent instrument to measure dry state (~0% RH) diameter of human hair fibers.

Vibrascopic Method

The vibrascope [88,89] is a device that applies an oscillatory force of known frequency to a filament under tension. The fiber cross-sectional area (A) in centimeters squared (cm^2) may be computed from the lowest (natural) frequency (f) in cycles per second that produces mechanical resonance. The tension (T) on the fiber is in dynes, the fiber length (L) is in centimeters, and its density is 1.32 g/cm^3:

$$f_i = \frac{1}{2L}\sqrt{T/1.32A}$$

The fiber diameter, radius, volume, and surface area may then be calculated as outlined previously.

This method assumes that the fiber is a homogeneous filament and provides an average diameter. Nonetheless, it has been shown to be in close agreement with microscopic measurements of the diameter of nylon fibers [88]. This method, like the linear density method, is an excellent averaging technique and is quicker than the microscopic methods.

Micrometer Caliper Method

This method works well for hard fibers like steel, tungsten, and glass, but not as well for softer fibers like human hair. It is a crude but fast way to approximate hair fiber diameter, but because hair fibers yield to low compressional forces, and these are difficult to control, this technique provides low values for fiber diameter and a large variance.

Sieving Hair Fibers

Busch [90] has used fine-mesh sieves to separate fine hair fibers from a bundle for further characterizations. It is conceivable that further separations of hair fibers may be achieved via sieving.

Laser Beam Diffraction Method

Brancik and Datyner [91] have described the diffraction of monochromatic light from a laser to measure the diameter of single wool fibers in liquids. Busch [90] has used a laser beam diffraction system with robotic control for characterizing hair fiber diameter and shape for a large number of hairs.

TABLE 8–10. Percent weight and volume changes versus RH.

% RH	Absorption		
	Volume increase (%) [94]	Weight increase (%) [29]	Weight increase (%) [56]
0	0	0	0
8.5	—	3.9	—
10	5.7	—	—
20	—	—	7.6
40	12.2	10.2	—
60	16.3	—	—
63	—	14.8	—
86	—	22.6	—
90	24.6	—	—
100	32.1	31.2	31.1

Centrifuge Method

The centrifuge method [56,92,93] involves treating a known weight of parallel fibers (400–800 mg, ~20 cm long) with a liquid, centrifuging to remove excess liquid between the fibers, and reweighing. This is a good averaging technique for multiple fibers and is well suited to follow reactions with hair fibers by measuring the percentage weight gain of liquid imbibed. This method may be used to approximate fiber volume changes in aqueous systems, because weight gains at different relative humidities correspond relatively well to volume increases (Table 8–10). From volume changes, cross-sectional area, diameter, length, surface area, and other dimensional changes may be computed (see the section entitled Linear Density Method).

Variation in Fiber Curvature, Cross-Sectional Shape, and Dimensions

Fiber curvature and cross-sectional shape variations of human scalp hair are controlled genetically. Therefore, racial information on hair characteristics is useful. Table 8–11 describes the three major races according to percentages in the world and the United States, and Table 8–12 summarizes the general characteristics of the scalp hair of these three races. Even though doubt has been cast upon the theory of three major races from which all of mankind originates, this classification is still useful for describing variation in certain characteristics of human scalp hair.

Randebrook [97] suggests that human hair from the scalp varies from 40 to 120 μm in diameter. Most of our work with Caucasian hair (the finest in diameter of the three major races) shows variation from 50 to 90 μm, in accord with Yin et al. [84]. Bogaty's review [98] of the anthropological

TABLE 8–11. Population percentages of the three major racial groups.

	Percent of Earth's Population [95]	Percent of United States Population [96]
Caucasoid (principally of European ancestry)	56	86.9
Negroid (black races of Africa, Melanesia, and Papua)	10	11.5
Mongoloid (Sinetics, Mongols, American Indians, and Eskimos)	34	<1.6

TABLE 8–12. Hair fiber characteristics by race.[a]

	Fiber Characteristics [95, 99]			
Race	Thickness	Curvature	Cross-sectional shape	Color
Caucasoid	Fine	Straight to curly	Nearly round to slightly oval	Blond to dark-brown
Negroid	Coarse	Wavy to woolly	Slightly oval to elliptical	Brown-black to black
Mongoloid	Coarse	Straight to wavy	Nearly round to slightly oval	Dark-brown to brown-black

[a] See also data in Table 8–13 and Figure 8–9.

literature, comparing hair from Caucasian adults and children, is in this same range, except for infant hair (15 individuals, age 2), which averaged only 31 μm in diameter. Because this is an average, smaller fibers were obviously present.

Table 8–13 summarizes data from Steggarda and Seibert [100], comparing scalp hair from the three major races. These data support the general conclusions on racial variation summarized in Table 8–12 and show that hair from Caucasians is finer than hair of Negroids or Mongoloids (about 25% finer in mean diameter).

Rows 9 and 10 of Table 8–13 were obtained by multiplying rows 7 and 8 by 1.273. This "normalization" is required to bring the mean diameter for Caucasian hair to 70 μm, in accord with our data and those of others [6,101]. Row 10 represents an estimate of the diameters of 95% of the fibers from the populations of the data by Steggarda and Seibert. It is a good estimate of the variability of the mean fiber diameter of human scalp hair (29–125 μm), and is in reasonable agreement with the estimate by Randebrook [97] of 40 to 120 μm.

Table 8–13 also shows that hair from Negroids has a greater deviation from circularity (average D_1/D_2 of 1.75) than that of the other two races (Figure 8–13). The hair from Mongoloids is the most circular (average

TABLE 8–13. Variation in cross-sectional dimensions of human scalp hair by race.[a]

	Racial group			
		Mongoloid		
	Negroid	Navajo	Maya	Caucasoid
Number of total hairs	873	1002	986	858
Number of individuals[b]	10	10	10	10
Average maximum D_1 in micrometers	90.62	78.76	79.53	63.93
Average minimum D_2 in micrometers	51.70	61.96	64.90	47.28
Ratio D_1/D_2	1.75	1.27	1.23	1.35
Area $\times 10^6$ cm^2	40.06	40.65	41.05	24.11
Mean D, assuming circularity (μm)	71	72	72	55
Estimate of 95% of fiber population (μm)	36–94	26–98	43–93	21–75
Factored mean D (μm)[b]	90	92	92	70
Factored estimate of 95% of fiber population	46–120	36–125	55–119	29–96

[a] Factor of 1.273 used to obtain items 9 and 10. This provides 70 μm for mean diameter for Caucasian hair, consistent with our data and Yin et al. [84].
[b] Subject ages ranged from 9 to 19 years; 5 males and females per group. Hair taken from occipital region of scalp. (From data of Steggarda and Seibert [100].)

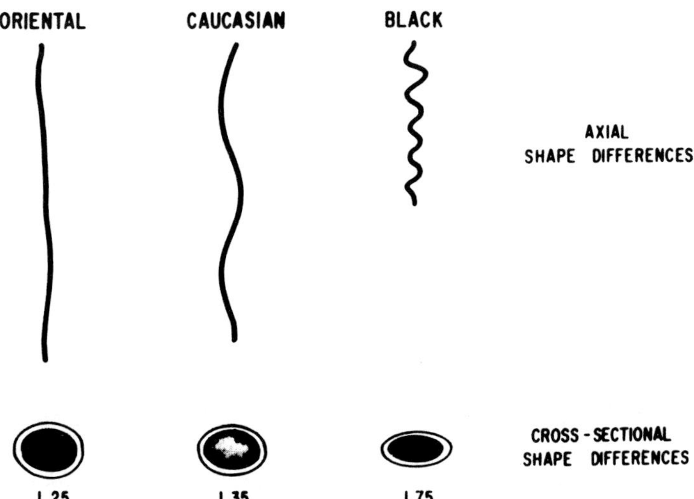

FIGURE 8–13. Hair of different racial origins.

D_1/D_2 of 1.25); for Caucasians, this ratio is 1.35. This suggests that the calculated circular diameter for Mongoloid and Caucasian hair averages approximately 11% and 15%, respectively, from the major and minor axes of noncircular fibers. Therefore, in most circumstances, the assumption of circularity is an acceptable approximation; however, this deviation averages approximately 38% for Negroid hair, and therefore, the assumption of circularity for Negroid hair is generally not an acceptable approximation.

The size and shape of hair fibers also vary along the fiber length, providing a plus for the averaging techniques for determining fiber cross-sectional dimensions. These variations have been attributed to growth patterns as well as to environmental effects [102].

Curvature

The longitudinal fiber shape (curvature) of Table 8–12 is also racially related (Figure 8–13), and its cause is not fully understood. Mercer [103] has suggested an asymmetric keratinization of the fiber during growth. The shape of the folloicle has also been suggested as responsible for the longitudinal shape of hair fibers. For more details, see Chapter 1.

Curvature is perhaps the most important fiber characteristic for styling, combing, and other aspects of hair behavior. Permanent waves and hair relaxers function primarily by changing fiber curvature to either a curlier or a straighter form. When curvature is low (the hair is straight), other fiber properties play a more important role in hair behavior. However, when curvature is large (hair is very curly), this fiber property dominates other fiber properties and controls hair effects.

Hardy [104] first described a technique for measuring average fiber curvature, and this technique has been used to a limited extent in forensic science [105]. Robbins and Reich [3] have also described a method to characterize hair fiber curvature from three simple meaurements on single hairs. These three measurements are the taut length (L_T), the curled length (L_c), and the number of wave crests of the hair (N). L_T is measured with a 1-g weight attached to the end of the hair (small alligator clip). L_c is measured with the fiber hanging freely. Each fiber is then viewed as a simple sine wave counting from crest to crest for N. The fiber curvature C may be calculated from this expression:

$$C = N/(L_c/L_T)$$

Influence of Relative Humidity on Dimensions of Hair

Stam et al. [94], using a microscopic method, measured the changes in the length and diameter of human hair at different RHs and calculated changes in cross-sectional area and volume (Table 8–14). These data show an increase in fiber length and a much larger increase in diameter with increasing RH. In fact, the length changes of hair fibers are nearly logarithmic between 20% and 100% RH and serve as the basis for hair hygrometers. In this device, hair fibers are mounted under tension so that one end is

TABLE 8–14. Dimensional changes of human hair versus RH [94].

	Moisture absorption			
% RH	Increase in diameter (%)	Increase in length (%)	Increase in cross-sectional area (%)	Increase in volume (%)
0	0	0	0	0
10	2.3	0.56	4.7	5.7
40	5.1	1.29	10.5	12.2
60	6.9	1.53	14.3	16.3
90	10.6	1.72	22.3	24.6
100	13.9	1.86	29.7	32.1

fixed and the other end connected to levers that cause a pointer to move over a relative humidity scale as the fibers increase and decrease in length with changing humidities.

Although the swelling behavior of keratin fibers is usually explained in terms of their microfibril-matrix components, Swift [106] and others have provided evidence that the nonkeratin portions of hair are also important to fiber swelling. For example, Swift demonstrated by microscopic studies involving the penetration of fluorescent labeled proteins in water that a large-order swelling occurs in the nonkeratin regions of hair. The diametral swelling of hair by water from the dry state is usually given as about 14% to 16%. On the other hand, X-ray diffraction measurement of intermicrofibrillar separation distances indicates that swelling of only 5.5% occurs [107] in these regions.

Swift therefore proposed that the difference can be explained by the large-order swelling that occurs in the nonkeratin portions of hair, primarily the cell membrane complex and the endocuticle of hair. For additional details, see Chapter 1.

Equally interesting is the hysteresis in the moisture absorption-desorption curves. At any given RH between 0% and 100%, the hair contains less water when it is absorbed from the dry state than when it undergoes desorption from the wet state (Table 8–15). This same phenomenon was observed earlier by Chamberlain and Speakman [28] for human hair and by Speakman [25] for wool fiber (Table 8–16).

Stam et al. [94] also found that tension on fibers influences their dimensions (moisture content) at different RHs. For example, stretching hair fibers below 60% RH provides less swelling than for unstretched hair, and stretching at RH above 60% provides more swelling than unstretched hair.

Table 8–16 also shows that moisture loss (desorption) and regains (absorption) for human hair and wool fiber are virtually identical. It has been suggested [22] that the moisture regains of human hair and wool fiber are virtually identical up to about 90% RH, where they diverge; wool has a regain of 33%, and hair about 31%. However, even if divergence does oc-

TABLE 8–15. Absorption versus desorption of moisture [94].

% RH	Absorption (% increase in volume)	Desorption (% increase in volume)
0	0	0
10	5.7	6.8
40	12.2	13.0
60	16.3	17.3
90	24.6	25.1
100	32.1	32.1

TABLE 8–16. Water content of human hair and wool fiber versus RH.

	Human hair (weight gain)[a]		Wool (weight gain)[a]	
RH (%)	Absorption (%)	Desorption (%)	Absorption (%)	Desorption (%)
0	0	0	0	0
8.5	3.9	5.1	—	—
35	—	—	8.4	9.7
40.4	10.2	12.0	—	—
63	14.8	16.7	14.3	—
86	22.6	23.3	—	—
100	31.2	31.2	31.9	31.9

[a] Data from Chamberlain and Speckman [28].

cur, the actual differences in moisture regain by these two fibers is indeed small.

The rate of moisture regain is much slower from water vapor than from liquid water (diffusion-controlled reaction). Liquid water at room temperature will penetrate the hair in less than 15 min and in less than 5 min at 92°F [93], whereas 18 to 24 h are required for single fibers to equilibrate in a humid atmosphere, with even longer times required for a fiber assembly [108].

Variation of Fiber Surface Area with Diameter

For a given weight of hair, the fiber surface area is inversely proportional to the fiber diameter. Table 8–17 shows how the calculated fiber surface area varies with diameter for 1 g of hair (assuming it is a right cylinder).

Fiber Dimensions and pH

The swelling of human hair in aqueous solutions after 24 h or longer at different pH values exhibits four distinct parts [52,81] (see Figure 8–9):

1. A minimum in swelling from pH 4 to 9.
2. Above pH 10, a large increase in swelling.
3. pH 3 to 1, a slight increase in swelling.
4. Below pH 1, a slight decrease in swelling.

TABLE 8–17. Fiber diameter and surface area.

Fiber diameter (μm)	Calculated[a] surface area (cm^2 for 1 g hair)	Calculated relative surface area
40	758	3.0
80	379	1.5
120	253	1.0

[a] For 1 g of hair, assuming it is a right cylinder.

The minimum in swelling is consistent with the observation of Steinhardt and Harris [101]. In the absence of added electrolyte, there is no combination of wool fiber with mineral acid or alkali from pH 5 to 10. This is in the vicinity of the isoionic point of hair. The large increase in swelling above pH 10 is largely caused by ionization of diacidic amino acid residues in the hair and partly by keratin hydrolysis. The increase in swelling from pH 3 to 1 results from the combination of acid with the dibasic amino acids. Breuer and Prichard [57] attribute the decrease in swelling below pH 1 to an irreversible structural change.

Ehrhardt [93] observed alkaline swelling after only 5 min of reaction with hair at 95°F. Under these conditions, a similar effect was not observed using 0.1 N hydrochloric acid. The swelling in acid below pH 3 is slightly less than that of wool fiber [85] and has been attributed to the higher cross-link density of human hair [26,85].

Solvents and Swelling

Acetonitrile, triethylphosphate, and glycerol swell hair to a lesser extent than does water [56,92]. In fact, glycerol is sometimes used in tensile or torsional testing to approximate 0% RH [20] because of the lack of swelling by this solvent. Dimethyl formamide, ethylene glycol monomethylether, and diethylene glycol monomethylether swell hair similar to water, although the rate of swelling by these solvents is less than the rate of swelling by water [56,92]. Glacial acetic acid and formic acid swell hair to a much greater extent than water [56,92]. Formamide and urea (aqueous solutions of urea) also produce swelling beyond that of water, probably by promoting greater cleavage of hydrogen bonds [56,92].

Amines such as ethyl amine, at 25% in water or higher concentrations, swell hair to a greater extent than water. These solutions rupture amide linkages and ultimately disintegrate (dissolve) the hair (after several days) [56,92]. Concentrated solutions of alkali halides (25%–60%), after several days, produce extensive dimensional changes to hair [56,92]. Barnard and White [85] found extensive swelling by potassium iodide, sodium bromide, lithium bromide, and lithium chloride solutions, but not for sodium chloride.

Hair Swelling and Cosmetic Treatments

Shansky [109] developed a special microscopic cell to follow hair swelling and used this system for studying permanent-wave reactions with hair. With this system, he found that the reduction step produces an increase in diametral swelling, increasing with increasing rupture of disulfide bonds [109,110]. Shansky also found that rinsing with water after reduction produces additional swelling, and he attributed this to osmotic forces, because there is a lower salt concentration outside the fibers at this stage of the process. Neutralization then reverses swelling; that is, swelling decreases with the reformation of disulfide cross-linkages [109,111].

Eckstrom [87] described a moving boundary between swollen and unswollen fiber as the reducing front penetrates the hair during permanent waving and in depilation. Keil [112] noted a similar moving boundary in his studies on permanent waving, using the polarizing microscope. Wickett [113] has shown that for reaction of sodium thioglycolate with hair, above pH 10, moving boundary kinetics occurs, and diffusion of mercaptan into the hair is rate controlling. However, for reaction at lower pH values (below pH 9), the type of reaction depends on the reactivity of the particular individual's hair. For one individual whose hair was highly reactive, the reaction followed pseudo-first-order kinetics and therefore was reaction controlled; however, for another individual with difficult-to-wave hair the reaction exhibited moving boundary kinetics.

Powers and Barnett [110] found that a large excess of reducing solution at pH 10 (for reduction times of 15 min or longer) produced swelling in excess of 200% and destroyed the hair. They also found that the amount of swelling increased with increasing pH from 8 to 10, which was undoubtedly related to the increasing rate of disulfide rupture with increasing pH. Finally, these scientists indicated that the effectiveness of the neutralization step in permanent waving could be assessed by relating it to the swelling action of hair in water. Therefore, hair swelling assessments have been extremely valuable for providing information about chemical alterations to human hair, relevant to permanent-waving reactions.

Swelling Test for Hair Damage

Klemm et al. [114] have described a swelling test to assess hair damage by permanent waves and bleaches. This test consists of measuring diametral swelling in a series of three solutions: the first, water for 10 min; the second, 60% lithium bromide for 60 min; and the third, water for 10 min. From an empirical equation, numerical values of swelling behavior may be calculated. This method distinguishes between single and multiple permanent waves and between single and multiple bleaches on hair. The use of swelling tests to assess hair damage are similar to tensile tests in revealing damage to the cortex, however, they are less sensitive for revealing damage to the cuticle and to the cell membrane complex.

Friction

Friction is the force that resists motion when one body slides over another. The classical laws of friction were formulated by Leonardo da Vinci and later by Amontons, with whom they are generally associated [99].

The frictional force necessary to slide one surface over another is proportional to the normal load pressing the two surfaces together (W).

$$\text{Frictional force} = \mu W$$

The proportionality constant (μ) is called the coefficient of friction, and the frictional force is independent of the area of contact. The force necessary to initiate movement determines the coefficient of static friction (μ_s), and the force necessary to maintain movement determines the coefficient of kinetic friction (μ_k). μ_k is generally less than μ_s.

These laws of friction apply to dry, unlubricated surfaces and to boundary lubricants (very thin solid or nonfluid films separating the surfaces) but not to hydrodynamic lubricants [115], which are fluid layers that separate the moving surfaces (e.g., engine-lubricating oils). Generally, lipid on the surface of the hair provides a reduction in friction. This is an experimental variable of concern to control, which may be done by careful cleansing of the test surfaces or testing fibers in surfactant solutions. Several theories attempt to explain friction; for discussion, see the book by Howell et al. [115], *Friction in Textiles.*

Two of the most important variables relevant to friction on hair are relative humidity (or moisture content of the hair) and the normal load (W) pressing the two surfaces together. It is advantageous to consider friction on hair in terms of two conditions for relative humidity and two conditions of load, thus forming a 2 × 2 matrix:

LOAD	Low RH (60% RH)	High RH (in water)
High (g)	Dry High load	Wet High load
Low (mg)	Dry Low load	Wet Low load

Relative Humidity

The dry high-load condition attempts to simulate hair friction conditions relevant to dry combing, whereas the wet high-load condition is relevant to wet combing. On the other hand, the dry low-load condition is relevant to those critical hair on hair interactions involved in style retention and hair body. Robbins [116] has described three different experimental conditions to characterize hair fiber friction by these three relevant humidity load conditions; these methodologies are summarized in the next section.

Methods for Measuring Friction on Hair Fibers

Two different capstan methods (a fiber over a rod) have been used to measure frictional coefficients of single hair fibers [117,118] at high load both in air (low RH) and in aqueous media.

The apparatus used by Schwartz and Knowles [117] involves draping a fiber with equal weights on each end over a cylinder. One weight is placed on a torsion balance to measure the frictional forces developed as the cylinder is moved against the fiber.

The method of Scott and Robbins [118] involves attaching the root end of a hair fiber to the load cell of an Instron tensile tester. The fiber is weighted at the tip end and partially wrapped around two mandrels (these may be rotated however test sensitivity increases when the mandrels are stationary). The mandrels are attached to the crosshead, and as they move against the fiber, the frictional tension is recorded.

In a capstan system, the coefficient of kinetic friction (μ_k) may be calculated from the following expression:

$$\mu_k = \frac{1}{\phi} \ln \frac{T_2}{T_1}$$

Where ϕ is the angle of wrap in radians, T_2 is the tension after passing over the rod, and T_1 is the tension before passing over the rod. This equation assumes that friction is independent of load, which is valid for the load ranges used in these two studies of frictional effects on human hair. Scott and Robbins used a 1-g load (high load, on the tip end of the hair) small enough so that the total load did not exceed the fiber yield point.

Different rubbing speeds of about 10 in. per minute or higher do not appreciably change the frictional coefficient [118]. However, Robbins [116] showed that friction increases with decreasing rubbing speed at about 0.5 in./min and appears to level near 0.05 to 0.02 in./min, probably approaching static friction. Furthermore, greater differences may be demonstrated between treatments, especially on dry hair at low rubbing speeds. Therefore, the preferred laboratory conditions for simulating the actions involved in wet combing of hair involve a load of approximately 1 g/fiber, with the fiber immersed in water, at a low rubbing speed in the vicinity of 0.02 in./min. For simulating dry combing, the preferred condi-

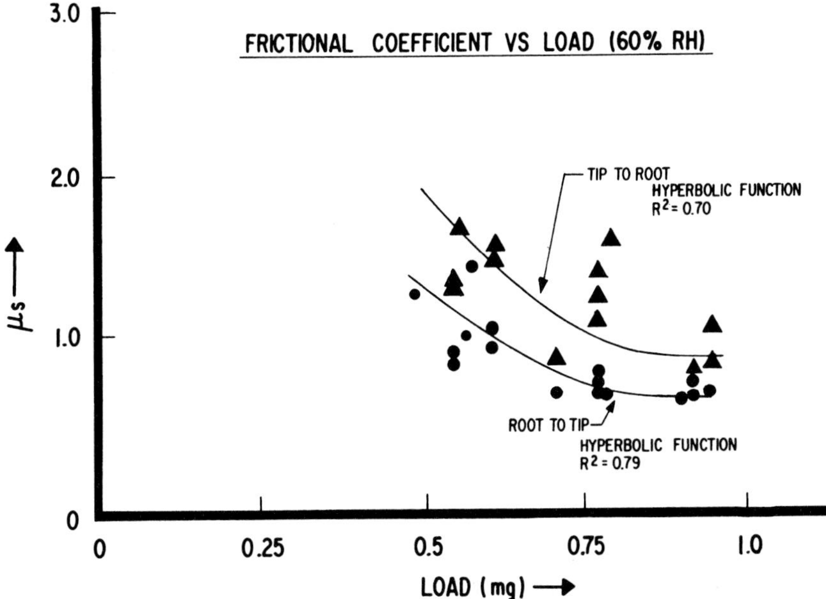

FIGURE 8–14. Directional nature of the frictional coefficient of human hair and its variation with load (at 60% RH).

tions are a load of approximately 1 g/fiber, near 60% RH, and a low rubbing speed of approximately 0.02 in./min.

Wrap angle produces significant differences in friction, as indicated by the capstan equation, suggesting that the frictional coefficient increases with increasing wrap angle. The two frictional methods presented measure coefficients of kinetic but not static friction. Because Amonton's laws state that static friction is generally higher than kinetic friction, there is probably a directional similarity.

Robbins [116] has also developed a procedure for determining dry static friction at low load (in the milligram range) by modification of the incline plane fiber loop method of Howell and Mazur [119]. This procedure attempts to measure those intimate fiber–fiber interactions associated with hair body and style retention by determining the angle of slip for a small hair loop sitting on two parallel hair fibers. Above a 1-mg load, the frictional coefficient decreases very slowly with increasing load. However, below 1-mg load, the frictional coefficient increases abruptly with decreasing load (Figure 8–14). The diameter of the fiber loop (~2 cm) affects the frictional coefficient. This effect presumably results from scale distortion as the loop becomes progressively smaller. Larger loops are recommended (~5 cm in diameter), providing a load of 1 to 2 mg depending on fiber thickness.

At low load (milligram range), the fiber system is sensitive to cohesive forces from thick layers of product deposits, because the fiber system has difficulty ploughing through thick deposits. For example, a pomade-type combing aid was evaluated and shown to dramatically increase the apparent frictional coefficient at low load because of the cohesive forces of the viscous pomade. However, when this same pomade was tested by a high-load dry friction method, the frictional coefficient decreased relative to untreated control fibers. These effects suggest that this pomade will hold fibers of an assembly in place better, because fiber movement at low-load forces will be resisted; however, when high-load forces are applied as in combing, the movement of the comb against the hair will be facilitated by the product, which will act as a lubricant to high-load forces. This experiment verifies the need for both low- and high-load friction methods to develop a more complete understanding of hair product behavior. Other approaches to measure fiber friction are described in the books by Howell et al. [115] and Meredith and Hearle [120].

Factors that Influence Hair Fiber Friction

Relative Humidity and Friction

As indicated, wet friction for human hair is higher than dry friction (Table 8–18). In addition, both static and kinetic friction and the differential friction effect increase with increasing RH. These same phenomena have been observed for wool fiber [121,122].

Fiber Diameter

Scott and Robbins [118] found that high-load friction is independent of hair fiber diameter, in agreement with theory and with the results of Fishman et al. [123]. However, others [124] have reported a slight increase in friction with wool fiber diameter. For low-load friction, this variable would be difficult to test, because as fiber diameter increases, the load also changes appreciably.

TABLE 8–18. Wet and dry frictional coefficients for human hair on different comb substrates (high-load condition) [117].

	μ_k		
	Hard Rubber	Nylon	Aluminum
Dry	0.19	0.14	0.12
Wet	0.38	0.22	0.18

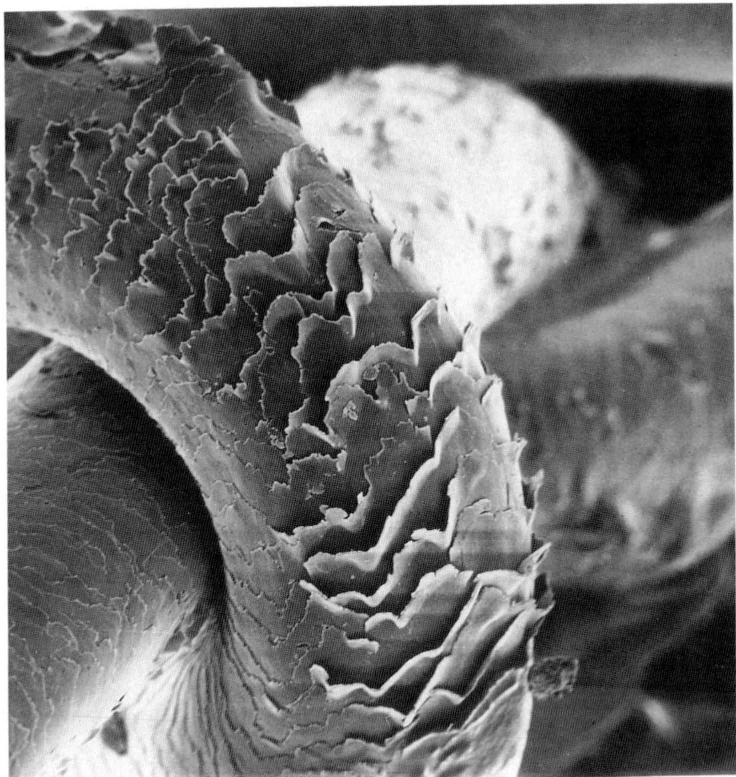

FIGURE 8–15. Scanning electron micrograph of a knotted hair fiber illustrates the cuticle cell surface structure.

Directional Friction Effect

As with most animal fibers, human hair shows a directional friction effect; that is, it is easier to move a surface over hair in a root-to-tip direction than in a tip-to-root direction (Fig. 8–15; see also Fig. 8–14). This friction effect is useful for orienting hair fibers for experimentation. For example, simply take a hair fiber between the thumb and forefinger, and gently rub back and forth along the fiber axis. The fiber will move in the direction of the root end. Moistening the fingers makes this effect even more apparent, because the differential friction effect increases with increasing moisture at the fiber surface.

Table 8–19 illustrates this effect when rubbing hair fibers against a hard rubber surface in dilute shampoo and creme rinse solutions. Note that the differential friction effect is greater in this creme rinse solution than in the

TABLE 8–19. Directional friction effect for human hair, determined at high load [108].

	High cleaning shampoo (μ_k)	Creme rinse (μ_k)
Rubbing root to tip (μ_1)	0.425	0.293
Rubbing tip to root (μ_2)	0.546	0.475
Directional friction effect (DFE)	0.285	0.621

$$\text{DFE} = \frac{\mu_2 - \mu_1}{\mu_1}$$

test shampoo. This is because the test creme rinse lowered the root-to-tip coefficient more than the tip-to-root factor. Scott and Robbins [118] and Schwartz and Knowles [117] examined the root to tip rubbing in more detail, since it appears to be more closely related to combing.

Swift and Bews [125] suggested that dry wool fiber does not display a differential friction effect (absolutely dry wool against a glass surface). Others [121] have demonstrated a differential friction effect (DFE) from dry wool against horn, and for human hair against hard rubber [116]. The debate centers around whether absolutely dry keratin fibers display a DFE. The facts are that the DFE does become smaller with decreasing RH, but it is purely academic whether or not the DFE disappears completely at absolute zero RH, which in itself is academic. Nevertheless, Swift and Bews [125] have proposed the following explanation to account for the decreasing DFE with decreasing water content in animal hairs.

The two major layers of cuticle cells, the exocuticle and endocuticle, may be expected to behave differently with respect to moisture regain. The exocuticle, because of its high cross-link density, should not swell so readily as the endocuticle, with its paucity of cross-links and high density of ionic groups. Swelling of the endocuticle on regain could convert it to a gel-like structure, and this could contribute to the DFE in animal hairs [115,120,122,126].

Mandrel Composition and Fiber Friction

Both Scott and Robbins [118] and Schwartz and Knowles [117] found a wide variation in the coefficient of kinetic friction for rubbing human hair fibers against different mandrel compositions. Some of Schwartz's results are summarized in Table 8–18. Most interesting are the relatively low values on aluminum, which suggests a benefit for aluminum combs. However, more recently, Hambidge and Wolfram [127] have shown that the frictional characteristics of combing materials are not a very important factor in hair combing. This work suggests that hair on hair friction is more important to combing ease than hair on comb friction.

Experimentally, it is generally easier to test friction of hair against another substrate than to test hair-on-hair friction. It is also highly likely that the relative frictional effects of hair against rubber or another substrate will correlate well with hair-on-hair friction, as evidenced by the results showing lower dry than wet frictional coefficients for all substrates (see Table 8–18).

Temperature and Fiber Friction

Scott and Robbins [118] found that temperature changes from 75° to 110°F produced virtually no changes in the high-load frictional coefficient.

Treatment Effects on Hair Fiber Friction

Bleaching and Hair Fiber Friction

Scott and Robbins [118] have shown that bleaching hair increases fiber friction, and that friction increases with increasing bleach damage. This same effect has been observed both at high load and at low load.

The results of Table 8–20 (see "in shampoo" heading) illustrate this effect, and those in "creme rinse" illustrate the strong effect of creme rinse on bleached hair. These results suggest the stronger interaction of bleached hair with cationic creme rinse ingredients that has been shown [128] and are consistent with the higher concentration of ionized cysteic acid groups in the fibers [36]. The influence of bleaching and its frictional effects on the cosmetic properties of hair are discussed in the last part of this chapter.

Permanent Waving and Hair Fiber Friction

Permanent waving also increases hair fiber friction [117], and its influence on hair properties is discussed in the last section of this chapter.

TABLE 8–20. Effect of bleaching on hair fiber friction [118] (high load).

	μ_k in shampoo	μ_k in creme rinse
Unmodified hair	0.249	0.220
1 × mild bleach	0.342	0.190
3 × mild bleach	0.427	0.193

TABLE 8–21. Frictional coefficients by shampoos [118] (high load).

	Hair on hard rubber (μ_k)
High cleaning shampoo	0.342
Experimental high conditioning shampoo	0.155

Shampoos and Hair Fiber Friction

Table 8–21 summarizes some of the data of Scott and Robbins [118] for the wet frictional coefficients of shampoos at high load. As one might expect, the coefficient of friction for the high conditioning shampoo is substantially lower than that of the high cleaning shampoo. This effect suggests easier wet combing, which has been verified.

Creme Rinse Ingredients and Hair Fiber Friction

From frictional behavior versus concentration, Scott and Robbins [118] have suggested three types of cationic creme rinse ingredients (Table 8–22). For CTAB, the frictional coefficient decreases with concentration to a "simulated rinse" (0%), in which it increases. SBDAC illustrates another type of behavior, where friction decreases with concentration; however, at 0% it remains low. The third type of behavior is illustrated by both IQ and DDAC, where the frictional coefficient is low under all test conditions. These data suggest a point of superiority for these ingredients over CTAB. The increase in the frictional coefficient with concentration for SBDAC is unexpected. These results indicate the necessity for thorough rinsing with this ingredient for optimum creme rinse performance.

Scott and Robbins [118] tested CTAB above and below its critical micelle concentration, using salt to promote micelle formation. No significant difference was found in the frictional coefficient, suggesting that micellar sorption versus molecular sorption does not affect the frictional coefficient for systems of this type.

For additional discussion of creme rinses and the effects of reduced fiber friction on hair properties, see the last section of this chapter (Hair Assembly Properties).

TABLE 8–22. Influence of cationic concentration and rinsing effects on hair friction [118].

	Concentration		
Ingredient	0.1% μ_k	0.01% μ_k	0%[a] μ_k
Cetyl trimethyl ammonium bromide (CTAB)	0.390	0.298	0.537
Stearyl benzyl dimethyl ammonium chloride (SBDAC)	0.450	0.394	0.298
Distearyl dimethyl ammonium chloride (DDAC)	0.171	—	0.298
Imidazolinium quaternary (IQ)	0.188	0.169	0.166

[a] This point was determined after the higher concentration (0.1%) simply by changing solution to deionized water, to simulate rinsing.

Hair Assembly Properties

The following section defines and summarizes the literature and some of the author's views on the more important cosmetic assembly characteristics of human hair including, hair conditioning, flyaway hair, luster or shine, combing ease, hair body, style retention, and manageability. Hair feel assessments, including dryness-oiliness, clean-feel, conditioned feel, etc., are important assessments waiting to be defined in a more meaningful way.

Over the past several years those three elusive cosmetic terms—hair conditioning, hair body, and manageability—have been defined in ways that have permitted progress to be made in understanding these important hair properties and in stimulating new methods for their measurement.

The most useful working definition for hair conditioning is based on the action or function performed by a hair conditioner; thus, a hair conditioner is an ingredient or a product that when applied to hair in its recommended use, procedure, and concentration improves the combability relative to appropriate controls. In the case of an ingredient, the control should be a product containing all ingredients except the potential active. In the case of a product, the control should be a cleansing medium such as 12% sodium laureth-2 sulfate in water.

This definition does not say that combability is the only property of a hair conditioner, but it defines combability as the "acid test" or the "price of entry" for a hair conditioner. Market research studies with consumers and current scientific literature are consistent with the above definition. The advantage of this definition is that it permits ease of combing, a method that can be used for reproducible measurement, to be used to study and to improve hair products for hair conditioning (see the section on ease of combing in this chapter).

Hair body has been defined by Robbins and Scott [1] as thickness or apparent volume of a hair assembly, involving sight and touch for assessment. This definition does not lend itself to one simple method for measurement; however, it does help to improve our understanding of this important hair property and has permitted several scientific tests to be proposed that correlate with hair body. For a more complete discussion, see the section on hair body later in this chapter.

Manageability involves the ease of arranging hair in place and its temporary ability to stay in place. Therefore, the suggestion has been made [129] to consider using three types of manageability, rather than the single elusive term manageability, to permit measurement and scientific evaluation For a more complete discussion, see the section on hair manageability later in this chapter.

The foregoing definitions of hair conditioning, hair body, and manageability do not provide a single method to serve to measure these properties.

However, they do permit a better understanding between scientist and consumer and a better overall understanding of these properties. The definitions therefore permit us to move forward to improve these important hair properties for consumers, because better methods for measurement will result from the greater understanding provided by definitions that bridge the gap between consumer and scientist. It is not only in terms of definitions of important assembly properties that progress has been made, however, but also in the understanding of hair assembly behavior.

Robbins and Reich [3] determined empirical relationships between combing ease and the fiber properties of curvature, friction, stiffness and diameter and demonstrated that a hair assembly property such as combing ease can be defined by a few fundamental single-fiber properties. In this study, a quadratic relationship was found for curvature, but linear relationships were found for the other fiber properties. This means that when the hair is wavy to straight, small changes in curvature are of lesser importance and the other fiber properties are very important to combing behavior. However, as the hair becomes very curly, the effect of hair curl becomes more important while the effects of the other fiber properties are of lesser importance. Robbins subsequently suggested that a quadratic curvature effect exists for all hair assembly characteristics such as hair body/limpness, flyaway hair, hair manageability and its components, curl or style retention, etc. This means that for all assembly properties of hair the effect of curvature on that property increases with increasing curvature and that the effects for all other properties are constant. Therefore, the effect of curvature increases with increasing curliness and the other properties become of decreasing importance. Thus, for all hair assembly properties, at a high enough curvature, the amount of curliness of the fibers will dominate assembly behavior and the principal way to improve the assembly behavior is to change the curliness.

Flyaway Hair: Static Charge

When a comb is brought into contact with hair, charges of opposite sign but equal magnitude are generated on both the comb and the hair surface because of the difference in the affinity of each of these materials for electrons (electrochemical potential). The resultant static electricity on the hair produces the cosmetic condition of "flyaway" that makes the hair difficult to manage. Static electricity, which consists of electrons or ions that are not moving, results from rubbing and from pressure during combing and brushing. An electric charge from friction is called triboelectricity; that from pressure is called piezoelectricity.

After combing or after separation of the comb from the hair, the dissipation of the static charge is governed by the conductivity (reciprocal of resistivity) of the fibers or their electrical resistance. In general, materials with a high electrical resistance (e.g., human hair, wool, silk, or nylon) are more prone to static buildup than are lower resistance materials like cotton and rayon [130]. Therefore, the phenomenon of static flyaway is concerned with three conditions: static charge generation on the fibers, conductivity or the removal of static charge from the hair, and hair type. The primary factors involved in static charge generation are (1) the difference in electrochemical potentials of the two surfaces, and (2) the rubbing forces involved (e.g., the use of lubricants can reduce rubbing forces and therefore provide less charge generation) [131]. Jachowicz et al. [132] have shown that the conductivity of the hair surface is also important to static flyaway. These scientists have demonstrated that long-chain quaternary ammonium salts increase the conductivity of the hair surface in addition to decreasing hair fiber friction. For these two reasons, long-chain quats are excellent antistatic agents.

Hair type is also relevant to the condition of flyaway hair. For example, hair fiber curvature is critical to flyaway. Robbins has demonstrated that straight hair has a greater tendency to flyaway than very curly hair, even when a similar level of static charge has been built up on the fibers. The large number of fiber entanglements in curly hair inhibit the separation of hairs in static flyaway, and this effect becomes greater as the hair fiber curvature increases. Thus, the quadratic effect for curvature and flyaway must exist resulting in a larger rather than a constant effect on flyaway with increasing curvature. This means that when hair is straight to wavy, chargeability, fiber friction, and other fiber properties are highly important to flyaway hair; however, when the hair is very curly, curvature dominates flyaway and these other fiber properties are of lesser importance.

Methods Relevant to Static Flyaway

Mills et al. [133] have described two techniques for estimating static charge on human hair. The "ballooning method" actually estimates static flyaway and consists of acclimating tresses at a desired humidity in a chamber (generally for 24 h) and then combing the hair in a controlled manner and estimating the relative amount of static charge by the amount of ballooning or separation of the fibers of the tress. The second method of Mills et al. [133] consists of acclimating tresses, combing in a controlled manner with a special comb containing a bare copper wire in its back that leads through an insulated holder to an oscillograph. The charge on the comb is measured, and theoretically it is equal and opposite to that on the hair.

Similar, yet more sophisticated approaches have been described by Barber and Posner [134], Lunn and Evans [131], and Jachowicz et al. [132]. Lunn and Evans [131] have described methods for measuring charge generation on hair tresses, charge mobility on the hair, and charge distribution along hair fibers. These methods involve combing hair tresses to induce the charge. Jachowicz et al. [132] have developed a useful method to measure both charge generation and charge decay by rubbing hair fibers against different rubbing elements under controlled conditions. Jachowicz et al. have also measured charge density along the length of tresses and have shown that the electrical field concentrated near the fiber tips causes flyaway hair.

The measurement of static charge on textile fibers, fabrics, and yarns has also received a considerable amount of scientific attention and is relevant to this same subject on human hair. For an introduction into this subject, the reader is referred to the books by Meredith and Hearle [135] and Morton and Hearle [130]. The measurement of electrical resistance (reciprocal of conductance) of fibers is also fundamental to their static electrification and is described by Hersh [136] for human hair and other fibers and by Meredith and Hearle [135] for textile fibers.

Triboelectric Series

Attempts have been made [137,138] to classify materials according to "triboelectric series"; that is, to list materials in an order so that the higher one on the list will be positively charged and the lower one negatively charged when any two of the materials are rubbed together. In theory, triboelectric classifications should be useful, because the relative affinity for electrons of each of the materials in contact (electrochemical potential) is very important to the charge developed [132]. However, Morton and Hearle [130] claim that such series are generally self-inconsistent and inconsistent with respect to each other.

Factors That Influence Static Charge

Moisture Content, Resistance, and Static Charge

The moisture content of human hair probably provides a larger influence on static charge than does any other variable, and its direct action is on the electrical resistance (conductance) of hair; that is, increasing moisture in hair decreases its electrical resistance [136], and therefore increasing moisture increases the conductivity of the fiber surface so that it is less prone to develop a static charge.

The electrical resistances of wool and human hair have been shown to be very similar at 85% RH (Table 8–23), and their resistances are similar from 52% to 85% RH [136]. Because their moisture binding RH relationships from 0% to 100% RH are virtually identical (see Table 8–16), their resistance–RH relationships from 0% to 100% RH must also be very similar.

TABLE 8–23. Electrical resistance of wool and human hair at 85% RH [136].

Wool	4.2 to 7.4×10^{12} ohms
Human hair	10.0 to 17.0×10^{12} ohms

TABLE 8–24. Effect of RH on the static charge developed on human hair.[a]

Treatment	% RH 27	51	76.5
Experimental shampoo	14.3	10.8	3.3
Creme rinse	2.5	1.5	0.4

[a] Data are relative pip heights from oscillograph recordings of the charge developed on the comb (from Mills et al. [133]).

The resistance of wool fiber has been shown to vary by a factor of approximately 10^5 from 10% to 90% RH and by a much larger factor from 0% to 100% RH [130].

Confirming this relationship of resistance–RH and static charge for human hair is the effect of RH on static charge shown by Mills et al. [133] (Table 8–24). These data show a progressive decrease in the static charge developed on the hair with increasing RH, for each of two different types of treatment. For both treatments, the principal effect of RH is to increase the water content of the hair, which decreases its electrical resistance, making the fiber a better conducting system and therefore less capable of retaining a static charge.

Temperature and Static Charge

The resistance of keratin fibers generally decreases as temperature increases. An increase of 10°C will produce approximately a fivefold decrease in resistance [133]. Therefore, the effect of temperature on conductivity does not appear to be the cause of greater flyaway from hot combing as compared to room temperature combing.

Impurities and Static Charge

For hygroscopic fibers like wool fiber and human hair, the resistance can be influenced by electrolyte content. For example, the addition of potassium chloride lowers the resistance of wool, whereas washing in distilled water can increase wool's resistance [133] by removal of electrolyte from the hair.

Rubbing Velocity and Static Charge

The studies of Hersh [136] with textile fibers suggest that the amount of static charge generation is virtually independent of rubbing speed, when rubbing high-resistance fibers against high-resistance fibers. However, Cunningham and Montgomery [139] have shown an increase in static charge with increasing rubbing velocity when rubbing high-resistance fibers against metal fibers. Rubbing velocities in both these studies were approximately 1 to 30 cm/s.

Work of Combing and Static Charge

Qualitative experiments after tresses are shampooed and combed show that the amount of ballooning depends on the amount of combing. As already indicated, Lunn and Evans [131] concluded that the primary way that long-chain quaternary ammonium salts reduce static buildup is by decreasing interfiber friction, which in turn reduces the work of combing. Therefore, less static charge develops during grooming, and one way to reduce static buildup is to reduce the work of combing, that is, to make the hair easier to comb [134].

Sign of the Charge

The sign of the static charge that develops on both human hair and wool when rubbed against similar fibers has been shown to be related to the direction of rubbing [132,136,140]. If the fibers are oriented in the same direction and one fiber is removed from the bundle by pulling it out by its root end, a positive charge develops on this fiber. If a fiber is removed from the bundle by pulling it by its tip end, a negative charge develops on the fiber. If the fiber that is removed is oriented opposite to the other fibers of the bundle, or if it contains no scales, no charge develops on it. Although this effect is not fully understood, it has been attributed to the heterogenicity of the scales. The points of the scale edges have been suggested to have a different triboelectric nature from the main scale surfaces and rubbing a fiber root to tip rubs primarily against scale surfaces, whereas rubbing tip to root rubs mainly against scale edges [136].

An equally interesting effect on the sign of the static charge in hair has also been found to relate to the nature of surface deposits or treatments [130,132]. After washing hair tresses with an anionic shampoo and then combing, a large amount of static ballooning was observed. Flyaway was considerably less for tresses treated with a cationic creme rinse followed by water rinsing and drying. However, if tresses were treated with a cationic creme rinse and not rinsed or only lightly rinsed with water, a relatively large amount of ballooning was apparent. The charged fibers from this treatment were attracted to (not repelled from) charged fibers from the anionic shampoo treatment, indicating opposite signs of static electricity. This different sign effect was confirmed by oscillographic measurements. Related effects have been described by Jachowicz et al. [132].

For discussion and theoretical explanations and controversies of the static electrification of fibers, see the book by Morton and Hearle [130], the thesis by Hersh [136], and the paper by Jachowicz et al. [132]

Effect of Treatments on Static Charge

Table 8–25 describes the effects of shampoos (high cleaning) compared with creme rinses on the development of static charge on hair. The

Table 8–25. Effect of creme rinse on static charge in hair.

Shampoo[a] [121]	Creme rinse[a] [121]	% RH	Approximate R_S for wool[b] [133]
11.5–14.3	2.0 to 3.8	27	10^{12} at 27% RH
2.2–3.3	0.4 to 0.5	77	10^{8} at 77% RH

[a] Data are relative pip heights from oscillographic recordings of charge on the comb.
[b] R_S, resistance in ohms between the ends of a specimen 1 cm long and of 1 g mass (ohm-gm/cm^2).

amount of flyaway, which is related to both the chargeability (charge generation and conductivity combined) of the hair fibers and to the work of combing. The lower static values (see Table 8–25) for both creme rinse and shampoo treatments as a function of RH result primarily from a decrease in the electrical resistance of the hair, and the lower work of combing, as well as the increased moisture content (see Tables 8–10 and 8–25). When the resistance drops below a certain value near 10^8 ohm-g/cm^2, the charge can apparently spread more readily over the entire hair to adjacent surfaces and dissipate into the air, thus not exceeding the charge density required to cause noticeable flyaway.

Jachowicz et al. [132] found that the adsorption of long-chain quaternary ammonium compounds, cationic polymers, and polymer–detergent complexes decreases the electrochemical potential and increases the conductivity of hair. Of course, long-chain quaternary ammonium compounds also decrease the rubbing energy by a lubricating action. This lubrication decreases the charge generated, and Lunn and Evans [131] suggest that lubrication is the major reason for the antistatic effect of creme rinses. Jachowicz et al. [132] suggest that the increase in surface conductivity by quats in creme rinses is also important to the effectiveness of these ingredients as antistats. Therefore, the lower static charge for creme rinse versus shampoo treatments (see Table 8–25) is a result of lower frictional drag during combing as well as an increase in the conductivity of the hair. The change in resistance (of keratin fibers) from 10^{12} to 10^8 ohm-gm/cm^{2}that occurs between 22% and 77% RH (see Table 8–25) approximates the required decrease in resistance for a creme rinse effect at 27% RH, assuming that a large part of the antistatic effect by creme rinses is due to an increase in conductivity.

Jachowicz et al. [132] also modified hair fibers by reduction, bleaching, and oxidation dyes and found only a small difference in triboelectric charging versus chemically unaltered hair and no increase in surface conductivity by these treatments.

Luster or Hair Shine

Our consumer research suggests that hair shine is a cosmetic term more meaningful to consumers than luster. The word luster is used more frequently in scientific works on textile materials. In most of our work, the objective was to develop methods to correlate with the consumers' subjective assessment of hair shine. Therefore, the term hair shine is used most frequently in this discussion.

Perhaps the most important factor for maximizing hair shine is to align the fibers as parallel as possible for the desired hair style, and obviously fiber alignment is related to fiber curvature. The more curly the hair, the poorer the fiber alignment and therefore the poorer the shine of the hair. In fact, when the hair is very curly, the greatest impact on shine occurs with treatments to maintain or improve fiber alignment. This is of course done by reducing curvature and keeping the fibers aligned by fixatives or oils/greases or reducing the amount of curliness; reduction in curliness with improved alignment provides the best results. Thus, it would seem as if a decreasing effect (quadratic curvature) for fiber curvature and shine exists, while approximately constant effects exist for shine and other important fiber properties as for combing ease, hair body/limpness and flyaway hair.

When hair is illuminated, the incident light may be reflected at the sur-

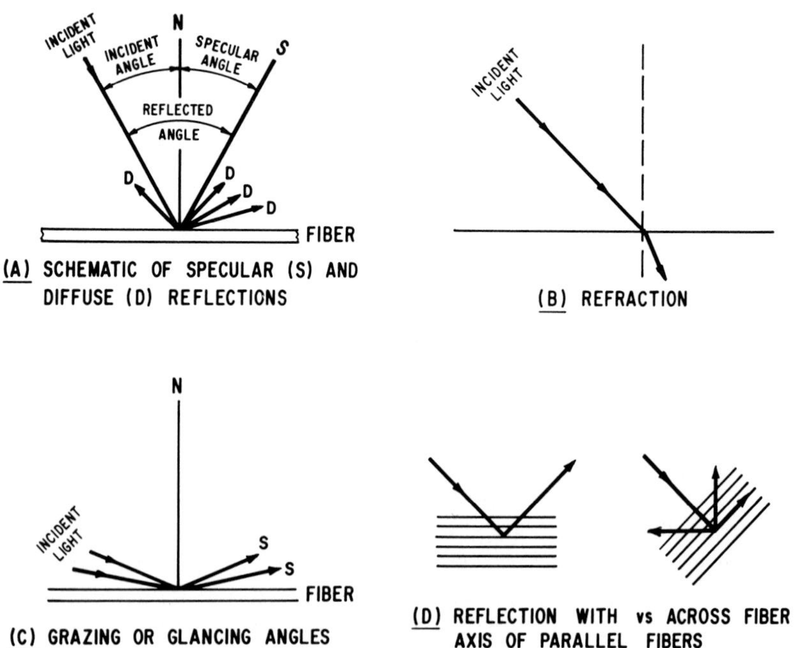

FIGURE 8–16. Some parameters important to hair luster.

face or refracted (bent) (Figure 8–16). It may enter the fiber and be absorbed (by pigment), or it may reemerge, usually after hitting the rear wall of the fiber, where it is partly reflected and refracted again. The phenomenon of light scattering is the main subject of this section, and we shall see how reflection of light may either enhance or reduce hair shine, how refraction of light can reduce hair shine, and how absorption of light may enhance it.

When incident light strikes a surface, it may be reflected specularly (S), when the angle of reflectance equals the incident angle, or it may be reflected diffusely (D), at angles other than the incident angle (scattered) (see Figure 8–16). Unless one examines a perfect mirror, a combination of both specular and diffuse reflection takes place.

Light striking hair in a root-to-tip direction at an incident angle of 30°, provides a specular reflection of 24° (large peak of Figure 8–17), rather

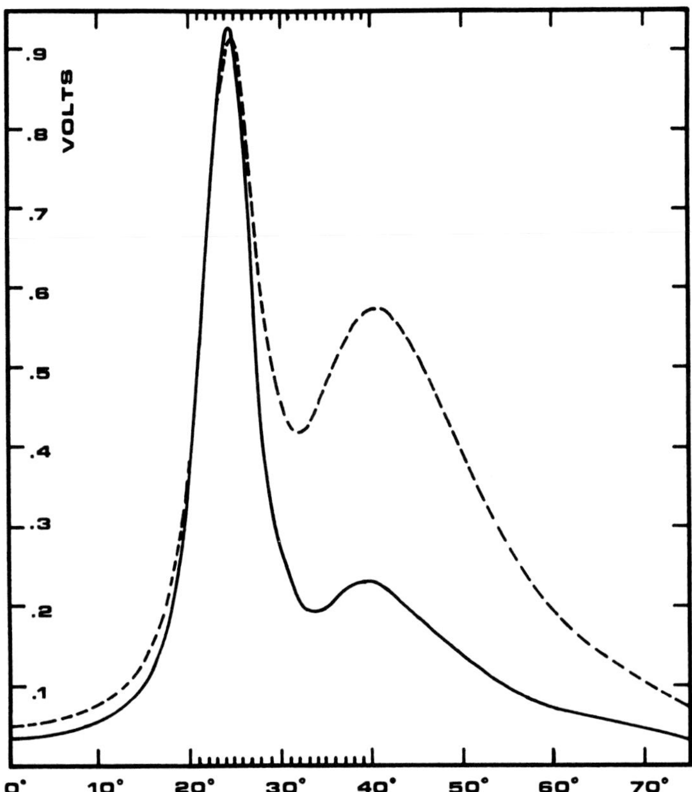

FIGURE 8–17. Light-scattering curves from dark brown hair (*solid line*) and blond hair (*dashed line*) [144]. (Reprinted with permission of the *Journal of the Society of Cosmetic Chemists.*)

than 30°. The specular reflectance may be estimated from light-scattering curves (Figure 8–17), either from the specular peak height or by measuring the area of the specular peak. If the axis of the hair is 30° relative to the incident light, then the scale angle must be 3° ($30 - 2 \times$ scale angle $=$ 24°). The second peak of Figure 8–17 occurs at approximately 40°, and represents light that has entered the fiber and is reflected from the back wall of the hair. This second peak is much larger for blond hair (dashed line) than for dark brown hair (solid line) because of the greater absorption of light in the cortex by the melanin pigment of dark hair.

Diffuse scattering may be estimated from light-scattering curves by drawing a line between the light intensities (voltages of Figure 8–17) at 0° and 75° and measuring the area under the line. A light scattering curve with an incident light of 30° striking the fiber in a tip to root direction, shows the back wall reflection near 15° and the specular reflection at 36°, once again suggesting a 3° scale angle.

Some ratio of specular to diffuse reflectance is generally accepted as a measure of luster for fibers and yarns [141–143]. Ward and Benerito [141] have shown that the ratio of specular to diffuse reflectance for cotton fibers correlates with visual luster assessments. Fourt [142] suggested a contrast ratio for evaluating luster of wool fabric, using a ratio of specular reflectance at a 45° angle of incidence to diffuse reflectance at 0°. Stamm et al. [144] suggested the function

$$\text{Hair shine} = (S - D)/S$$

where $S =$ specular reflectance and $D =$ diffuse reflectance.

Hair Shine Methods

Hair shine may be evaluated subjectively on tresses or on heads of hair (preferably on half-heads), or it may be evaluated instrumentally. The subjective evaluation of hair shine on tresses can be very consistent and reliable, provided care is taken to consistently align the fibers of the tresses, to control lighting, to use multiple observers to evaluate tresses, and to use replication. Six to nine tresses seems to be optimum for any single test. Therefore, only two to three treatments can be evaluated in any single test, and the data should be analyzed by a nonparametric procedure such as the Friedman test [145].

Similar care should be taken for subjective analysis of shine on live heads. The primary variables to control here are lighting and hair alignment, which can be controlled to a limited degree by parting the hair in the center of the head and blow-drying the hair while combing it straight after treatment. The optimum system is to treat hair on heads and to take 30 to 40 fibers per side after treatment and to evaluate the fibers instrumentally for shine by light scattering.

Several different instrumental methods have been used to evaluate the shine of human hair. The method of Thompson and Mills [143] measures reflectance from an assembly of hair fibers; the methods of Stamm et al. [144] and Reich and Robbins [146] measure light scattered from either single hairs or from a parallel array of taut hair fibers. The method of Schebece and Scott [147] measures scattering of light from single hair fibers.

The method of Thompson and Mills [143] illuminates a small tress of hair after the fibers have been carefully aligned over a cylinder 4 in. in diameter. A photocell is placed at an angle of 160° to the incident beam, and at varying distances from the hair sample, which is rotated to give a maximum reading. The light source is at a fixed distance (10 in.) from the hair sample. From these parameters, both the specular (S) and diffuse (D) reflectance can be calculated using the following expression, after taking readings at more than one sample to photocell distance (X). Cos C is the cosine of 80° in this case.

$$\text{Intensity of light reflected} = S \frac{D \cos C}{X^2}$$

The limitations of fiber alignment, the inability to vary incident angle or to scan all reflected light from the surface, with the Thompson and Mills method are solved by the goniophotometric methods of Stamm et al. [144], Schebece and Scott [147], and Reich and Robbins [146]. In these methods, either a single hair or multiple hair fibers are stretched and mounted on a removable frame and placed in a light-scattering goniophotometer to permit recording the intensity of reflected light versus angle of observation. From a strip chart recording of intensity of reflected light versus angle of observation (see Figure 8–16), the specular reflectance and diffuse reflectance are obtained. From these parameters, an objective measure of hair shine may be calculated (see earlier discussion).

Fiber Alignment, Orientation, and Hair Shine

When hair fibers of an assembly are aligned in parallel, maximum specular (mirror) reflectance can be obtained with minimum scattering [148]. The problem of consistently aligning the fibers of a tress or head exactly parallel or comparably perhaps produces the largest variance in shine evaluations and therefore interferes with the ability to see small changes in hair shine; thus, the obvious advantage of the instrumental methods that evaluate either single hairs or a parallel array of hairs [144–147].

In a single hair or in an assembly, shine is always more apparent along the fiber axis than across it [143] (see Figure 8–16). When the incident light is in the scale direction, reflectance is at a maximum, and scattering is less than in the "against-scale" direction [144]. The shine of the surface may appear different when illuminated and viewed from grazing angles as opposed to larger angles where highlights are more prominant (see Figure 8–16).

TABLE 8–26. Effects of shampoos on hair shine (specular/diffuse scattering).[a]

Step 1: Oily hair from scalp	0.411
Step 2: Wash with commercial soap-containing shampoo	0.466
Step 3: Wash with commercial TEALS-based shampoo	0.538

[a] Washing and rinsing in 100-ppm hardness water. (Data provided in private communication by F. Schebece [149].)

Fiber Color or Dyeing and Shine

Dark hair often appears shinier than lighter hair because part of the light is reflected at the fiber surface, and part enters the fiber and is scattered by reflecting off irregularities of the interior. When light reemerges, the diffuse component is increased. If the fiber is colored or dyed, some of this diffuse component is absorbed before reemerging, reducing it, and making the fiber appear shinier. This effect, depicted in Figure 8–17, compares light-scattering curves for a dark brown hair and a blond hair fiber and has already been described.

Shampoos, Sebum, and Hair Shine

Schebece [149], using a light-scattering technique, showed that sebum delusters hair and that soap-containing shampoos also diminish hair shine. The soap effect can only be seen when hardness is present but can be detected in hardnesses as low as 80 ppm (Table 8–26). Thompson and Mills [143] found a similar effect with soap-containing shampoos at 300 ppm of hardness. More recently we have found such effects with soap-containing shampoos at 60 to 80 ppm of hardness.

These data clearly show that sebum dulls hair and that soap deposits from shampoos also deluster hair. Similar dulling effects may be observed after shampooing hair with a shampoo containing cationic polymer ingredients [149]. These deposits are generally not uniform on the hair surface, and they increase the diffuse scattering; however, decreases in the specular component may also be seen with increasing deposition of conditioning ingredients.

Hair Sprays and Hair Shine

Several commercial hair sprays were examined by Schebece for hair luster before and after soaking fibers in different concentrations of the product concentrates. In all cases, the specular/diffuse reflectance ratios decreased (from 2% to 14%) depending on the concentration and type of resin used. In addition, combing and other physical manipulations of the resin on the hair produced cracks in the hair spray resins, increasing the diffuse scattering and further dulling the hair.

Permanent Waves and Visual Assessment of Hair Luster

Dark-brown hair tresses, after treatment with a commercial home wave, were visually assessed by panelists as slightly less shiny than the untreated control. A color shift to a slightly lighter shade was also noted. This shine change is believed to result from an actual dulling of the fibers rather than curvature or alignment changes.

Oxidation of Hair and Visual Assessment of Hair Shine

A group of five subjects were treated half-head style with a surface oxidative treatment. This treatment oxidized the hair surface but did not penetrate to the cortex; therefore, no pigment was oxidized and no color change occurred. Evaluations for shine were made by both subjects and independent observers through 1 month after treatment. Independent observers all agreed that the treated side was less shiny than the untreated side. However, only one subject could detect this delustering effect on her own hair. These results suggest that oxidative treatments, including hair bleaches, dull the hair. In addition, independent observations are more sensitive to hair shine changes on heads than is self-evaluation of hair shine.

Abrasion of Hair and Hair Shine

Schebece [149] abraded hair fibers against hair fibers or a smooth Bakelite surface in one experiment (Figure 8–18) and in another experiment against a 50-μm tungsten wire. The data show that abrasion decreases hair shine (specular/diffuse reflectance), and this dulling effect increases with increasing abrasion. In certain instances, the total amount of reflected light and the specular component both increased; the specular/diffuse ratio was always decreased. These results suggest that teasing (back-combing) hair and other abrasive actions such as vigorous combing or brushing can deluster the hair by breaking scale edges and creating more irregularities on the fiber surface (see Figures 1–17 and 1–18). These effects dull hair by increasing diffuse scattering.

Combing Ease

Combing ease may be defined as the ease of aligning fibers of an assembly with a comb so they are essentially parallel. It may be considered in terms of single-fiber properties or treated as an assembly property. The fiber properties that are relevant to combing ease are curvature, friction, stiffness, diameter or cross-sectional area, length, and cohesion. Table 8–27 describes how "changes" in these fiber properties affect combing ease. This table suggests that increasing fiber curvature, friction, or static charge will each make hair more difficult to comb, whereas increasing fiber stiffness, diameter, or cohesive forces will make hair easier to comb.

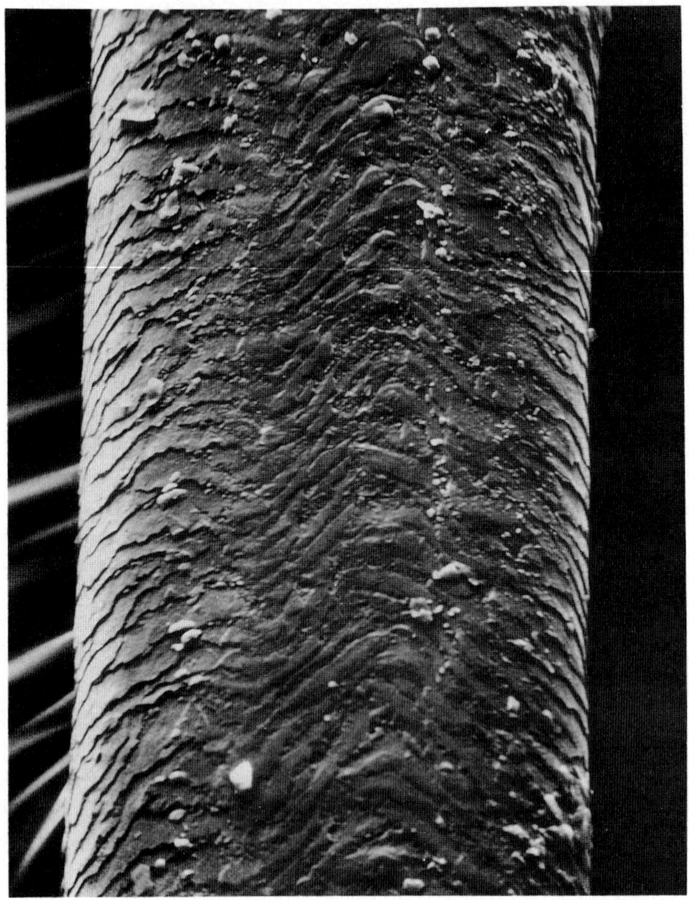

FIGURE 8–18. Scanning electron micrograph of hair fiber mechanically abraded by rubbing against another hair. Note cuticle abrasion and hair fragments.

Robbins and Reich [3] have described studies relating quantitative combing behavior to the single-fiber properties of curvature, friction, stiffness, and diameter for straight, wavy, and kinky hair, all treated with a shampoo detergent (sodium lauryl sulfate), a long-chain quaternary ammonium compound (stearalkonium chloride), a commercial pomade (made from mineral oil and petrolatum), and a hair bleach (peroxide/persulfate system).

The results of this study show that curvature has more of an impact on combing behavior than any other fiber property. There is a quadratic relationship between curvature and combing ease, but linear relationships between combing ease and other fiber properties. Thus, the curvature effect is small when the hair is relatively straight, but it increases with increasing curliness. Fiber friction and stiffness also contribute to combing be-

TABLE 8–27. Single-fiber properties in relationship to combing ease [3].[a]

Increase in these fiber properties makes combing easier	Increase in these fiber properties makes combing more difficult
Stiffness	Curvature[b]
Diameter[c]	Friction
Cohesion	Length
	Static charge (chargeability)[d]

[a] Note: Fiber length is not changed by cosmetic treatments.
[b] Relatively large effect; the effect increases with increasing curliness.
[c] Relatively small effect.
[d] Relatively small effect predicted.

havior, and the effect of these variables becomes more important as curvature decreases. These effects are most important with very straight hair. Fiber diameter is not as important as these other variables. Increasing fiber curvature or fiber friction makes combing more difficult as expected (see Table 8–27). However, increasing fiber stiffness results in lower combing forces. For pomades and other oily or wax-containing conditioning products, cohesive forces serve to lower combing loads. This effect probably occurs because these ingredients inhibit the formation of entanglements as the comb traverses through the hair. Thus, the cohesive effects by pomades on combing ease, in a sense, partially counteract the negative effects of curvature by inhibiting entanglements.

Methods to Evaluate Combing Ease

Qualitative combing of tresses in replicate and evaluation of the data by nonparametric statistics can be a powerful tool when properly applied, because it provides a fast, sensitive, and reproducible method. However, quantitative instrumental methods have also been useful [150,151]. Basically, these methods consist of attaching a tress or swatch of hair to a strain gauge (e.g., the load cell of an Instron tensile tester) and measuring the forces or work required to move a comb through the swatch under standardized conditions. An alternative approach is the raspiness method of Waggoner and Scott [152]. This method utilizes an electronic comb designed to pick up vibrational frequencies emitted as the comb teeth rub along the hair scales.

Treatment Effects on Combing Ease

Both permanent waving and bleaching make hair more difficult to comb [117,151]. Permanent waving increases fiber curvature and interfiber friction [117], primary factors that make hair more difficult to comb (see Table 8–27).

In the case of bleaching, the primary factor is the frictional increase [117,118]. There is no measurable curvature change in bleaching [3], and stiffness [3], diameter [3], and chargeability [132] changes are also negligible. Creme rinses [150,152] and some conditioner sets [150], in contrast to permanent waves and bleaches, make hair easier to comb by providing a decrease in interfiber friction [3,117,118]. Chargeability [132] may also decrease, which helps to improve dry combing. Pomades decrease fiber friction and increase cohesive forces between hairs. This cohesive effect helps to inhibit the formation of entanglements beneath the comb as it travels through the hair, and at the same time it helps to keep the fibers parallel after each comb stroke. Thus, cohesive forces make hair combing easier.

Shampoos offer a category with wide variability, since they can make hair either easier or more difficult to comb [118,153]. High-cleaning shampoos remove surface oils, increasing interfiber friction [118], making clean hair more difficult to comb than dirty hair. However, certain conditioning shampoos deposit ingredients onto the hair surface that can decrease fiber friction, making hair easier to comb.

Shampoo ingredients [132] can also alter the chargeability of the hair. For example, high-cleaning shampoos remove surface oils and deposit small amounts of anionic surfactant onto the hair, thus increasing chargeability, whereas some conditioning shampoos lubricate the hair surface, providing easier combing and at the same time decreasing chargeability, leading to less flyaway and easier dry combing. Changes in fiber stiffness, curvature, and diameter by current shampoos are negligible, and therefore changes in these properties are not relevant to combing effects by shampoos.

Hair Body

Hair body is defined in the textile trade as that compact, soft, or firm feel of textile stock or fabric [154]—a tactile property. With regard to human hair, body may be defined as thickness or apparent volume of a hair assembly, involving sight and touch for assessment, and is the inverse of hair limpness. The quality of liveliness or springiness [155] is also associated with hair body. Hough et al. [2] defined body as the structural strength and resiliency of a hair mass, which is consistent with these definitions.

Body is a complex property involving several single-fiber properties, including friction, stiffness, curvature, cross-sectional area or diameter, weight, and length (Table 8–28).

If we consider changes instead of absolute values for hair body, it permits us to neglect properties that do not change during treatment (e.g., density of hair population on the scalp and fiber length) and thus simpli-

TABLE 8–28. Relation of single-fiber properties to hair body [1,3].[a]

Increase in these fiber properties produces an increase in hair body	Increase in these fiber properties produces a decrease in hair body
Curvature[b]	Cohesion[b]
Friction	Weight
Stiffness	
Diameter	

[a] Note: Fiber length is not considered, since it is not changed by cosmetic treatments.
[b] Relatively large effect predicted.

fies the analysis. This approach suggests that if one makes the fibers stiffer, increases their diameter or curvature, or increases the frictional forces between the fibers, hair body will increase.

Hough et al. [2] used a related approach and concluded that five groups of fundamental parameters govern hair body:

1. Hair density on the scalp
2. Stiffness
3. Diameter
4. Fiber interactions
5. Curvature.

Further, increasing hair density on the scalp, stiffness, diameter, or curvature increases hair body. For fiber interactions, the effects depend on the nature of the interactions. Obviously, these two independent approaches are related, and both lead to the conclusion that hair body can be studied systematically through single-fiber properties.

Robbins has suggested that curvature has more of an impact on hair body/limpness than any other fiber property and has suggested a quadratic relationship between hair body and curvature similar to that for combing ease and curvature and essentially constant effects for body and other fiber properties. Further, he has shown that straight hair exhibits limp behavior with one third the amount of synthetic sebum on it compared to kinky hair that exhibits no detectable limpness. This experimental result is consistent with a quadratic relationship between hair body and curvature.

Methods to Evaluate Hair Body

Several laboratory methods have been described to characterize hair body [2,84,156–160]. The Tolgyesi omega loop method [84] examines structural strength of a hair assembly, emphasizing the bending properties of an assembly more than the frictional properties of the fibers. One method that emphasizes frictional behavior but attempts to measure the bulk of hair fibers is the cylinder method of Scott and Robbins [157]. This method involves randomly dropping 1- to 2-in. hair fibers into a graduated cylinder. Fibers with higher interfiber friction tend to provide a larger volume and therefore more body. The method of Clarke et al. uses image analysis to approximate the volume of hair tresses [160].

The Textile Research Institute (TRI) method [156] involves transverse compression of a bundle of fibers to determine the compressibility and recovery behavior of hair tresses. Garcia and Wolfram [158] considered the force or work necessary to pull hair tresses through a Teflon ring. Here the net force is a combination of fiber bending and fiber–fiber and fiber–ring friction. But it is not clear with either of these two methods which parameter or combination of parameters correlates best with hair body. TRI suggests that their method is intended to measure the "tactile component of hair body as it is reflected in the resistance which hair offers to compression."

Another approach, by Robbins and Crawford [159], is a modification of the Garcia–Wolfram method. In this system, the bulk of a hair assembly is assessed by measuring the work required to pull the tress through a succession of very thin (0.076-cm) circular templates of decreasing diameter. A plot of work values versus circle diameter is extrapolated to zero work to obtain the "maximum tress diameter," This parameter appears to be a measure of the bulk of hair assemblies. Blankenburg [161] of the German Wool Research Institute has independently developed a similar method using this principle.

Treatment Effects and Hair Body

Permanent waves and hair bleaches are both known to increase hair body, although the mechanisms of action of these products are different. Permanent waving increases hair body by increasing both fiber curvature and fiber friction [117]. Bleaching hair does not increase hair fiber curvature, but it can increase interfiber friction substantially [117,118], and both of these effects are primary factors that increase hair body (see Table 8–28).

Creme rinses are purported to make hair limp, and limpness is the inverse of hair body. This effect is probably caused by the creme rinse ingredients decreasing interfiber friction [117,118].

Conditioner sets are an interesting category, because some of these products increase hair body and at the same time make the hair comb easier. Such effects probably arise because these products reduce high-load frictional forces when the hair is being combed wet (or even when dry), and on drying they increase the low-load friction and or cohesive forces between fibers. Certain hair sprays behave in this same manner. Increases to the "apparent" fiber diameter are also possible for conditioner sets and hair sprays when they are combed thoroughly and are distributed well throughout the hair.

Shampoos vary in their effects on hair body. High-cleaning shampoos increase the body of dirty hair by removing sebaceous soil and other oily soils. These effects decrease cohesive forces between fibers and increases interfiber friction. On the other hand, continued use of conditioning shampoos can lead to limp hair by reducing interfiber friction and increasing cohesive effects between hairs. These effects are similar to the action of creme rinses. This limpness effect is greatest for straight hair and least for very curly to kinky hair.

Style Retention

Style retention may be defined as the ability of hair to stay in place after styling. It is time dependent and includes curl retention, wave retention, and straightness retention. Style retention may be described in terms of single-fiber properties or it may be treated as an assembly property. It is most important to permanent waves and to hair sprays as well as to conditioner sets or wave sets and to mousses.

Some of the more important single-fiber properties that are related to style retention are curvature, friction, stiffness, cohesion, and weight [1,3]. Table 8–29 suggests that increasing fiber stiffness and weight will decrease style retention. However, for straight hair styles, increasing weight improves style retention. The role of weight is straightforward; that of stiffness is more subtle. Hair fibers are generally water set to produce an optimum curvature, which usually differs from the natural curvature of the hair. Changes in humidity promote deterioration of the water set and a gradual change toward the hair's natural curvature [27] (see Chapter 3). Only frictional and cohesive forces tend to hold the assembly in the desired style. The stiffer the fibers, the more readily they tend to overcome these frictional forces and the desired style. We thus conclude that increasing fiber stiffness decreases style retention.

TABLE 8–29. How single-fiber properties relate to style retention [1].

Increase in these fiber properties produces a decrease in style retention	Increase in these fiber properties produces an increase in style retention
Curvature[a]	Friction (low load)
Stiffness	Cohesion
Weight	Curvature[a]

[a] A curvature increase can increase or decrease style retention (see text).

The maximum fiber curvature consistent with the desired styling will produce the optimum number of entanglements for that style. If style retention is still not adequate, for example, for straighter hair styles, then reinforcement of the fiber–fiber interactions with hair fixatives will be necessary. Generally, an increase in fiber curvature will increase style retention. However, if the desired style is considerably straighter than the natural curvature, then a decrease in fiber curvature will be necessary to improve style retention. The converse holds for styles that are curlier than the natural curvature of the hair.

Methods Relevant to Style Retention

Most companies involved in hair products have developed procedures for evaluating curl or wave retention of hair assemblies (tresses, wigs, or heads). The laboratory procedures vary, depending on the type of product to be evaluated. Differences in application of hair sprays and permanent waves and the difference in performance characteristics require different procedures. The extent of quantitation may also vary. For example, Sante et al. [162] described the evaluation of mercaptans as potential permanent-waving agents by simply treating tresses wound on curlers and qualitative evaluation of the hair for "curl power" after removal from the curlers. Other, more quantitative approaches involve calculation of percent curl retention [163] or percent waving efficiency [164,165].

Basically, curl retention procedures involve treatment of a tress of known length, winding it on a mandrel or a rod or around pegs, and drying or conditioning it. After removing the hair from the rod, either length measurements versus time (under controlled humidity) are taken to use to arrive at percent curl retention [163], or it is placed on water and the change in elongation provides a means to obtain percent waving efficiency for cold wave lotions [164,165].

The curl strength method of Stravrakas et al. [166] is an interesting alternative. This method involves treating hair, curling it on rods, conditioning it at high humidities, and determining the resistance to the deformation of curls and waves formed by the hair treatment, using an Instron tensile tester or a similar strain gauge device. This method provides a large variance; however, it can distinguish between cold waves, hair sprays, and water-set hair.

For hair sprays or hair fixatives, several techniques have been described to evaluate curl-holding ability under conditions of controlled humidity [167,168]. One novel approach [167] involves the rate of untwisting of tresses treated with the hair fixative. This method correlates with curl retention tests, yet it allows for faster evaluation of treatments.

A single-fiber method for determining curl retention or straightening has been described by Robbins [169]. This method offers advantages when used in conjunction with a fiber assembly method (tress method), because it eliminates fiber–fiber interactions from curvature and stiffness changes. This method consists of uniformly winding a single hair fiber of known length onto a glass rod, treating it with water or product, and conditioning the hair at a controlled humidity (generally 50%–65% RH). The fiber is then removed from the rod and conditioned at different RHs, and changes in length with time are measured with a cathetometer. (See Chapter 3 for a discussion of some of the results with this method regarding water-setting hair.)

Style Retention and Hair Treatments

Both permanent waving and relaxing of hair improve style retention. Obviously, the type and size of rollers (curlers) in permanent waving are important in determining the final result, and they should be consistent with the curvature required for the desired hair style. Permanent waving also increases interfiber friction [117], which also contributes to improve style retention.

Bleaching increases interfiber friction [117,118], which helps to improve style retention. Crème rinse ingredients [118], in contrast to bleaches, provide a decrease in interfiber friction, which generally decreases style retention, except for some straight hair styles.

Conditioner sets, setting lotions, mousses, and hair sprays all increase the interfiber attractive forces that help to improve style retention. These effects are greater than that of the weight increase of these ingredients, which would tend to decrease style retention.

High-cleaning shampoos, when used on hair containing sebaceous soil, remove the surface oils and thereby reduce the cohesive bonding between fibers. These products thus provide for an improvement in style retention for "dry look" hair styles. Certain conditioning shampoos provide for a decrease in fiber friction and an increase in cohesive bonding and thereby provide for a limp look and a decrease in style retention for most styles.

Hair Manageability

Manageability is the ease of arranging hair in place and its temporary ability to stay in place; long-term effects on the hair fiber assembly are not considered for this property. Manageability is an even more complex consumer assessment than hair body. It is such an inclusive term that it cannot be measured by one single procedure. Therefore, Robbins et al. [129] recommended considering this important cosmetic property in terms of its component fiber assembly properties, that is, as different types of manageability that can be more readily visualized and measured. Manageability is concerned with arranging hair in place (combing/brushing), keeping hair in place (style retention during styling), and flyaway hair. Therefore, Robbins et al. [129] suggested these three types of manageability, rather than the single elusive term manageability, to permit measurement and scientific evaluation: style arrangement manageability (combing/brushing and arranging hair), style retention manageability (style retention when styling), and flyaway manageability.

Existing tests to evaluate hair combing ease, style retention, and flyaway hair or static charge may then be used to evaluate these different types of manageability. Further, each type of manageability may be expressed in terms of a ratio of control/treatment values.

Using this approach, those single fiber properties as defined in Table 8–27 that improve combing ease will also improve style arrangement manageability, while those properties of Table 8–29 that improve style retention will also improve style retention manageability. Similarly, those parameters that decrease static charge on hair, as described in the previous section on static charge, will improve flyaway hair manageability.

Therefore, increasing fiber curvature, static charge, or high-load friction will decrease style arrangement manageability, and fiber curvature plays a dominant role when the hair is very curly (see the section on Combing Ease). On the other hand, increasing low-load friction or cohesive forces between fibers will improve style retention manageability, and a change in fiber curvature can either increase or decrease style retention manageability. For example, if the treatment changes the curvature so that it is either too straight or too curly for the desired style, it will make the hair less manageable. However, if the treatment makes the hair curvature more consistent with the desired style, then that type of change will improve style retention manageability.

Because fiber length is not changed by cosmetic treatments, length is not relevant for this type of analysis that considers changes by cosmetic treatments. For flyaway hair manageability, any change that increases static charge on the hair will increase static ballooning of hair and thus will decrease flyaway hair manageability. When the hair is straight to wavy, flyaway hair can be a real problem; however, when the hair is very curly to kinky, flyaway is of little concern (see the section on flyaway hair for the explaination).

Treatment Effects and Hair Manageability

Because style arrangement manageability and style retention manageability often are opposed, the decision as to whether a given effect will improve manageability for any particular person's hair depends on which of these two components is desired more. And because different persons will attach different importance to the three different types of manageability, it is very difficult to talk about overall manageability. It is therefore preferable to discuss each of the three types of manageability separately rather than to try to arrive at a composite for this important cosmetic property.

Permanent waving can increase style retention manageability by helping to provide an amount of hair fiber curvature that is more consistent with the desired hair styling. Permanent waving also increases low-load friction, which also improves style retention manageability. Permanent waves are generally used to increase fiber curvature, which increases the number of possible entanglements and decreases style arrangement manageability. Increasing fiber friction also decreases style arrangement ease. Most satisfied users of permanent waves are more concerned with the staying-in-place component or style retention manageability. Therefore, they feel that their hair is more manageable after a permanent wave. Those dissatisfied with the style arrangement manageability of a permanent wave can decrease fiber friction with a creme rinse and thereby improve style arrangement manageability.

Creme rinse ingredients decrease high-load friction [117,118], which makes the hair easier to comb and improves style arrangement manageability. At the same time, creme rinse ingredients decrease fiber chargeability. This latter effect, in combination with combing ease, decreases the propensity of the fibers to fly away and thus improves flyaway manageability. Most satisfied users of creme rinses are more concerned with style arrangement and flyaway manageability than with style retention manageability. Therefore, they conclude that their hair is more manageable after using a creme rinse.

Satisfied users of conditioner sets, setting lotions, pomades, and hair sprays are generally more concerned with style retention manageability, and therefore the increase in interfiber cohesive forces from these products more than offsets any decrease in style arrangement manageability. However, some of these products (e.g., some pomade products) offer decreased high-load friction [116] and therefore offer improvements in both style arrangement and style retention manageability at the same time for certain hair styles, such as "wet look" or greasy hair styles.

High-cleaning shampoos remove sebaceous oils and decrease cohesive forces between the fibers, thereby making the hair more conducive to "dry look" hair styles. These products increase frictional forces relative to dirty hair. These effects aid style retention manageability for "dry look" styles. Satisfied users of these shampoos, who do not use other hair products, are generally more concerned with these factors than with style arrangement manageability. Users of high-cleaning shampoos plus creme rinses or high-conditioning shampoos are generally very concerned with improving style arrangement manageability and flyaway manageability for a "dry look" type of hair style.

In summary, all important consumer assessments of hair such as flyaway hair, combing ease, hair shine, hair body, style retention and manageability are evaluated as assembly properties rather than as single fiber properties. To optimize hair behavior and to objectively measure small changes that are relevant to these consumer assessments, quantitative measurements and a better understanding of the relationships between fiber properties and assembly behavior is necessary.

Over the last decade, considerable progress has been made in these areas of hair care primarily through a better understanding of how fiber properties (hair types) relate to hair assembly behavior. For example, for flyaway hair, combing ease, hair body and manageability, fiber curvature is dominant when the hair is curly to kinky. When the hair is straight to wavy, however, other fiber properties such as friction, stiffness, diameter, chargeability and cohesive forces play a more significant role.

The future progress for hair care products is largely dependent on a better understanding and better measurements for these fiber and assembly properties.

References

1. Robbins, C.R.; Scott, G.V. J. Soc. Cosmet. Chem. 29:783 (1978).
2. Hough, P.; Huey, H. & Tolgyesi, W. J. Soc. Cosmet. Chem. 27:571 (1976).
3. Robbins, C.R.; Reich, C. J. Soc. Cosmet. Chem. 37:141 (1986).
4. International Dictionary of Physics and Electronics. Van Nostrand Reinhold, New York (1956).
5. Elliot, A. Text. Res. J. 22:783 (1952).
6. Wolfram, L.J.; Lindemann, M. J. Soc. Cosmet. Chem. 22:839 (1971).
7. Scott, G.V. Private communication.
8. Robbins, C.R.; Crawford, R. J. Soc. Cosmet. Chem. 42:59 (1991).
9. Weigmann, H.D. Analysis and quantification of hair damage. Progress Report No. 2, TRI Princeton, Princeton NJ, Nov. 1991.
10. Simpson, W. J. Text. Inst. 51T:675 (1965).
11. Huck, P.; Baddiel, C. J. Soc. Cosmet. Chem. 22:401 (1971).
12. Hamburger, W.; et al. Proc. Sci. Sect. Toilet Goods Assoc. No. 14 (Dec. 1950).
13. Henderson, G.H.; et. al. J. Soc. Cosmet. Chem. 29:449 (1978).
14. Kamath, Y.; Weigmann, H.D. J. Appl. Poly. Sci. 27:3809 (1982).
15. Kamath, Y.; Weigmann, H.D. J. Soc. Cosmet. Chem. 35:21 (1984).
16. Deem, D.; Rieger, M. J. Soc. Cosmet. Chem. 19:410 (1968).
17. Harris, M.; et al. J. Res. Natl. Bur. Stand. 29:73 (1942).
18. Sikorski, J.; Woods, H. Proc. Leeds. Philos. Lit. Soc. (Sci. Sect.) 5:313 (1950).
19. Rebenfeld, L.; Dansizer, C. Text. Res. J. 33:458 (1963).
20. Speakman, J. J. Text. Inst. 37T:102 (1947).
21. Sookne, A.; Harris, M. J. Res. Natl. Bur. Stand. 19:535 (1937).
22. Wolfram, L.J.; Lennhoff, M. J. Text. Inst. 57T:591 (1966).
23. Beyak, R.; et al. J. Soc. Cosmet. Chem. 20:615 (1969).
24. Feughelman, M.; Robinson, M. Text. Res. J. 37:441 (1967).
25. Speakman, J. Proc. R. Soc. 103B:377 (1928).
26. Menkart, J. J. Soc. Cosmet. Chem. 17:769 (1966).

27. Breuer, M.M. J. Soc. Cosmet. Chem. 23:447 (1972).
28. Chamberlain, N.; Speakman, J. Z. Electrochem. 37:374 (1931).
29. Speakman, J. Trans. Faraday Soc. 25:92 (1929).
30. Robbins, C.R.; Scott, G.V. J. Soc. Cosmet. Chem. 21:639 (1970).
31. Rebenfeld, L.; et al. J. Soc. Cosmet. Chem. 17:525 (1966).
32. Astbury, W.; Street, A. Philos. Trans. Proc. R. Soc. 230A:75 (1931).
33. Crawford, R.J. Private communication.
34. Humphries, W.; et al. J. Soc. Cosmet. Chem. 23:359 (1972).
35. Alexander, P.; et al. Biochem. J. 49:129 (1951).
36. Alexander, P.; et al. Wool, Its Chemistry and Physics, 2nd Ed., pp. 61–65. Franklin Publishing, NJ (1963).
37. Harris, M.; Brown, A. Symposium on Fibrous Proteins, p. 203. Publ. Soc. Dyers Col., Bradford, UK (1946).
38. Garson, J.C.; et al. Int. J. Cosmet. Sci. 2:231 (1980).
39. Robbins, C.R.; Kelly, C. J. Soc. Cosmet. Chem. 20:555 (1969).
40. Edman, W.; Marti, M. J. Soc. Cosmet. Chem. 12:133 (1961).
41. Robbins, C.R. J. Soc. Cosmet. Chem. 22:339 (1971).
42. Randebrook, R.; Eckert, L. Fette Seifen Anstrichm. 67:775 (1965).
43. Heilingotter, R. Am. Perfumer 66:17 (1955).
44. Freytag, H. J. Soc. Cosmet. Chem. 15:667 (1964).
45. Hamburger, W.J.; Morgan, H. Proc. Sci. Sect. Toilet Goods Association, No.18 (Dec. 1952).
46. Kubu, E.; Montgomery, D. Text. Res. J. 22:778 (1952).
47. Heilingotter, R.; Komarony, R. Am. Perfumer 71:31 (1958).
48. Whitman, R. Proc. Sci. Sect. Toilet Goods Assoc, No. 18:27 (1952).
49. Brown, J. J. Soc. Cosmet. Chem. 18:225 (1967).
50. Cook, M. Drug Cosmet. Ind. 99(5):53 (1966).
51. Wall, F.E. In Cosmetics, Science and Technology, Sagarin, E., ed., Ch. 21, Interscience, New York (1957).
52. Crawford, R.J. Private communication.
53. Zahn, H.; et al. Fette Seifen Anstrichm. 70(10):757 (1968).
54. Scott, G.V. Private communication.
55. Speakman, J.; Scott, J. Trans. Faraday Soc. 30:539 (1934).
56. Valko, E.; Barnett, G. J. Soc. Cosmet. Chem. 3:108 (1952).
57. Breuer, M.; Prichard, D. J. Soc. Cosmet. Chem. 18:643 (1967).
58. Beyak, R.; et al. J. Soc. Cosmet. Chem. 22:667 (1971).
59. Robbins, C.R.; Kelly, C. Text. Res. J. 40:891 (1970).
60. Harris, M.; Smith, A. J. Res. Natl. Bur. Stand. 20:563 (1938).
61. Korastoff, E. Br. J. Dermatol. 83:27 (1970).
62. Swanbeck, G.; et al. J. Invest. Dermatol. 54:248 (1970).
63. Wilson, J.T. Int. Symp. Forensic Hair Comparisons, Quantico, VA (June 1985)
64. Anzuino, G.; Robbins, C.R. J. Soc. Cosmet. Chem. 22:179 (1971).
65. Hirsch, F. J. Soc. Cosmet. Chem. 11:26 (1960).
66. Robinson, M.S.; Rigby, B.J. Text. Res. J. 55:597 (1985).
67. Brown, A.; et al. Text. Res. J. 20:51 (1950).
68. Scott, G.V.; Robbins, C.R. Text. Res. J. 39:975 (1969).
69. Scott, G.V.; Robbins, C.R. J. Soc. Cosmet. Chem. 29:469 (1978).

70. Gutherie, J.; et al. J. Text. Inst. 59T:912 (1954).
71. Morton, W.; Hearle, J.W.S. Physical Properties of Textile Fibers, Ch. 17. Butterworths, London (1962).
72. Hearle, J.W.S.; Peters, R. Moisture in Textiles, p. 173. Butterworths, London (1960).
73. Meredith, R. J. Text. Inst. 45T:489 (1954).
74. Goodings, A. Text. Res. J. 38:123 (1968).
75. Morton, W.; Permenyer, F. J. Text. Inst. 40T:371 (1949).
76. Mitchell, T.; Feughelman, M. Text. Res. J. 30:662 (1960).
77. Bogaty, H. J. Soc. Cosmet. Chem. 18:575 (1967).
78. Wolfram, L.J.; Albrecht, L. J. Soc. Cosmet. Chem. 36:87 (1985).
79. Meredith, R.; Hearle, J.W.S. Physical Methods of Investigating Textiles, Ch. 8.3. Interscience, New York (1959).
80. Scott, G.V. Private communication.
81. Abbott, N.; Goodings, A. J. Text. Inst. 40T:232 (1949).
82. King, A. J. Text. Inst. 17T:53 (1926).
83. Hearle, J.W.S.; Peters, R. Moisture in Textiles, p. 144. Butterworths, London (1960).
84. Yin, N.; et al. J. Soc. Cosmet. Chem. 28:139 (1977).
85. Barnard, W.; White, H. Text. Res. J. 24:695 (1954).
86. White, H.; Stam, P. Text. Res. J. 19:136 (1949).
87. Eckstrom, M. J. Soc. Cosmet. Chem. 2:244 (1951).
88. Montgomery, D.; Milloway, W. Text. Res. J. 22:729 (1952).
89. Dart, S.; Peterson, L. Text. Res. J. 19:89 (1949).
90. Busch, P. 3rd Int. Hair Science Symp., Syburg, W. Germany (Nov. 1984).
91. Brancik, J.; Daytner, A. Text. Res. J. 47:662 (1977).
92. Barnett, G. The swelling of hair in aqueous solutions and mixed solvents. M.S. Thesis, Polytechnic Institute of Brooklyn, NY (1952).
93. Ehrhardt, H. Private communication.
94. Stam, R.; et al. Text. Res. J. 22:448 (1952).
95. World Book Encyclopedia. Field Enterprises Educational Corp., Chicago (1969).
96. Statistical Abstracts of the United States (1976).
97. Randebrook, R. J. Soc. Cosmet. Chem. 15:691 (1964).
98. Bogaty, H. J. Soc. Cosmet. Chem. 20:159 (1969).
99. Encyclopedia Britannica. Wm. Benton, Chicago (1973).
100. Steggarda, M.; Seibert, H. J. Hered. 32:315 (1941).
101. Steinhardt, J.; Harris, M. J. Res. Natl. Bur. Stand. 24:335 (1940).
102. Seibert, H.; Steggarda, M. J. Hered. 33:302 (1942).
103. Mercer, E.H. Biochim. Biophys. Acta 3:161 (1949).
104. Hardy, D. Am. J. Phys. Anthropol. 39:7 (1973).
105. Bailey, J.G.; Schliebe, S.A. Int. Symp. Forensic Hair Comparisons, Quantico, VA (June 1985).
106. Swift, J.A. 8th Int. Hair Science Symp. DWI, Kiel, Germany (Sept. 1992).
107. Spei, M.; Zahn, H. Melliand Textilber. 60(7):523 (1979).
108. Crawford, R.J. Private communication.
109. Shansky, A. J. Soc. Cosmet. Chem. 14:427 (1963).
110. Powers, D.; Barnett, G. J. Soc. Cosmet. Chem. 4:92 (1953).
111. Reed, R.; et al. J. Soc. Cosmet. Chem. 1:109 (1948).

112. Keil, F. J. Soc. Cosmet. Chem. 11:543 (1960).
113. Wickett, R. J. Soc. Cosmet. Chem. 34:301 (1983).
114. Klemm, E.; et al. Proc. Sci. Sect. Toilet Goods Association 43:7, (1965).
115. Howell, H.; et al. Friction in Textiles. Butterworths, London (1959).
116. Robbins, C.R. 3rd Int. Hair Sci. Symp. DWI, Syburg, Germany (1984).
117. Schwartz, A.; Knowles, D. J. Soc. Cosmet. Chem. 14:455 (1963).
118. Scott, G.V.; Robbins, C.R. J. Soc. Cosmet. Chem. 31:179 (1980).
119. Howell, H.G.; Mazur, J. J. Text. Inst. 43T:59 (1952).
120. Meredith, R.; Hearle, J.W.S. Physical Methods of Investigating Textiles, Ch. 11. Interscience, New York (1959).
121. King, G. J. Text. Inst. 41T:135 (1950).
122. Wool Research, Vol. 2, Physical Properties of Wool Fibers and Fabrics, Ch. 8. Wool Industries Research Association, Leeds, UK (1955).
123. Fishman, D.; et al. Text. Res. J. 18:475 (1948).
124. Martin, A.; Mittleman, R. J. Text. Inst. 37T:269 (1946).
125. Swift, J.; Bews, B. J. Soc. Cosmet. Chem. 27:289 (1976).
126. Alexander, P.; et al. In Wool, Its Chemistry and Physics, 2nd Ed., pp. 25–46. Franklin Publishing, NJ (1963).
127. Hambidge, A.; Wolfram, L.J. 3rd Int. Hair Sci. Symp., Syburg, W. Germany, German Wool Res. Inst. Publ. Abstracts (1984).
128. Scott, G.V.; et al. J. Soc. Cosmet. Chem. 20:135 (1969).
129. Robbins, C.R.; et al, J. Soc. Cosmet. Chem. 37:489 (1986).
130. Morton, W.; Hearle, J.W.S. Physical Properties of Textile Fibers, Ch. 21. Butterworths, London (1962).
131. Lunn, A.; Evans, R. J. Soc. Cosmet. Chem. 28:549 (1977).
132. Jachowicz, J.; et al. J. Soc. Cosmet. Chem. 36:189 (1985).
133. Mills, C.; et al. J. Soc. Cosmet. Chem. 7:466 (1956).
134. Barber, R.; Posner, A. J. Soc. Cosmet. Chem. 10:236 (1959).
135. Meredith, R.; Hearle, J.W.S. Physical Methods of Investigating Textiles, Ch. 13.3. Interscience, New York (1959).
136. Hersh, S. Static electrification of fibrous materials. Ph.D. Thesis, Princeton University, Princeton, NJ (1954).
137. Henry, P. Br. J. Appl. Phys. (Suppl. 2):531 (1953).
138. Hersh, S.; Montgomery, D. Text. Res. J. 25:279 (1955).
139. Cunningham, R.; Montgomery, D. Text. Res. J. 28:971 (1958).
140. Martin, J. J. Soc. Dyers Col. 60:325 (1944).
141. Ward, T.; Benerito, R. Text. Res. J. 35:271 (1965).
142. Fourt, L. Text. Res. J. 36:915 (1966).
143. Thompson, W.; Mills, C. Proc. Sci. Sect. Toilet Goods Assoc., No. 15 (May 1951).
144. Stamm, R.F.; et al. J. Soc. Cosmet. Chem. 28:571 (1977).
145. Conover, W.J. Practical Nonparametric Statistics, pp. 299–308. Wiley, New York (1980).
146. Reich, C.; Robbins, C. J. Soc. Cosmet. Chem. 44:221 (1993).
147. Schebece, F.; Scott, G.V. Private communication.
148. Meredith, R.; Hearle, J.W.S. Physical Methods of Investigating Textiles, Ch. 12. Interscience, New York (1959).
149. Schebece, F. Private communication.
150. Garcia, M.; Diaz, J. J. Soc. Cosmet. Chem. 27:379 (1976).

151. Newman, W.; et al. J. Soc. Cosmet. Chem. 24:773 (1973).
152. Waggoner, W.; Scott, G.V. J. Soc. Cosmet. Chem. 17:171 (1966).
153. Ross, L. Private communication.
154. Harris, M. Handbook of Text. Fibers, 1st Ed.. Harris Research Labs, Washington, DC (1954).
155. Modern Beauty Shop Magazine (Dec. 1957).
156. Studies of the Modification of Human Hair Properties by Surface Treatments, Phase II—Hair Assembly Behavior. Progress Report No. 7, Text. Research Institute, Princeton, NJ (March 15:1978).
157. Robbins, C.R. Chemical and Physical Behavior of Human Hair, 1st Ed., p. 199. Van Nostrand Reinhold, New York (1979).
158. Garcia, M.L.; Wolfram, L.J. 10th IFSCC Congress, Sydney, Australia (1978).
159. Robbins, C.; Crawford, R.J. J. Soc. Cosmet. Chem. 35:369 (1984).
160. Clarke, J.; Robbins, C.R.; Reich, C. J. Soc. Cosmet. Chem. 42:341 (1991).
161. Blankenburg, G. Private communication.
162. Sante, R.; et al. J. Soc. Cosmet. Chem. 4:270 (1953).
163. Scott, G.V. Colgate Palmolive Curl Retention Methods Internal Report (1967).
164. Kirby, D.H. Proc. Sci. Sect. Toilet Goods Assoc. 26:12, (1956).
165. Watson, P.C. U.S. patent 2,936,543 (May 27, 1958).
166. Stavrakas, E.J.; et al. Proc. Sci. Sect. Toilet Goods Assoc. 31:36 (1959).
167. Ganslow, S.; Koehler, F.T. J. Soc. Cosmet. Chem. 29:65 (1978).
168. Reed, A.B., Jr.; Bronfein, I. Drug Cosmet. Ind. 94:178 (1964).
169. Robbins, C.R. J. Soc. Cosmet. Chem. 34:227 (1983).

Appendix

Physicochemical Constants

0°C = 273.16 K
R gas constant = 1.987cal deg^{-1} $mole^{-1}$
Standard gravity = 980.665 cm sec^{-2}
Faraday's constant = f = 23,062 cal (volt equivalent)$^{-1}$
Avogadro's constant = N = 6.0238×10^{23} molecules $mole^{-1}$
Density of hair = 1.32 g cm^{-3}
Refractive index of hair :
    Epsilon = 1.56 (light parallel to fiber axis)
    Omega = 1.55 (light perpendicular to fiber axis)
Elastic modulus :
    Stretching = 3.89×10^{10} dynes/cm^2
    Bending = 3.79×10^{10} dynes/cm^2
    Twisting = 0.89×10^{10} dynes/cm^2 *

* Bogaty, H.J. Soc. Cosmet. Chem. 18:575 (1967).

Approximate Diameter or width of a Few Keratin Fibers

Fiber	Diameter or width (micrometers)[a]
Human scalp hair	
Terminal hair	30–120
Vellus	< 4
Wool fiber	
Fine wool	17–33 [1]
Coarse wool	33–42 [1]
Horse hair	
Mane	50–150 [1]
Tail	75–280 [1]
Cat whisker (near skin)	~ 450
Porcupine quill	can be >1,000

[a] Sources: [1] Harris Handbook of Textile Fibers, Harris, M., ed. Harris Research Labs., Washington, DC (1954).

Units of Linear Measure

Unit	Abbreviation or symbol	Quantity
Meter	m[a]	
Centimeter	cm	10^{-2} m
Millimeter	mm	10^{-3} m
Micrometer[b]	μm	10^{-6} m
Nanometer	nm	10^{-9} m
Angstrom	Å	10^{-10} m

[a] The arbitrarily chosen standard of length of the metric system. It is the distance between two marks on a platinum-iridium bar kept at constant temperature at the International Bureau of Weights and Measures near Paris. For conversion to the English system, 1 meter equals 39.37 inches, 1 centimeter equals 39.37×10^{-2} inches, etc.

[b] The word micron is sometimes used for micrometer.

Index

M

N